AF794647

Air's Appearance

Air's Appearance

Literary Atmosphere in British Fiction, 1660–1794

JAYNE ELIZABETH LEWIS

The University of Chicago Press
Chicago and London

Jayne Elizabeth Lewis is professor of English at the University of California, Irvine, and the author, most recently, of *Mary Queen of Scots: Romance and Nation.*

The University of Chicago Press, Chicago 60637
The University of Chicago Press, Ltd., London
© 2012 by The University of Chicago
All rights reserved. Published 2012.
Printed in the United States of America

21 20 19 18 17 16 15 14 13 12   1 2 3 4 5

ISBN-13: 978-0-226-47669-8 (cloth)
ISBN-13: 978-0-226-47671-1 (e-book)
ISBN-10: 0-226-47669-3 (cloth)
ISBN-10: 0-226-47671-5 (e-book)

Library of Congress Cataloging-in-Publication Data

Lewis, Jayne Elizabeth.
Air's appearance : literary atmosphere in British fiction, 1660–1794 / Jayne Elizabeth Lewis.
pages. cm.
Includes bibliographical references and index.
ISBN 978-0-226-47669-8 (cloth: alkaline paper)
ISBN 0-226-47669-3 (cloth: alkaline paper)
ISBN 978-0-226-47671-1 (e-book)
ISBN 0-226-47671-5 (e-book)
1. Weather in literature. 2. English fiction—17th century—History and criticism. 3. English fiction—18th century—History and criticism. I. Title.
PR769.L49 2012
823.009'36—dc23

2012002383

♾ This paper meets the requirements of ANSI/NISO Z39.48–1992 (Permanence of Paper).

Contents

Figures

Acknowledgments

This book took shape in and near Los Angeles, a beloved city justly famous for its air. Here a vibrant human community has been my work's informing atmosphere, and I am grateful to colleagues past and present at the University of California, Los Angeles, and the University of California, Irvine: Jami Bartlett, Liz DeLoughry, Vivian Folkenflik, Richard Godden, Andy Kelly, Claire McEachern, Anne Mellor, Felicity Nussbaum, Hugh Roberts, Debora Shuger, Victoria Silver, Ann Van Sant, and Rob Watson. I have particularly benefited from the intellectual generosity of Emily Anderson, Noelle Chao, Helen Deutsch, Corrinne Harol, Sarah Kareem, Justine Pizzo, and Arden Reed, and, most especially, from my fellow traveler Jonathan Grossman, for his rigor, candor, and unfaltering companionship through "the atmosphere story." Heartfelt thanks for Laura O'Connor's lucent camaraderie, irrepressible faith, and unusual insight into the book, and for Jerry Christensen's generous and searching responses to many of its pages. I owe a special debt to Robert Folkenflik for his faithful erudition. Several graduate students at Irvine—Craig Carson, Maia Krause, Jennifer Liou, Emily Liu, Michelle Neely, John Nieman, and Eric Peterson—have taught me as much as I have them. Until his death, Richard Kroll was my vital first reader. I still hear his voice. Susan Davis shared undertakings. I have had two lifeguards: James McMichael's profound capacity as a poet, reader, and friend, inspired, heartened, and helped me; Michelle Latiolais showed me what fiction is supposed to feel like.

In many ways, I have written a book about belief, but I have learned the most about that from those who have believed in it. Of these, Hilary Schor, my curious heroine and beloved reader, comes first. I am also grateful to Deidre Lynch, John Richetti, Elisa Tamarkin, James Turner, and Cynthia

Wall, all scholars whose style and substance have sustained my own work and whose interest in this project, often expressed in very material ways, has buoyed it. Rick Barney, James Chandler, Mark Dorrian, Tim Erwin, Sophie Gee, Nancy Glazener, Jan Golinksi, Anna Kornbluh, Celeste Langan, Kimberly Latta, James Olney, Kenneth Olwig, Elliott Visconsi, Roxanne Wheeler, Matt Wickman, and Ruth Yeazell all shared their work or responded to mine in ways that truly mattered. Rachel Gamby and Eldon and Stella Lewis stayed present.

Many thanks to the staffs of the William Andrews Clark Memorial Library, the Huntington Library, the Horace Walpole Library, and the British Library, as well as to the Andrew Mellon Foundation. I am grateful to audiences at Pomona, UCLA, UC–Irvine, UC–Santa Barbara, the William Andrews Clark Memorial Library, the University of Chicago, Yale University, the Horace Walpole Library, and the Southern California Eighteenth-Century Group. Parts of the book have appeared elsewhere in print. Chapter 4 expands "The Atmospheres of *Robinson Crusoe*," in *Defoe's Footprints: Essays in Honour of Maximillian E. Novak*, ed. Robert M. Maniquis and Carl Fisher (Toronto: University of Toronto Press, 2009), 35–54. © University of Toronto Press, 2009. Reprinted with permission of the publisher. Chapter 5 was first published as "Spectral Currencies in the Air of Reality: Defoe's *Journal of the Plague Year* and the History of Apparitions" in *Representations* 87 (2004): 82–101. A small part of chapter 6 was published as "The Air of *Tom Jones*; or, What Rose from the Novel." This article originally appeared in *The Eighteenth Century: Theory and Interpretation* 52, no. 3–4 (Fall 2011/Winter 2012): 303–22, published by the University of Pennsylvania Press. I am grateful for kind permission to reproduce sections of these essays. I most sincerely thank my anonymous readers at the University of Chicago Press; my heroically vigilant and resourceful copy editor, Kathy Swain; Mark Reschke for his help in the final stages of the publication process; and my meticulous, responsive, and beneficent editors, Alan Thomas and Randolph Petilos.

And now for the other Thomases. Within my family circle, Janet Thomas, let me borrow the building blocks of chemistry. Brook Thomas did his level best to keep me afloat. Peter Thomas is my sunshine and my air. This book is for him.

Introduction

> We must remember that there is such a thing as atmosphere.
>
> VIRGINIA WOOLF, *On Being Ill* (1926)

> It is very necessary for a critic to remember about the atmosphere.
>
> WILLIAM EMPSON, *Seven Types of Ambiguity* (1930)

Why on earth would we need to be reminded about something as vital as (the) atmosphere? This book hunts for the key to a notoriously elusive dimension of literary experience in the library of eighteenth-century English fiction. I link the rise of atmosphere as an aesthetic program within that fiction to contemporary conceptions of the air as they were articulated in the interpenetrating spheres of natural philosophy, supernaturalism, and sociability in the period.

A truism in the history of science is that air became visible as an object of knowledge after the founding of the Royal Society in 1660, in concert with both the decline of supernatural accounts of what goes on in the air and the elevation of distinctively modern forms of sociability.[1] A truism among literary historians is that "atmosphere" became visible as a feature of imaginative literature during the gothic craze of the 1790s.[2] Because technically air is not visible, neither of these truisms can be strictly true. What can be true is that both English natural philosophy and English fiction assumed their seemingly modern shapes within the visual medium of writing. Over the long eighteenth century, this medium openly aspired to the ubiquity, liberty, and commonality of air. But writing of the period also seemed to absorb many of the occult powers, despotic and inscrutable, previously identified with the air, even as it orchestrated a model of personal sociability based on the assumption and manipulation of "an" air. As it considers how English letters created the impression that air can be seen, then looks at how individual writers experimented both with and within that impression, *Air's Appearance* weaves seemingly disparate conceptions of air into a genealogy of modern literary "atmosphere."

But just what is it we mean when we say that a literary work has "an atmosphere"? What might this term, and the presumably common experience that it conjures, have to do with "the" atmosphere? Such questions are, so to speak, in the air right now, not least because we're worried: about how long either literature or the air about us might last in its present form; about the nature, degree, and significance of our contribution to the shelters that both provide. It doesn't help that it's so hard to say exactly what either "a" literary atmosphere or "the" atmosphere about us actually is. With respect to books, we can speak of mood and voice (the German *Stimmung*) or parse novelistic weather reports—"fog everywhere"; "it was a dark and stormy night"—without arriving at much of a determination.[3] One of British modernism's most discerning readers of eighteenth-century literature, William Empson, supposed that literary atmosphere might come down to "an undifferentiated mode of being" that is "conveyed in some unknown and fundamental way as a by-product of meaning." But in the end even Empson threw up his hands. What was atmosphere? "Analysis cannot hope to do anything but ignore it," he concluded (or did not), and "criticism can only state that it is there."[4] As for the so-called sphere of vapors that physically surrounds us, and on which we so manifestly and entirely depend, Karl Marx was surely right to find that "after the discovery of the component gases of the air, the atmosphere remained unaltered."[5] We are, in other words, every bit as bewildered as we were before we knew anything at all about the air. Fog everywhere, indeed.

I take a broadly historical approach to the question of what literary atmosphere might entail, and to that of what it might in turn tell us about our notional, physical, and media environments, particularly as these interact with those who experience them. Eighteenth-century chemistry discovered—or at least named—the component gases of the air. The following pages propose that, while the idea, and experience, of "a" literary atmosphere seems to be a metaphorical extension of "the" physical atmosphere, as known to and by natural philosophers from Robert Boyle to Joseph Priestley, these two kinds of air actually developed interdependently. In English print, the word "atmosphere" first appeared in 1638, between the covers of John Wilkins's *Discovery of a World in the Moone*, a work soon, and repeatedly, reprinted as *A Discourse concerning a New World and Another Planet*. Here, on the strength of its analogy to "that Part of the Air which is nearest to the Earth [and] is of a thicker Substance than the other," polymath Wilkins less defined than (obviously) projected "Atmo-sphera" as an "Orb of gross, Vaporous Air immediately encompassing the Body of the Moon."[6] A few years later, Wilkins's fascinating treatise on cryptography, *Mercury, or the Secret and Swift Messenger* (1641), would spend pages picturing such immediately encompassing orbs as they might be mediated through the manifestly terrestrial bodies of letters. Wilkins wanted to show his wartime readers how to "writ[e] by the sunbeams, upon a wall or front of a house" and how to scribble on a glass "and then oppos[e] the glasse against the full Moone" so that "the letters would appeare through it."[7]

It would be hard to find a more literal instance of literary atmosphere, or at least of the evocative and enveloping spaces that can seem to thicken between lexical lines. And Wilkins instances something more: his atmospheres suggest that such lines are also informed by those same intervening spaces. In this sense, Wilkins's atmospheric writing captures the inevitably mediated nature of what surrounds us. Furthermore, it prompts us to imagine that one precipitating condition of sense experience might be its formal encoding. Wilkins's life overlapped with Newton's, but here he anticipates the intuitions of scientists writing after Einstein. "Our sensory experience forms a cryptogram," the Cambridge astrophysicist Arthur Stanley Eddington offered at almost exactly the same time that Empson went in search of literary atmosphere, "and the scientist is a Baconian enthusiast engaged in deciphering the cryptogram."[8] It is in early conceptions of "Air" and "Atmosphere" that we encounter this "cryptogram" at both its most apparent and its most "encompassing."

For all their devotion to the material and quantifiable properties of air, that is, a distinguished line of English pneumatic chemists—from Robert Boyle, Edmund Bohun, and Stephen Hales up to Joseph Priestley—were also compulsive writers and self-conscious literary stylists who consciously moved within "Orbs" of recurrent words.[9] At the same time, all had to grapple with their subject's invisibility, its orbicular ubiquity, and the pressure exerted by its persistent strong connotations, indeed its charismatic encryption, within folkloric and faith traditions. These properties of "the" air give figurative language in general—but especially language explicitly realized in literary space—an active and dynamic role in air's conceptualization and investigation.

It is certainly nothing new to see the languages of nature that would eventually be called science in light of their constitutive figures; even the historian of science will identify tropes from metaphor to personification and analogy as volatile agents of natural knowledge.[10] Considered as its original medium, however, air sustains a specifically metonymic relationship to human language, one that persists, deepens, and becomes capable of self-reflection as language assumes written form. Working within a notional environment created through the transparent layering of lexical elements, English aerial philosophers of the late seventeenth and eighteenth centuries built into the seemingly omnipresent medium of writing a propensity to give rise, at least apparently, to its own effectual surroundings. This propensity was dramatized whenever it came to articulating air's "spring" (a feat that Boyle's layered prose attempted with suggestive difficulty), to citing the elements that compose what we call "the air," or to capturing air's mediation through what was now "the Atmosphere."

According to John Arbuthnot—physician, satirist, asthmatic, and indefatigable chronicler of *The Effect of Air on Human Bodies* (1733)—"atmosphere" happens when air absorbs, but also displays, "the Steams, Effluvia, and all the Abrasions of Bodies on the Surface of the Earth."[11] In the ever-greater Britain that he knew, air's metonymic affinity with compositional forms that were, after all, designed to make intangibles apparent and effectual potentially licensed many different kinds of writing as principal investigators of "the" atmosphere. Indeed, the more widely diffused, the more superficial, and the more practiced in the mechanics of producing effects without supplying any immediate cause, the better. Arbuthnot did not bother with it, but it is at this point that popular fiction began to gain an edge over other kinds of writing about the air. For whereas the figural dimension of scientific or protoscientific writing has sometimes seemed to complicate and even undermine its truth claims, air grants writing that presents itself as essentially, not accidentally, figurative—that presents itself as, in fact, fiction—a claim to literal truth.

In this book, I see popular fiction's edge as a creative analyst of atmosphere to have been honed outside the formalizing precincts of the novel in two works of singular relevance to eighteenth-century writers of fiction, John Milton's *Paradise Lost* (1667–74) and Alexander Pope's *The Rape of the Lock* (1712–14). Both of these poems were also intimately engaged with contemporary speculation about the air, especially as this included a heightened conception of air as a social form (Pope's preoccupation) and as a self-limiting medium of knowledge (Milton's). Later, I suggest, novelists as disparate as Daniel Defoe, Henry Fielding, Charlotte Lennox, and Ann Radcliffe went on to appropriate and enhance natural philosophy's reflexive, and potentially interactive, manner of accounting for fundamentally unknowable environments. It is their literary experiments, not primarily those of the chemists, physicians, and weather journalists whom I consider along the way, that I most want to understand. The reader who seeks a history of science, or a conventionally interdisciplinary accounting of what was said about the air in this period, should look elsewhere.[12] Attending instead to how things were said in intimate relation with the air, I draw on the root meaning of the word "atmosphere"—"sphere of vapors"—to suggest that, for many eighteenth-century English writers, atmosphere *was* appearance: the making apparent, if not strictly visible, of an otherwise invisible medium. One thing that atmosphere makes visible is the inevitably mediated nature of what we call experience. In my reading, this renders atmosphere, especially in its literary form, the ironic shadow of the Enlightenment's vaunted faith in the self-evident. Mist creeps into the poetry and literary romance of the 1790s not to defy the Enlightenment assumption that the world is immediately available to the senses. Rather, it arises from the conceptual sphere within which that assumption itself took shape.

Two significant accidents motivate my emphasis on anglophone writers centered in or near London. One is the accident of geography that shrouded the British Isles in vapors so thick and shapely as to compel those who inhaled them into complex reflection on and through them. This apparently indigenous and natural phenomenon mixed with supposedly modern forms of urbanity and manufacture in the metropolis (London fog was mostly smog, and the topic of considerable gentlemanly banter). But there is also the accident of history that made English–its printed version conceived as the material form of a tongue that once "floated in the breath"—the lingua franca of so many convenient lies and inconvenient truths about the air over the course of the eighteenth century.[13] An obvious backstory is the denigration of oral (airborne) rivals to English literary culture in the period and the suppression of nonvisual forms of communication and knowledge (sound, smell, and touch, for example) in the interest of their ocular counterparts.[14] Such stories have been well told by other scholars. I assume but also complicate them insofar as I reconceptualize English literature in terms of its expressed affinities—its antic and mimetic continuity or, to borrow a phrase from French chemistry of the period, its *rapport*—with air and specifically with the modes of expression and experience that air compels. Here it is transference as opposed to reference, and metonymy more than metaphor or analogy, that create the impression of volume and depth, of interiority to a layered and influential world, that we still assume whenever we speak of "atmosphere."

One arc of my story inevitably follows the familiar curve of eighteenth-century natural philosophy, now legible to many in terms of what Bruno Latour terms its "hylozoism," its acquisition of authority from the interweaving of human and nonhuman agents and actors within a broad web of communication.[15] This arc makes its way upward from Boyle's evident fixing of air as a "physico-mechanical" body in the 1660s and 1670s. It drops some hundred years later into Priestley's teasing of "common air" into many separate "factitious" ones and his isolation of carbon dioxide, nitrogen, and (to some extent) oxygen as chemical building blocks of the former. Such edifying abstractions of the air have often looked like classic expressions of what once was anachronistically called the new science. Embedded in credible human communities, they were legitimated by all the technology that modernity could muster, from air pumps to aeoliphiles. But these same abstractions, however believable, proved treacherous allies of the period's most visible knowledge projects. Ruthless parodies of Boyle's experiments considered that air's, well, thinness potentially laid that of his empirical method bare.

Meanwhile, air belongs to everyone, as we still recognize when we speak of ideas being "in the air" or of the "castles in the air" that presumably any imagination can build. Such promiscuity may explain why the study of the air in the eighteenth century never achieved the social prestige or in-house legitimacy that other branches of natural philosophy could claim. But then why should it have? In this instance, the object of knowledge refuses to behave like one; the atmosphere cannot be stood outside of, so it can never be completely objectified, fixed, or, as a result, intellectually possessed. *Air's Appearance* tracks such inconveniences through the writing of Boyle and Priestley, who precisely *as* writers adapted to them by retaining ancient—or perhaps in Latour's term "amodern"—conceptions of the air within what often presented themselves as novel forms of denotation and reference.[16] On the way from Boyle to Priestley, I also consider the writing of middle-modern physicians (John Arbuthnot, George Cheyne, and John Floyer) and meteorologists avant la lettre (Thomas Short, Edmund Bohun, and Bernard Annely), who grasped their elusive subject matter through specific graphic and linguistic innovations. Instead of producing a strictly modern and distinctly English account of the air, however, their innovations often identified "air" with other times (including antiquity), other places, and even other worlds.

The possibility of other worlds brings us to those now deemed the supernaturalists of the day. A standard view holds that over the course of the eighteenth century, broad-based belief in such aeriform beings as fairies and apparitions vanished into thin air. Along with the "Prince of the Powers of the Air," the devil himself, those creatures are still seen to have been either vanquished by a purely physico-mechanical understanding of what air is and does or transformed into secular magics ranging from popular entertainment to repression psychology.[17] Recently, Simon During has cast the eighteenth-century overhaul of belief in the supernatural as a complicated, because knowing, submission to a new "language game," albeit one now split between "interiorized and literary magic" and "technological and exteriorized show business" (During, *Modern Enchantments*, 48, 50). While my own consideration of English surmise about things in the air converges with During's, I engage contemporary investigations of air's occult properties primarily as a lively and persistent form of writing that charted the readerly experience of being inside a self-contained world made up of mediating forms. Ongoing debates about what Defoe called the "Reality of Apparitions" devised notional environments that consisted solely and manifestly of such forms. In these self-consciously graphic venues, the transference of meaning is as primary as reference to objects, and mentalities are connected within a circle of implicit instructions about how to be in mediated environments. In eighteenth-century writing about the aerial phenomenon of apparitions, ambient phenomena thus sustain intermediate modes of belief. At the very least, they support beliefs founded on a rounded impression of immediacy that is known to be, quite possibly, illusory.

The paradox of mental life inside this shaken globe returns us to the problem, if no less the promise, of eighteenth-century English fiction. That fiction was subtly theorized from within by writers well aware that its visual embodiment in books made it uniquely perceptible as a factitious object even as it is somehow experienced as all-encompassing. At the same time, the habits of mind that make the nothing of fiction seem like something were precisely those that made "atmosphere" ever more apparent in an increasingly literate culture. So I float the hypothesis that what Empson would call a "tangibly intangible" sense of being in a mediated world took shape in dynamic exchange with a sense of atmosphere that owes as much to indigenous and borrowed versions of antiquity—and to the immediate, atemporal experience of the body in a sensed environment—as to new technologies of communication and knowledge. In calculating such debts, I hope to offer a new perspective on what fifty years of literary criticism has termed the "rise of the novel." The tenor of this metaphor has been questioned endlessly, but the vehicle itself never. To propose that a signature genre of modernity rose as a principal investigator of "the" air is, however, to take at face value a word ("rise") that traditionally has mainly implied this way of writing's steady social climb, its ascent to visibility as a "real" class of writing.[18] Here, I see eighteenth-century English fiction, in the particular (and possibly misnamed) form of the novel, to have risen into a reverberant cloud of unknowing that that fiction itself seemed to create.

The atmospheric techniques at work (and play) in eighteenth-century fiction also conditioned readers to comprehend the forms that defined their own experience of any given world. Attention to these techniques can help to provide a language for a dimension of literary experience for which we now possess no stable critical vocabulary. "Atmosphere" here stands in contrast to such novelistic staples as character, plot, or even style, all subjects of authoritative study.[19] It makes sense to develop a vocabulary of literary atmosphere from the ground up (or perhaps from the sky down) by looking to a cultural moment when the novel was, in more than one sense, in the air. Yet it is also in this period that "air"—by definition elusive—proves most difficult to parse, thanks not least to the dominant materialist idioms of the day. Alongside the writers featured here, I provisionally stabilize "air" by looking at how it mediates between material and immaterial, literal and figurative, registers of reality and by trying to remember that "air" may not meaningfully exist except through human interactions with it. Because these interactions inevitably involve the imagination—specifically the linguistic imagination—we are left less with a single unified concept of the air than with a set of associations that give rise to a "novel" order of experience. This potentially yields a new understanding of the novel form itself, even as it makes for the kind of analysis, frustrating to some, that drifts from the world of solid objects into the nebulous reaches of fictionality and language.

All the same, several recent studies of how books talk about "things" emphasize those things' disappearance as such talk transpires; the logical next step, taken here, moves in the direction of the surrounding conditions into which (shaped) things disappear.[20] What are such things like once we understand them to be "in the air," subject to our suspended convictions about them? May not those backgrounds be shapes themselves, negative spaces registered in both literature and literary criticism as atmosphere? With Eddington—a Cambridge don while Empson was studying mathematics there—I thus seek "a knowledge which is neither of actors nor of actions, but of which the actors and actions are a vehicle" (Eddington, *New Pathways*, 256). Eddington's perceptions, like Empson's, will sometimes surface in the following pages; both men not only breathed Cambridge's air in the early 1930s but were preoccupied with what awareness of "the atmosphere"—and so of whatever clouds knowledge—allows a person to know. Both developed evocative vocabularies to capture this paradox. For instance, noting the "lines" that the terrestrial observer will see "imprinted on the light" of interstellar nebulae, Eddington concluded that these are "traces of a familiar enough substance. It is oxygen and nitrogen—or, if you like, common air" (Eddington, *New Pathways*, 205). Eddington's atmospheric lines that the observer sees through—or (if you like) sees without knowing she sees—perhaps differ little from the lunar cryptograms once traced by John Wilkins, himself master of Trinity College, Cambridge.

As it is realized in "atmosphere," I follow "common air" through several signature fictions of the long eighteenth century, teaming each with a body of nonfictional writing about the air. The trail admittedly goes cold once we reach Ann Radcliffe's gothic fiction of the 1790s. I find not that English fiction—like English aerial philosophy and, if bizarrely, English demonology—*had* to become what we now call "gothic" but that it had been so all along. At the very least, when seen in light of their transactions with other literary treatments of the air, none of the works of interest here demands to be read in terms of the social, psychological, or even cultural materialism usually supposed to distinguish the literature of the period. They do not necessarily aspire to the status of things, nor do they automatically order themselves in relation to them.

Instead, the literary works that I discuss all test, in lexical space, what Hubert Damisch captures in his elusive but tantalizing "theory of /cloud/." Damisch's eye was drawn to the visual arts, roving from Correggio's Renaissance frescoes to the "hieroglyphs of breath" that mark medieval Chinese painting, from Constable's naturalistic cloud studies to Cézanne's late deliberate vapidities. While all these works depict some form of "cloud," for Damisch depiction is not the point: a cloud *is*, in its essence, not a thing but rather "matter aspiring to form." As such, it reveals "an entirely original mode of organizing, articulating, and defining or designating, the substratum, the surface upon which the picture is painted," and it does so in such a way as to open space within that surface.[21] As Mark Dorrian has noted, cloud (or "/cloud/") in Damisch's sense must also be form aspiring to matter; Dorrian himself goes on to imagine what an architecture based on perpetual equivocation between these two states might resemble.[22]

Literary fiction is not visual art. It is not architecture. Yet it may give rise to surprisingly many of the effects that Damisch traces. In the eighteenth century, these effects were sometimes rehearsed in nonfictional writing about clouds and other aspects of atmosphere. This made it possible for writers of literary fiction to appropriate them. Because of their exceptional intelligibility within prosaic writing, I have not pursued atmospheric effects through the poetry of the period, though that poetry's pictorial bent might seem to be more obviously of a piece with Damisch's "/cloud/" and though its sound schemes immediately involve it with air.[23] To other scholars I have left such projects as the English georgic apotheosized in James Thomson's massive unto crushing *The Seasons* (1727–42), James Macpherson's feigned Ossian poems of the 1760s, and William Cowper's *The Task* (1785), whose preoccupation with winter in Mary Favret's breathtaking reading breeds the same sense of distanced immersion with which eighteenth-century Britons were coming to know war.[24] The two poems foregrounded here—Milton's *Paradise Lost* and Pope's *The Rape of the Lock*—are approached not as poetry but in terms of their explicitly fictionalized accounts of imaginative engagement with the air. In both cases, this engagement is modeled by female figures (Eve and

Belinda), as it is in Lennox's *The Female Quixote* (1752) and Radcliffe's *The Mysteries of Udolpho* (1794) as well. In other chapters, even Defoe's *Robinson Crusoe* (1719) and the Benjamin Partridge who accompanies its eponymous hero through the pages of Fielding's *Tom Jones* (1749) emerge as prototypes of the female reader of atmospheric fiction. These literary figures, like the typographical contexts within which they appear, foreground the experience of subjection to inscrutable environments. Inseparable from the forms that at once communicate them and make them visible, such environments were most happily elucidated through synthetic and associative mental practice.

In the eighteenth century, that practice, like the forms of experience that it precipitated, was often identified with the feminine and with the air. This identification is immediately visible in Joseph Wright of Derby's oft-analyzed painting of 1768, *Experiment on a Bird in the Air Pump* (fig. 1). Wright famously captures the midcentury vogue for the kind of exhibition science that turned laboratory experiment into bourgeois parlor trick. His ghostly witnesses to the illuminating effects of air's expulsion from an avian body divide along gender lines. If anything, the distinction between male and female beholders is what deepens and organizes the image into the doubled spherical form

FIGURE 1. Joseph Wright of Derby, *An Experiment on a Bird in the Air Pump* (1768). Oil on canvas, 183 × 244 cm. National Gallery, London. Photograph: © National Gallery, London/Art Resource, New York.

that opens inward to receive its beholder. The bodies of Wright's stricken, enraptured, or stoical female viewers assume attitudes that complement, and so realize, the shape of the abstracted cockatiel above them. In this manner a notionally common medium—air—becomes perceptible through, and indeed with, the paint medium. Visibly impressed by air, Wright's dynamically posed female subjects make it apparent as a condition of formal experience, in this case that of the airless sphere that contains everyone in the scene. It is the curved, compressed space *around* the air pump, not the space inside it, that is Wright's true interest, so much so that experience itself, aesthetic and otherwise, finally seems to be the property as much of the surrounding area as of any individual figure in it. Their formal subjection to this space is what allows its female inhabitants to model, occasion, and illuminate the experience of interiority to a sphere of mediating forms.

By contrast, virtually all the pneumatic chemists, physicians, and supernaturalists I ponder are male. Along with the pointing magician virtuoso who manipulates the illuminated pump in Wright's painting, they therefore occupy relatively higher rungs of the period's social (and knowledge) hierarchy. Yet just as an atmospheric reading practice levels the playing field of prestige between fiction and natural philosophy, so does it make visible the otherwise invisible contingency, arbitrariness, and even transience of such hierarchies, absorbing the authority of the pointing finger into the ambient circumstance that makes it visible. As the performance theorist Jean-Louis Barrault wrote in reference to postwar French theater, a pointing finger may be seen as the "representative organ of touch"; it "pierces through things and puts us in contact with the unknown or the invisible."[25] So Milton's insubordinate Eve naturally turns out to be more wedded to the air than she is to Adam. So, too, approaching Sweden in the summer of 1795, the radical feminist Mary Wollstonecraft could find herself unaccountably "warmed with the impression" that the "scenes" there "made upon" her. "Every cloud that flitted on the horizon was hailed as a liberator," Wollstonecraft wrote. But there was a cost, as there is to any investment in atmospheric forms: "Like most of the prospects sketched by hope, it dissolved under the eye into disappointment."[26] Wollstonecraft's sentence shifts into abstraction ("disappointment") in order to arrest the process of dissolution. But in this manner, she also manages to transfer the experience of "every cloud" to the intimate sphere of literary experience.

The following pages strive to map a hitherto uncharted dimension of that sphere. A slave to what Kenneth R. Olwig evocatively terms the "imagined scholarly discipline of aerography," I follow chapter 1's overview of eighteenth-century aerographies with a consideration, in chapter 2, of Boyle, usually hailed as the father of modern pneumatic chemistry.[27] While Boyle's famed *New Experiments, Physico-Mechanical, Touching the Spring of the Air* (1660) set out to instill such cold facts as "the spring and weight of the air," they also limn a countering romance of atmosphere, one that rose in part from Boyle's own studiously literary representations and in part from the speech impediment that made him more aware of air's mediatory properties than of its "physico-mechanical" ones. Boyle's hesitation between these modalities binds the aristocratic chemist to his middle-class Puritan contemporary, John Milton. When read as a spiritually ordered study of atmosphere's emergence from pure air, *Paradise Lost* reveals its kinship with Boyle's figural method.

Chapter 3 turns to a debased version of Milton's epic, Alexander Pope's *The Rape of the Lock* (1712–14), whose setting almost exclusively in drawing room and boudoir fails to defend it from the plagues of wind, lightning, and "Clouds of titillating Dust." I derive *The Rape of the Lock*'s influential aesthetic program from contemporary complaints about bad air—including the pathologies of inspiration, such as asthma, from which a growing proportion of the English population seemed to suffer—and from emergent models of sociability that posited the human body as the producer of formal airs. Chapters 4 and 5 gauge the atmospheres that develop between the covers of, respectively, the most canonical and the most baffling of Defoe's fictions, *Robinson Crusoe* (1719) and *A Journal of the Plague Year* (1722). I identify the weather-centered journal that Crusoe keeps while cast away with contemporary weather writing, then look at the "Shapes and Figures, Representations and Appearances" hallucinated collectively and individually throughout Defoe's *Journal.* These bind the *Journal* to the apparition narratives that flooded the English book market both in the days of the historical plague and in Defoe's own. I find such narratives to have mutually shaped contemporary experience of aerial figures that range from airborne microbes to specters to the spoken word.

Meanwhile, Enlightenment reconceptualizations of what the seventeenth-century demonologist Joseph Glanvill called "the *Immaterial Hemisphere*" demanded some reckoning with air's supernatural potential.[28] Prompted perhaps by its spirit of wit, I find midcentury comic fiction so to have reckoned. Chapters 6 and 7 approach Henry Fielding's *Tom Jones* (1749) and Charlotte Lennox's *The Female Quixote* (1752) as enlightened redactions of nonempirical aerographies of the period. Precisely as such redactions, these novels developed a modern paradigm of what we call ambience. Lennox's Lady Arabella is figured as a medium diffused through what Fielding, with respect to Lennox's novel, described as "the Coldness of the Narration." Fielding's own

Benjamin Partridge unites the "air" of his masterpiece *Tom Jones* with the air created in the astrological almanacs to be found everywhere in the period.

Chapters 8 and 9 bring Joseph Priestley's aerial investigations of the 1770s and 1780s into conversation with Ann Radcliffe's groundbreaking gothic romance, *The Mysteries of Udolpho* (1794). I treat Priestley as a linguist as well as a pneumatic chemist. His model of the sublime and his sense of literary language as a system of differences corresponding to the air lent Radcliffe the equipment necessary both to produce and to probe the atmospheric properties of the page. Radcliffe's writing in its turn taught readers such as Jane Austen's Catherine Morland to believe that purely graphic forms can chill the blood, tingle the spine, and make the "hair stan[d] on end the whole time."[29] Such expressions evoke the spring of the figural into the literal—of factitious object into influential environment—that makes literary atmosphere whatever it is that it is. English air studies of the long eighteenth century both analyzed that spring and reproduced it, thereby making new worlds fit for human habitation. So, despite its later reputation as a kind of coercive obscurity, my "atmosphere" turns out to be no enemy of the Enlightenment project, instead bearing witness to that project's creative, inclusive, and autocritical potential. It is because its visibly shifting shapes most fully realize that potential that, at the end of the day, this particular book believes in fiction.

1

Rounds of Air

AIR. Any thing light or uncertain.

SAMUEL JOHNSON, *A Dictionary of the English Language* (1755)

As eighteenth-century writers often remarked, the atmosphere of the British Isles is novel at every turn. "In many parts of the world, wet and dry are regularly expected at certain periods," Samuel Johnson's *Idler* complained in 1758, "but in our island, every man goes to sleep, unable to guess whether he shall behold in the morning, a bright or cloudy atmosphere."[1] Such volatility marks a moment in time as well as a point in space, since in Johnson's lifetime Britain passed out of the so-called Little Ice Age. Floods alternated with droughts, hard frosts, and apocalyptic storms; temperatures fluctuated so wildly that consecutive summers could yield the hottest and coldest temperatures on record. In the late seventeenth century, dust storms in East Anglia brought John Evelyn to mind of "sands in the Deserts of Libya"; yet on several occasions between 1690 and 1728, Eskimos kayaked up Scotland's river Don.[2] Johnson's words thus seem to domesticate the unsettling sense that British insular "atmosphere" is itself only insofar as it never quite is.

Air should be a more stable matter. But what, come to think of it, *is* air? The question teased the material- and mechanical-minded of the day; the definition of air that we find in Johnson's *Dictionary of the English Language* (1755) tells us why it did. "AIR," begins the *Dictionary*, confidently enough:

> AIR. n.s. [*air*, Fr. *aer*, Lat.] 1. The element encompassing the terraqueous globe. 2. The state of the air; or the air considered with regard to health. 3. Air in motion; a small gentle wind. 4. Scent; vapour. 5. Blast; pestilential vapour. 6. Any thing light or uncertain; that is as light as air. 7. The open weather; air unconfined. 8. Vent; utterance; emission into the air. 9. Publication; exposure to the publick view and knowledge. 10. Intelligence; information. This is not now in use. 11. Musick, whether light or serious; sound; air modulated. 12. Poetry; a song. 13. The mien, or manner, of the person; the look. 14. An affected

> or laboured manner or gesture; as, a lofty *air*, a gay *air*. 15. Appearance. 16. [In horsemanship.] *Airs* denote the artificial or practised motions of a managed horse.[3]

No one has seen air at any time. It has this in common with God, if not the other elements. And to this invisibility, Johnson's "AIR" owes its own uncertainty, its suspense among so many different meanings that, in the end, all we "see" in his definition are the words that work to make air apparent. Its head possibly turned by its descent from one of the flightier romance languages, Johnson's "AIR" signifies both matter and "an affected or laboured manner," no more or less an "element" (or a solid "state") than pure "appearance." In English an air might be an "utterance" or a piece of "musick" . . . but then again it might just be a "look." A transient, invisible "scent," it is also a lasting "publication" held up for "exposure to the publick view." Air, "light or serious," can be either a "small gentle wind" or "a blast." Even though "knowledge" is one thing that it conveys, "AIR" is, as Johnson's sixth definition admits, always "uncertain." "Any thing light or uncertain" is air.

The twenty-first-century scholar who sets out to ascertain much at all about the air that appears in eighteenth-century anglophone documents such as Johnson's *Dictionary* thus sets herself a task at least as foolhardy as any the lexicographer undertook. This is not just because air is everywhere and can be almost anything but because, "light and uncertain," it is so easily mistaken for nothing itself. That was often its fate in the literate culture that informed Johnson's Britain, a culture whose certainties apparently depended on its skepticism of anything that did not take visibly substantial form. Those who treated uncertain air as a material body subject to empirical analysis came in for ridicule from figures as various as Shadwell, Hogarth and Swift; of course, those who treated it as an immaterial one reaped their share of scorn, often at the same hands. A burgeoning literary marketplace supported both schools of derision. But here came an irony. Balanced just tenably between matter and not matter, air mimicked the material bought and sold in that marketplace—the writing, visible purveyor of things invisible, that had become a cutting-edge technology of national identity, cultural authority, and common knowledge. To make matters yet more interesting, as it found itself dispersed far and wide via the novel medium of print, writing in English set out to emulate, capture, contain, displace, and (at least notionally) replace the air. "Intelligence; information," Johnson writes, citing the tenth of air's sixteen possible meanings. "This is not now in use."

Johnson, it is well known, arrived at all of his definitions of English words by reading, taking notes, and vetting and arranging marks made by earlier

English writers; the irreducibly literary nature of this undertaking is what makes his *Dictionary* of a piece with the building of the English literary canon through such works as the comparably massive *Lives of the English Poets* (1779–81) and gives it ground common with both contemporary natural philosophy and the literature of the supernatural. In concert with both of these traditions, Johnson's *Dictionary* reached through the dense textual medium in order to "lash the wind"; its aim was to make at least one kind of air—the kind shaped and traded in English speech—apparent in the visible and material form of letters. "That fine matter" (as Johnson calls it) only laughs at this ambition's pretensions, its airs.[4]

Yet there is a twist. Haunted by the ephemerality of all sublunary things, including English and including air, Johnson's lugubrious preface to the *Dictionary* never cracks a smile. At the end of it, the lexicographer puts down his pen and sinks into "the gloom of solitude," resigning himself with "frigid tranquillity" to the wind of "empty sounds" that he has learned he cannot lash—that, if anything, lashes him (Johnson, Preface to *Dictionary*, 113). He is convinced that (by definition) he has done nothing of significance. Yet aside from *what* Johnson's preface says about the relationship between thin air and thick letters, its melancholy mood differently, if less directly, registers "uncertain" air. His contemporary George Cheyne associated the notorious "Moisture" and "Variableness" of insular air with the "Lowness of Spirits" that constituted the "ENGLISH MALADY."[5] Ill-defined and unknowable through conventions of reference, the shifting atmosphere in which English speech took shape permeates and yet seems to rise from Johnson's own melancholy writing. So is that writing informed by the very air that, at the referential level, has so obviously escaped it. Over time, it has been his air of frigid tranquility that has allowed Johnson's readers not just to analyze his hefty work of reference but to sense its presence. So does Raymond Williams register the "air of massive impersonality" that permeates Johnson's lexicography, and Freya Johnston note his "fear that what he has written will be received in a severely cold spirit."[6]

Almost against its will, Johnson's *Dictionary* suggests that to communicate through or in English is to engage with more than one kind of air and to do so in more ways than one. Still, in a last-ditch effort to take the guesswork out of "AIR," Johnson turns to the hymnodist Isaac Watts and his *Logick; or the Right Use of Reason in the Enquiry after Truth* (1725). No help, alas, is at hand: "If I were to tell what I mean by the Word *Air*," Watts wrote, and Johnson echoes under his first stab at defining air, "I might say, it is that thin Matter which we breathe in and breathe out continually; or it is that fluid Body in which the Birds fly a little above the Earth; or it is that invisible Matter which fills all Places near the Earth, or which immediately incompasses [*sic*] the Globe of Earth and Water." Watts already knew that "Logick" is defied by a "Matter" as "fluid," as "thin," and as "invisible" as air. So how do we know whether air is "matter" at all? In the case of both Johnson's *Dictionary* and Watts's *Logick* (in its eighth edition by 1745), it seems to become matter according to the variable "manner" in which it is articulated and received. Hence Watts declined to say what air was, coyly tendering only what he might have said "if I were to tell what I mean by the Word *Air*." In fact, in Watts's *Logick* the definition that Johnson's *Dictionary* uses to stabilize air appears merely as a way to show that "a Definition of a Name [is] only a Declaration in what Sense the Word is used."[7]

One moral, at least, seems clear: instead of asking what air is, these two writers propose, we may be better served by asking how we say anything about it at all. Because "the 'spirit' is afflicted with the curse of being burdened with matter," Karl Marx was to generalize, it "makes its appearance in the form of agitated layers of air, sounds, in short of language."[8] There is no subject matter that makes the "sounds," or indeed the rounds, forms and postures of human articulation more apparent than air does; in Watts's words, as echoed by Johnson, "that thin Matter" cannot be isolated from the schemes of those compelled to "breathe [it] in and breathe [it] out continually." Air can itself be grasped only peripherally, as it is transferred in specific languages from one such dependent to another. As Marx's example shows, this is scarcely a perception unique to English writers of Johnson's day. Aristotle's *Meteorologica* (350 BCE) begins with the confession that "our first difficulty concerns what we call the air"; in its quest for information about natural phenomena from clouds to comets, his book acknowledges air's unique, necessary, and difficult implication with "what we call it."[9] The Stagirite even implied a certain reciprocity between the way one would speak about the air (literally what one would call it) and what one holds it to be, which was, in Aristotle's view, a composition of the exhalations, hot or cold, moist or arid, that rise from particular patches of earth (Aristotle, *Meteorologica*, 29). To "see" air as made of exhalations is in turn to see that vocal expression must literally participate in its composition. When voice is transferred to writing (and especially when it seems to rise from it, and even just when we say it does), what transpires is best summarized in the parlance of postmodern culture's ethereal Facebook: it's complicated.

Air's relationship to the written word was especially complicated in Johnson's insular Britain, within whose English writing culture both modern meteorology and the modern novel took shape. Indeed, the complicity between these two nascent genres is surprising and, like the word "atmosphere" itself, can be traced back to the middle of the seventeenth century. During the English Civil War, Puritans and Royalists both floated competitive readings of atmospheric "appearances" from comets and cloud banks to miasmas and the northern lights. The air studies later sanctioned by the Royal Society thus grew out of literary evidence that visible aerial events support rival interpretations. Reading the skies became in itself a reflexive quasi-literary practice, at least insofar as suspense among competing systems of meaning was built into it. In an evocative study of "romantic weather," Arden Reed captures this sense of being in weather as being adrift in a system of signs that are "unstable, unpredictable, indeterminate," the preformal formalism of "something evermore about to be." Following Michel Serres, however, Reed deems a Newtonian eighteenth century to have been preoccupied with the "erasure" of this aspect of meteorology; "the Enlightenment response" to mutable air was apparently to bring it indoors as an object of study so as to derive uniform principles in "a visual form of reconciliation." In Reed's own view, it was then for the romantic poets, in effect, to unsee air.[10]

Not to put the romantics (or the romanticists) out of a job, but by the end of the eighteenth century many English aerial philosophers had nursed a deep sense of writing as a medium of reflexive investigation continuous with its aeriform object and indeed animated by it. True, representative figures such as John Pointer and Bernard Annely gave their books unimaginative titles, such as *A Rational Account of the Weather* (1738) and *A Theory of the Winds* (1729). But they also published them self-consciously in a medium whose appeal to the analytic eye was frankly disingenuous. For example, John Mills's *Essay on the Weather* (1770) voiced a commonplace: "The only model by which the changes of the wind can be traced with precision, undoubtedly is to keep regular registers of the weather, and mark every appearance in the heavens of the earth."[11] The "regular registe[r]" that Mills stereotypically envisions is easily seen as a fictional and experimental form within typographical possibility, one that produced a graphic cloud through which the environment itself was seen, if not precisely by the physical eye. Mills's own prescription for "regular registers" is conspicuously underwritten by the language of art. Explicit frames interacted with the aerial objects that they organized so as to delineate—and produce—a novel and common literary environment, one with its own claims to immediacy and its own rules of order.

These claims and rules turn out to have been more coherent with the aspirations of fiction than they were with those of poetry.[12] This is because eighteenth-century writers developed a theory and practice of fiction that held it to be an act of conspicuous pretense, one that incorporates the presumption of distance into its apparent claims to immediacy and retains the language of reference as pure form.[13] To eyes trained to make their way through the printed page according to such protocols, the skies themselves were legible as just such a fiction. "Sights have been seen, even in England," claims the anonymous *Wonderful History of all the Storms . . . that Have Happened in England* (1704), "Apparitions of fiery Armies, Castles, Towns, Monsters, and many other wonderful and dreadful Sights in the Air." Such reports, of course, potentially store the marvelous and the sublime in the air. But our *Wonderful History* actually looks another way. It is more interested in speculating that the splendors of the skies *may* be the projections of "certain Glasses and Instruments." Then again, they may "proceed from the Reflections of things on Earth, . . . being a mixture of Forms, as if a History was confusedly painted, and, the reflections of the Sun carrying the Images of Things on its refracted Beams to the Clouds, they impress themselves there." Here the analogy between a confusedly painted history and cloudy apparitions is less important than the "as if" that makes them both dependent on the medium at hand for their perception.[14]

It is through an autocritical, explicitly self-constituting British literary history that some sense of this dependency has been at once transmitted and preserved. There has never been a writer more informed by (or about) what she called Britain's "ordinary library" than Virginia Woolf, and it was Woolf who got at the intricate relationship between literary experience and its atmospheric counterpart. In 1926 the modern woman of letters found herself adrift in a slow-moving biography of four eighteenth-century sisters and braced herself for the long haul with the reflection that we common readers must "remind ourselves that there is such a thing as atmosphere." Woolf identified such atmospheric reading—at once detached and immersed—with the "interminable experiment with gold shafts and blue shadows" that one conducts while simply gazing up at the sky.[15] Atmospheric reading, at least in Woolf's book, never pretends to pin stable signs on some posited referent. Nor is it merely a lyric interlude. It keeps disappearing into writing, and writing itself is construed both as an activity and as the atemporal physical form to which that activity gives rise.

"Weather is a literary specialty," Mark Twain thus very accurately wrote in the preface to an 1892 novel, *The American Claimant*, set in predictably atmospheric England. "And no untrained hand can turn out a good article of it."[16] Even the trained hand is challenged. Thomas Uzzell's best-selling 1923 creative writing manual, *Narrative Technique*, warns that "the atmosphere story" is "a field of literature attempted by but few." Why? While "it is neces-

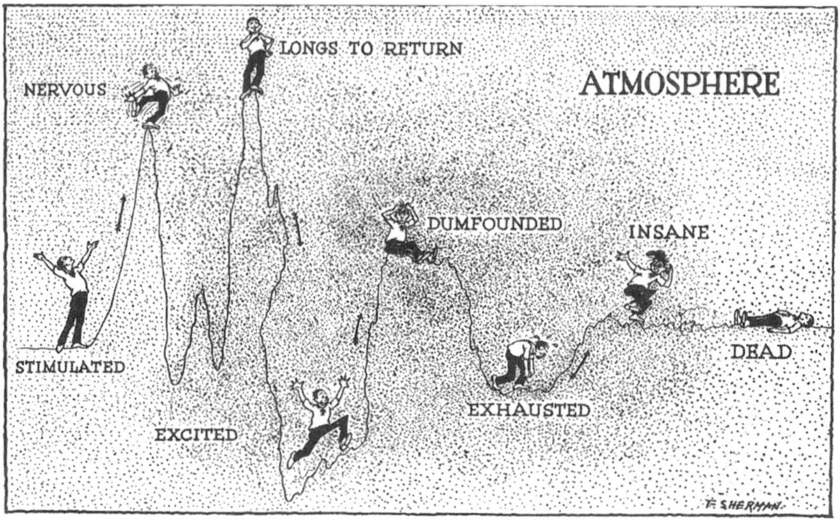

THE PATTERN OF THE ATMOSPHERE STORY

FIGURE 2. "The Pattern of the Atmosphere Story." From Thomas Uzzell, *Narrative Technique* (New York: Harcourt, Brace, 1923).

sary to make a powerful impression about the quality of the atmosphere," to this end "no mere description of a scene will suffice"; air must be registered through its effects on the "characters [that] have to be chosen to act for it." For Uzzell, these were both "human characters" and, as a diagram laying out "the pattern of the atmosphere story" made clear, graphic ones that both effected and represented the interplay of actions and effects (fig. 2).[17] And what was choosing the characters "to act for it"? It's not clear.

Uzzell's *Narrative Technique* takes us very far from the graphic atmospheres of the long eighteenth century. But it also returns us to them. In 1638 John Wilkins had drawn his readers into the possibility that "the Moon may be a World" literally through circles impressed on the page (fig. 3). One, for example, represented a sphere of "a contracted vapor" that even in the physical world would be obvious only through its own "appearance."[18] Another clergyman and natural philosopher, Joseph Glanvill, employed a strikingly similar graphic device to convey "the true Notion of a Spirit" (fig. 4). Here, a circle of letters was meant to help Glanvill's readers "suppose" themselves "environed with a Ring of Trumpeters." Glanvill's arrangement of graphic marks so as to convey the "circumsonant Clangor of these surrounding Trumpets" anticipates gothic romances of the later eighteenth century—works that made the sounds of other places and times present through the conspicuous agency of various forms of writing, from found manuscripts to alphabetical characters themselves.[19]

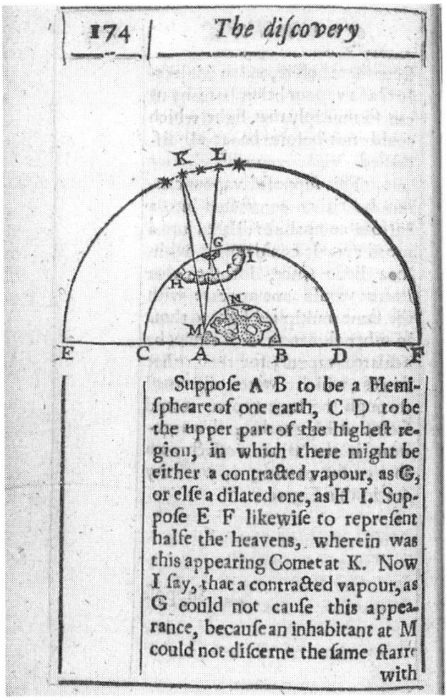

174 *The diſcovery*

Suppoſe A B to be a Hemi-ſpheare of one earth, C D to be the upper part of the higheſt region, in which there might be either a contracted vapour, as G, or elſe a dilated one, as H I. Suppoſe E F likewiſe to repreſent halfe the heavens, wherein was this appearing Comet at K. Now I ſay, that a contracted vapour, as G could not cauſe this appearance, becauſe an inhabitant at M could not diſcerne the ſame ſtarre with

FIGURE 3.
Detail from John Wilkins, *A Discourse concerning a New World, and Another Planet* (1638). Photograph: The William Andrews Clark Memorial Library, University of California, Los Angeles.

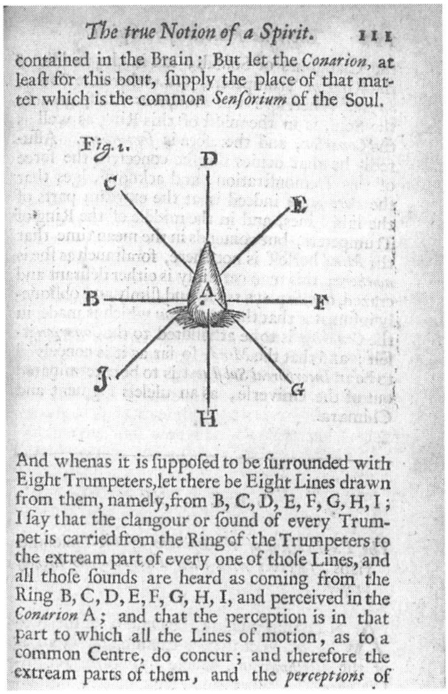

The true Notion of a Spirit. 111

contained in the Brain: But let the *Conarion*, at leaſt for this bout, ſupply the place of that matter which is the common *Senſorium* of the Soul.

And whenas it is ſuppoſed to be ſurrounded with Eight Trumpeters, let there be Eight Lines drawn from them, namely, from B, C, D, E, F, G, H, I; I ſay that the clangour or ſound of every Trumpet is carried from the Ring of the Trumpeters to the extream part of every one of thoſe Lines, and all thoſe ſounds are heard as coming from the Ring B, C, D, E, F, G, H, I, and perceived in the *Conarion* A; and that the perception is in that part to which all the Lines of motion, as to a common Centre, do concur; and therefore the extream parts of them, and the *perceptions* of

FIGURE 4.
Detail from Joseph Glanvill, *Saducismus Triumphatus* (1681). Photograph: The William Andrews Clark Memorial Library, University of California, Los Angeles.

Sciences of Air

Gothic romances not only anxiously track the movements of the air but visibly manufacture the impression of it. Technically they do so by way of the printed alphabet—via stereotype, taken in its most literal sense.[20] At the same time, gothic romances manufacture atmosphere thematically through such stereotypical fixtures as the rusting hinge, the darkening flesh, and the sputtering flame. These devices were often borrowed from English natural philosophies of the air as they had developed over the last 130 years. Two books have come to represent the end points of this period, Robert Boyle's *New Experiments, Physico-Mechanical, Touching the Spring of the Air* (1660) and Joseph Priestley's *Experiments and Observations on Different Kinds of Air* (1774). Boyle and Priestley also happen to be the subjects of two books that bracket the last twenty years in study *of* British air studies of the Enlightenment period: Steven Shapin and Simon Schaffer's massively influential *Leviathan and the Air Pump* (1989) and Steven Johnson's recent *The Invention of Air* (2008). Shapin and Schaffer's subject was Boyle; Johnson's is Priestley. The contrast between the books is instructive: Johnson's is popular, whereas Shapin and Schaffer's is academic; *The Invention of Air* is loose, synthetic, and even associative in its methodology, whereas *Leviathan and the Air Pump* contrives a hermetically sealed rhetorical analysis. As a writer, Johnson entirely trusts his chemical hero. Shapin and Schaffer are skeptical, to say the least: the famous final sentence of their exposé of Boyle's rhetoric of transparency—"Hobbes was right"—vindicates his most ruthless contemporary critic.[21]

Such differences correspond to differences between Boyle—the aristocratic "skeptical chymist" whose concept of experiment gave rise to an entire "literary technology" based on the spurious claims of virtual witnessing—and Priestley, the dissenting Unitarian minister who not only collaborated with everyone he met but eventually forsook the cramped quarters of an island off the coast of Europe for America's spacious skies.[22] But the fact that it was through writing in English that both Boyle and Priestley engaged the air helps if not to reconcile some of the apparent differences between them, then at least to recast those differences as exigencies of critical belief—as, indeed, conditions of interpretive community in our own day.

Between Shapin and Schaffer and Johnson lie several valuable studies of the ways that Britons living between Boyle and Priestley saw the air about them, mainly in the form of weather. Books by Mark Monmonier, Vladimir Janković, Jan Golinski, and Richard Hamblyn have focused on eighteenth-century meteorology not only in its provincial, pragmatic, traditional, and idiosyncratic iterations but also in its increasingly centralized institutional forms. All find that rational, quantitative conceptions of atmosphere that were centralized in the metropolis stabilized, and then established dominance, over the course of the eighteenth century. Both Janković and Golinski implicate the new institution of popular press coverage in this process. Both are ambivalent about a trend toward uniformity that seems to have suppressed indigenous, intimate, and authentic languages of the air—idiosyncratic weather journals, the homespun practical knowledge of farmers, and almanac traditions. Hamblyn, for his part, waxes downright elegiac; he presents Luke Howard's landmark 1803 essay *On the Modifications of Clouds* as having resuscitated, even as it codified, a lyric strain that British weather writing had submerged over the course of the eighteenth century.[23]

Here we sense air studies' natural kinship with poetry. But another, less recognized term that binds eighteenth-century fiction to contemporary natural philosophy is "history." Francis Bacon had hoped that one fruit of his experimental science might be a regular "history of the air"; he even wrote a tellingly brief *Historia ventorum* with the idea of including it in the magisterial *Instauratio Magna* (1620) that is so often seen to have set English empiricism on the path it would follow for the foreseeable future. In the end, though, Bacon foundered, for "the power and nature of the air . . . is entirely unknown" and the winds in which it is manifested are endlessly promiscuous, variable, and elusive. "They penetrate and bluster everywhere," Bacon wrote, "yet hide their head in clouds." Poets seemed to have more of a purchase on the wind than anyone else, and Bacon, after citing a few too many of them for his own comfort, at last puts down his pen. If one really wishes to experiment with it, he concludes, the best "approximation is the common games which depend on this wind."[24]

Defeat was not assured. On the other side of a print revolution sparked by civil war, a few game fellows of the newly minted Royal Society went back to the wind and made it a playground of sorts. Unlike Bacon, they willingly exploited air's ludic and elusive energy to realize writing as an environment in its own right and thus as an ironic foundation of knowledge. Edmund Bohun's 1671 *Discourse concerning the Origine and Properties of Wind* exulted in the fact that "the Body of the *Air*" is one of the "Appearances in nature, which may be rationally explicated severall ways" and made the plumbing of its "flatulent Emotions" an occasion for linguistic experiment. He giddily reminded his readers that if we want to know the air, we must be willing to be "tost to as many points of the Compasse, as *Columbus* or *Drake*; we must climb into the Regions of the Air."[25] In the next century, one of Bohun's nimblest and most prolific beneficiaries was the Sheffield physician Thomas Short, author of the popular "Short Syllabus of the Air." From its title forward, Short's two-volume *General Chronological History of the Air, Weather, Seasons, Meteors &c in Sundry Places and at Different Times* (1749) confronts the irony of its own task as it reins in a dizzying multiplicity of aerial phenomena by means of Enlightenment techniques of generality, chronology, and historical narration. Here, the act of compiling such a narration is a matter as much of amalgamating "sundry" pieces of writing as of coordinating diverse measurements of air's erratic performances at different places and times. Short's *History* collects "scraps of Histories [that] lay scattered in a vast multitude of Authors of different Designs and Professions," "Thousands of Volumes of Histories of the Effects of Different Seasons, Weather, and Constitutions of the Air." These, Short promises, can be organized into knowledge by an "exact judicious Review and tedious Rumination, . . . drawing from the Whole, such Deduction, Inference, Aphorisms or Conclusions, as might be of general Service" so as to predict "the next Season or Constitution."[26]

This self-consciously literary approach to unpredictable air builds a strong analogy between the disparate and heterogeneous documents through which it works and the "different . . . Constitutions of the Air" to be seen through them. More striking, though, Short's *History* discernibly takes shape within a nonhistorical environment that consists of nothing but mediating forms. Short's own writing less abstracts something from these forms than takes them for its own condition. For example, the details of Short's apparently historical account of the violent winds that ravaged England in the winter of 1709 come from "a short Abstract" from the journal of one conveniently named Reverend Say. Say "says he began his Journal . . . [so] as to Contradict some common, groundless, and superstitious Observations" (Short, *History*, 1:491). By the time we get to Short, though, we find ourselves very far off the "groun[d]" that Say ostensibly sought to establish; intervening are, in short, a history, an abstract, and a journal. Short neither hides this nor lets it get him down: it is part of his history of the air, and the transfer of information creates the frame through which air's own constitution is seemingly glimpsed.

In other British writing about the air, the air itself seems to supply such a frame. Edward Barlow's *Meteorological Essays* (1715) see the air ("the subject Matter of the Wind") by seeing through it: "In *Winter*, when the Air is chill with Cold, the Scene is chang'd."[27] Registering a dry spell, Barlow both sees and says that "the suppress'd Air began, as it were, to reel and stagger To and Fro under the Burthen; plainly appeared by the various and irregular Motions of some *wandring Clouds*, which the Barometer also resented [*sic*] by *Rising* and *Falling*" (Barlow, *Meteorological Essays*, 50). Barlow's "as it were," like the "as if" earlier noted, subtly introduces the visible agency of the writer, indeed of writ*ing*, into the picture of air and makes that part of its knowledge—knowledge that can thus turn back on the writing itself, rendering an apparent medium of information its legitimate, and uncannily vivacious, subject.

Historically speaking, it took air longer than earth, fire, or water to present itself as the element it still appeared to be at the turn of the eighteenth century. The ancient Greeks at first regarded air as a form of water, and early and medieval Christian writing knows it primarily both as a metaphor for spirit (*pneuma*) and as its medium. After Boyle, however, people began to behave as if they could see air, and hence they seemed to themselves to know more about it. They isolated it, measured it, manipulated it, and broke it down. They tried to decide what might be specifically English about the kind they breathed. They treated it as if it were a thing. Above all, they wrote about it. And yet they also depended on it, feared it, and made fun of those who did all of these things. Air slipped through all of these outstretched fingers: it was everywhere and nowhere, ultimately defying the ocular protocols that had come to define legitimate knowledge in the wake of the so-called scientific revolution. Partly through books, English readers (and the writers they often became) grew increasingly convinced that they were involved in air's manufacture. This only made it more distressing: "Since the Invention of firing so much Gun-powder," wrote Pointer anxiously in 1738, "I cannot but think the elements are much altered."[28]

Steven Johnson wonders why air itself should so suddenly have come to everyone's attention. He decides that it did so because the scientific revolution produced a technology for noticing it, a whole magazine of instruments from the air pump to the barometer to the thermometer: "In a way," Johnson muses, "the air pump had enabled the entire field of pneumatic chemistry in the seventeenth century by showing, indirectly, that there was something to study in the first place" (Johnson, *Invention of Air*, 67). But in what way? Many more people knew air through books than via Robert Boyle's celebrated air pump. The rise of air as an object of uneasy knowledge thus more meaningfully coincides with that of the ubiquitous print medium and with new ways of seeing "indirectly." This shift made it possible to see air as another such medium through the contiguous and not dissimilar one of writing. At the same time, writing's atemporal dimension becomes apparent as a condition of experience (thus, in Lockean terms, of knowledge) in its own right.

Patronized early on by the Royal Society, barometers, thermometers, and baroscopes are naturally ubiquitous in atmospheric writing. But their presence there is equivocal and subject to the kind of seeing indirectly through a common medium that making sense of print demands. In Barlow's *Meteorological Essays*, for instance, the barometer makes an appearance, but it isn't anything that makes air visible, even metaphorically. Rather, both air and the barometer register the shaping figural activity of the print medium. Barlow surveys the drought of 1710, through whose ominously persistent "Serenity" we find "the *Weather-Glass* keeping steadfast in its utmost Pitch . . . to shew that the Drought of the weather was then at a stand." Our author eventually conjures a baroscope "to illustrate all that I have said." This device—"a kind of Scale which weighs Air against Quicksilver"—also "shows the Ponderation of the Atmosphere at all times, and consequently how our Theory . . . corresponds with the rise and fall of the Quicksilver" (Barlow, *Meteorological Essays*, 92). In both of these instances, the barometer primarily makes visible what's been "said" and something about a "Theory."

Words alone could sometimes seem to measure air. The so-called Shepherd of Banbury was a staple of the bustling midcentury trade in British weather writing, his folksy voice the marketable production of any number of hack writers, some anonymous and others not. Drawn supposedly both from the "common Sayings of our Country People, and from old *English* Books of Husbandry," the shepherd's "rational" aphorisms concerning "the nature of Wind, Rain, Snow, &c." will trump your thermometer, hygrometer, or barometer every time.[29] The thermometer, after all, "measures exactly the Degrees of Heat, but the Air must be hot to such a Degree before it is discerned by this Instrument." As for the barometer, its quicksilver certainly "expresses the Alterations in [air's] Weight with wonderful Nicety, but then those Alterations are the Cause of this." And the hygrometer? It "measures Alterations but change must occur before these Alterations are visible" (Claridge, *Shepherd*, iv).

By contrast, our shepherd's observations are of quite another "Nature," one that claims more stably to convey that of air itself. This "Nature" is itself fully compatible with literary art: the shepherd is a pastoral fixture after all, and Virgil's eclogues are often cited. In the modern British case, the Shepherd of Banbury's adages allow words to *become* weather. In turn, analysis can double as a canny kind of meteorology, one that admits its own pretensions. Take, for example, the shepherd's words on one weather phenomenon: "MISTS: If they rise in low Ground and soon vanish > Fair Weather" (Claridge, *Shepherd*, 9). "This," pronounced one of his antiquarian "editors," "is a sure Sign, and very well-expressed, that is, very cleerly [*sic*]." But what is "this"? Technically speaking, it is the adage itself, which has been placed as an artifact at the top of the page.

The words here *are* the "MIST," the mediating form we see through and also see; for this reason they are also "very clee[r]." The shepherd's supposed editor captures the print medium's potential to constitute both a climate and an instrument of knowledge. He expects skepticism: "Why should we depend on any such Observations?" he rhetorically demands, vindicating not just the local knowledge of the illiterate shepherd but the authenticity of observations formulated as such (Claridge, *Shepherd*, 22). That the shepherd is unlettered is vital to the project of the medium that constructs his observations. The *Remarks on the Shepherd of Banbury's Rules* that John Mills appended to his 1770 *Essay on the Weather* note that "who the Shepherd of Banbury was, we know not, nor indeed have we any proof that the rules called his were penned by a real shepherd." These "rules" could still raise to legitimacy the "remarks of illiterate country people" that had been "despised or neglected" by natural philosophy.[30] Here they are restored to the status of self-evident facts. Likewise, the anonymous *Calendar of the Weather for the Year 1781* would concede that "what the poets on georgical subjects, early by repetition, stamped on the memory, have said, must of course make for a lasting impression." Such impressions, brought out in print, constituted their own forms of environmental knowledge, even as they visibly informed the environment that that knowledge seems to elucidate.[31]

Against this background, the baroscope, the barometer, and the weather glass all appear as aspects of the same instrument of ostensible vision, differently named. They materialize out of writing to make its techniques of making visible themselves apparent. Critical attention in our own day has been trained on the quasi-magical properties of their vital component, mercury; Terry Castle has illuminated its mysterious kinship with mythologies of communication traceable back to the Greek Hermes at least.[32] But, as it was back in Wilkins's day, mercury is also a "secret and swift Messenger" that immediately registers the modern lexical medium. So natural philosophy's "transactions" with the air might be imagined in terms other than those of the rise of an abstracted system. So imagining, we cannot accuse those same transactions of obfuscatory transparency, the designs of ideology. To the contrary, what we find in these treatments of the air might be described as transparent obfuscation. It is the spiritually and intellectually enlightening reminder that we see through a medium, one whose figures are inextricably mixed with our own ways of figuring things out.

For the record, the seemingly homely Shepherd of Banbury trailed clouds of glorious association with classical writing. Much eighteenth-century writing about the air did the same, not least because it had been a favorite topic in a range of key Greek and Roman texts, from Aristotle's *Meteorologica* to Lucretius's *De rerum natura.* It's clear, for example, how Hippocrates's *Waters, Airs, and Places* (400 BCE) shaped eighteenth-century *ideas* about air in the form of climate, and much has been written about the reception of his correlations of the states of bodies with those of the air in which they found themselves.[33] The Hippocratic model was still in place in Johnson's day and supported contemporary maps of the world's people that classified them characterologically, physiologically, and hygienically according to their climates; it also implied strict differences among climates themselves. John Arbuthnot posited that England's own perpetually unstable barometric pressure stimulated bodily fibers in such a way as to produce a people of exceptional energy, courage, and volatility (though they were nothing to the even more barometrically challenged Scots).[34] Hippocrates also opened an approach to the air through the human body that helps to explain why so many eighteenth-century analysts of the air were also physicians.[35]

The Hippocratic *idea* clearly broke down and scattered over the eighteenth century. Even climates that to us appear to be as radically different as those of India and England were not necessarily so perceived by 1800; the same airs were eventually seen to circulate in both, and differences of temperature, humidity, and even viscosity had to do with dispositions and arrangements of the air in relation to such topographical features as mountains or swamps:

the air on an Indian mountain and the air on an American mountain were at last conceived to be more similar than the air on an Indian mountain and the air in an Indian swamp.[36] At home, meanwhile, David Hume's essay "Of National Characters" (1742) set out to vanquish the climate thesis with the sword of the secularization thesis, declaring that native climate could hardly be said to shape collective character because that character is itself subject to dramatic change. "Our ancestors, a few years ago," Hume wrote, "were sunk in the most abject superstition, last century they were inflamed with the most furious enthusiasm, and are now settled into the most cool indifference with regard to religious matters, that is to be found in any nation of the world."[37]

What common character, traceable to the air its inhabitants breathed, could a modern nation possibly boast? Yet even as Hippocrates's deterministic and differentiated air begins to dissipate, Hume retains the Hippocratic *mode* as metaphor within the enlightened medium of his own written characters. Hippocrates's air matters, and persists, not because of *what* he said but through the figure of communication between interiors and exteriors that writing about the air developed. This figure, moreover, survived in and through books—books busy doing their own part to make different places seem the same insofar as the worldwide dispersal of the English print medium mediated the perception that they were not so different after all. In his skeptical philosophy, rotund Hume meanwhile loved to break down "the air of a paradox," an air, in his view, upheld by the complementary "dispositions" of philosophers who "greedily embrac'd" it and by readers thereby seized with "surprise and admiration."[38] In this manner, Hume preserved air as both a medium and an example of the mind's transition from speculation to experience. Small wonder that his famous thought experiments should ask the reader who would grasp abstractions such as the relationship between space and time to "suppose . . . a man to be supported in the air, and to be softly conveyed along by some invisible power" (Hume, *Treatise*, 1.2.5, 40) or to remember that "when I hold up my hand before me, and spread my fingers, they are separated as perfectly by the blue colour of the firmament, as they cou'd be by any visible object, which I cou'd place betwixt them" (Hume, *Treatise*, 1.2.5, 41).

A second classical source of eighteenth-century ideas about the air was Seneca. Golinski points out that modern meteorology doubled as moral philosophy to the extent that presumably it yielded stabilizing, or at least probable, knowledge conducive to the stoic value of *ataraxia*. In *Naturales Quaestiones* (63 CE) the Roman moralist proposed that all atmospheric "phenomena are involved in other phenomena" (*ali aliis ita implexa sint*) but held that these phenomena, even in their first appearance, are themselves accordingly read through the screen of air. In particular, Seneca invited his reader to consider the rings of light that seem to circle certain celestial bodies (the Greek *halo*, the Latin *corona*). Such rings appear when sunlight strikes the lower air: "Moisture and air, and in fact anything which is shaped by a blow, is forced into the same shape as that which shaped it. Now all light is round. Therefore, air also struck by light will go into this round formation." So although a corona might *seem* to encircle a faraway star, in fact "an image of this sort is formed very close to the earth," beyond whose immediate atmosphere "nothing of this sort can possibly happen because only tenuous ether exists there. For shapes are usually impressed only on dense and thick matter."[39] Furthermore, the atmospheric phenomena for which Seneca accounted are often themselves written as images—*simulacrae*, *ostensae*—whose status as replicas does not mean that they are not also things. This duality makes Senecan simulacra immediately relevant to the graphic signs that are themselves impressed on a "dense and thick" background, all of which leaves it hard to conceive of Senecan "*aer*," let alone its contents, without his writing about them. Seneca himself appears to have realized as much, for his *Naturales Quaestiones* are posed by way of the prior writings of Pliny and Aristotle and mediated through the cited poetry of Ovid and Virgil, among others.

Lucretius's *De rerum natura* (50 BCE) was yet a third conduit between classical and modern conceptions of the air. Its importance for the mechanistic and corpuscularian models of the new science has been evident for quite some time. Air for Lucretius consists in streaming and radiant particles that hover somewhere between materiality and immateriality. Daniel Tiffany's recent study of modern poetry shows how this suspension at once animated and automated the lyric mode, and Michel Serres exploited the rifts within Lucretius's universe to open a critique of the modern physics that relies on it.[40] Besides providing a treacherous foundation of modern knowledge and a guarantor of the possibility of lyric expression, however, Lucretius's air is also a dynamic figure of mediation. Objects cast off simulacra that travel through air but also *are* air; these are impressed on percipients who in turn cannot be said strictly to be seeing them. In the later seventeenth century, this model of communication gave Lucretius's air an animating affinity with the specific matter of the literary medium. This affinity could illuminate that medium as a material (if also figural) realm of participatory knowledge.[41]

In eighteenth-century Britain, literate attitudes toward Lucretius were ambivalent at best. Joseph Addison appears to have led the charge against that Roman's materialism, considered as a truth claim. "He was so pleased with the Matter of Fact which he could not have the Confidence to destroy, that he was forced to account for it by one of the most absurd philosophical Notions that ever was started," Addison's eidolon, Mr. Spectator, scoffed. The offending notion? "That the Surface of all Bodies are perpetually flying off from their respective Bodies, one after another, and that these Surfaces or thin Cases that included each other whilst they were linked in the Body like the Coats of an Onion are sometimes entire when they are separated from it, by which means are often beheld the Shapes and Shadow of Persons who are dead or absent."[42] For all his contempt for Lucretius's "Notion," Mr. Spectator here at once glimpses and conveys the potential of his air to figure new ways of transferring meaning, not least the one that has brought his own voice into apparent being. Lucretius's "philosophical Notion" seems "absurd" not because its content is not believable but because its form is all too easily so. That "Notion" thus at once resonates with and reveals the trick whereby Mr. Spectator's disembodied voice seems to—indeed does—materialize within a newly ubiquitous and particulated print medium. The new communicative practice latent in Lucretian air is potentially the realm of apparition that Hobbes once dismissed as the "Kingdom of Darknesse."

Other Powers of the Air

Mr. Spectator vets Lucretius's aerial "Notion" in an essay whose official topic is the probability of apparitions. This probability is tested less propositionally than atmospherically: the essay begins with Mr. Spectator detained "among Ruins of an old Abbey," listening to the cawing of rooks while falling night "heighten[s] the Horrors of the Place" (Addison, *Spectator*, 1:455). The question of whether it is possible to "behold the Shapes and Shadows of Persons who are dead or absent"—who are not, strictly speaking, there—is left up in the air. If anything, the periodical essay conduces to such a belief as it registers the effects of an atmosphere on an "absent" speaker, thereby conveying that speaker's shape and shadow to the modern reader.

Like the natural philosophy of its time, contemporary literature of the occult reflexively engaged the communicative element of air. Of course, one of the eighteenth century's major projects was discriminating between nature and the supernatural; the fact that one of these categories is now a proper common noun whereas the other hovers somewhere between a noun and an adjective tells us how things went. From a historical perspective, to acknowledge that a line had to be drawn between the explicable and the inexplicable is to acknowledge that in the period itself no such line clearly existed. Other geometrical figures—usually spheres and circles—better describe (i.e., encircle) their relationship then. Air has a special place in this conundrum, because of all objects of natural philosophy, it is (light excepted) the one that is the most difficult to detach from the marvelous, the occult, the sublime, and, indeed, the divine. As the Cambridge Platonist Henry More had put it in *The Immortality of the Soul* (1662), "Daemons and Separate Souls" find their "natural abode" in the "Air" and share an "Aetherial Congruity" with it.[43]

Demons, though, are fundamentally communicative entities. So is air. Its primary assignment—to convey—involves it with language in the manner that bedeviled Samuel Johnson. This, along with its invisibility, is why air was linked with the shadier side of what we now call the supernatural, whereas light retained its association with God and divinity (often repurposed as reason). A standard view in both Protestant and Roman Catholic theology was that Satan himself had fallen from heaven's light not to hell but into the lower atmosphere, where he lurks still as the source and agent of its most mystifying forms.[44] In Defoe's *Political History of the Devil* (1726), he is "the Prince of the Powers of the Air," though the novelist tails him principally through such books as Milton's *Paradise Lost*.

Works such as Defoe's *Political History* may be read as if they participated in the same sphere of inquiry that those of meteorology and chemistry did. Many writers comfortably approached the air from both perspectives. Boyle, the presumed father of the physico- mechanical model of air, funded investigations into the second sight and hobnobbed with the likes of the demonologist (and fellow of the Royal Society) Joseph Glanvill. But more important than the fact that many seemed to believe in air both as a material body fully explicable by the laws of nature and as the domain of apparitions and angels is the fact that both conceptions were conveyed through the same medium. Faith-based and folkloric conceptions of a populous, active air that did not necessarily obey the ordinary laws of space and time were often projected outside literature, left in the custody of the illiterate, and handed over to the second-sighted. But they were themselves grounded in literary transaction and derived their forms from the same habits of conviction that provided their content.

As chapter 5 considers, dozens of books were published on both sides of a lively debate about what Defoe called "the Reality of Apparitions." Elias Ashmole built up an occult library at Oxford, home also of the Invisible College that would become the Royal Society. The pragmatic John Evelyn wondered about the mysterious "mason word" that mystically conveyed sympathy from one man to another regardless of geographical distance, and Reginald Scot's *Discoverie of Witchcraft* (1584) was reprinted with the entirely credulous *Discourse of Devils and Spirits* (1695). Given so many equivocations, we need not belabor what anybody said about air's capacity to support supernatural phenomena or dwell on the occasional speculation that air might be irreducibly occult in its own right. In making common media visible, printed controversies about apparitions negotiated the terms of a shared investment in a communicative medium, ultimately projecting through it a common, reflexive, and remarkably durable sphere of belief.[45] As Steven Shapin has lately explained, "spirits" themselves were a notional hybrid blending immaterial and material accounts of reality in response to the later seventeenth-century demand for discursive compromise.[46]

When we switch our inquiry from the content of beliefs about the air to the question of what shape they took in a mixed environment, we find that these beliefs charted a sphere of experience uncompromised by the inevitability of mediating form. Occasionally this happened within a specifically ethnographic framework, and toward its appreciation no people of the air stand readier to assist us than the fairies. Indigenous fairy belief has been anatomized by Katharine Briggs, who, with respect to its upholders, tellingly figured it as "the air they breathed."[47] English-speaking believers are themselves believed to have fallen away over the course of the eighteenth century, leaving fairies to cling to the leaves of explicitly imaginative literature or to persist as the contemptible *sithe* of the Celtic fringe. In the *Secret Commonwealth of Fairies* (1692) by the Scottish Presbyterian Robert Kirk, "these siths or fairies" are "said to be off a middle nature betwixt man and Angell (as were daemons thought to be of old); of intelligent studious Spirits, and light changeable bodies (like those called Astrall) somewhat of the nature of a condens'd cloud, and best seen in twilight."[48] In Kirk's influential guide to what even then seemed to be a disbanding circle of local folk belief, fairies share the nature of the clouds that did survive enlightened scrutiny—that are, in fact, what enlightened belief sees.

Kirk slyly mentions that the fairies are also "said to have many pleasant Toyish Books. But the operation of these peeces only appears in som paroxysm of antic Corybantick jollity—as if ravish't and promted by a new Spirit entering into them at that instant, lighter and merrier than their own" (Kirk, *Secret Commonwealth*, 83). While it may be a conceit that tips Kirk's satiric hand, fairy reading recalls the Puritan poring over the pages of holy scripture even as it anticipates eighteenth-century satires on Methodists erotically invaded by the holy spirit; Kirk himself both translated the New Testament into Gaelic and mysteriously disappeared at the end of his life into what was rumored to be a fairy hill. For its part, Kirk's *Secret Commonwealth* less documents disappearing folk belief than condenses reflexive forms of perception and representation within a consolidated English literary practice. Kirk is careful to tether what is said of the fairies to their objective reality, reconciling both, however, only on the surface of his own pages. Indeed, "the Sight" itself, which presumably gave access to fairies, Kirk held to have been "modestly deduc'd" from "Holy Write" (Kirk, *Secret Commonwealth*, 102). This conception of the fairies both as clouds and as readers realizes what is "said to be" to make fairy lore a vital and encompassing reflection on the dynamics of belief in literate culture.

Here is where Kirk's *Secret Commonwealth* converges with books we would expect to be hostile to it. Robert Plot was a professor of chemistry at Oxford and a keeper of the Ashmolean; his natural histories of certain English counties systematically rewrite geographic spaces previously shaped through indigenous folk belief in terms that would be recognizable to the modern naturalist. But when it came to the supernatural populations with which local belief invested such places, it became clear that this was less to give them their true and natural contours than to transpose them into Plot's own book. Such books were in turn rendered conceptually habitable in hybrid notional forms that assimilated faerie signs to atmospheric ones. For example, in Plot's *Natural History of Staffordshire* (1686), that county is defined and conveyed partly through the so-called fairy circles of Staffordshire. These were grass impressions presumably left by the fairies because "all agre[e] (if to be believed) that their dances were always circular."[49] The natural historian measures the circles' diameter and finds them to correspond to the hoofprints of cattle that graze in circles. He eventually abandons the bovine hypothesis, however, and floats yet a third explanation of the fairy circles: they "must needs be the effects of Lightening, exploded from the Clouds most times in a circular manner . . . which though of a viscous sulphureous consistence, yet taking fire and violently breaking the *Cloud* wherein it was pent, must naturally expand it self every way obliquely" until it becomes a circle (Plot, *Natural History*, 15).

Lightning's "adventitious Sulphur" spreads outward like "a disease that creeps on" and "brushes the earth on the lower side, the upper part vanishing in the thin Air" (Plot, *Natural History*, 19). Figured in the expanding circle and uniquely fit to investigate the ways belief works, this new communicative form is one that Plot identifies with his own writing: "And thus much for these Circles, and enough too perhaps may the Reader say to break my promise so solemnly made in the first [paragraph] of the Chapter, *of not enlarging to the utmost compass of my theme*" (Plot, *Natural History*, 19). At this point, the circle of Plot's writing encompasses fairy circles, cow prints, and lightning, unifying literal and figural media within a novel airform. Here belief itself will be visible, if still common to all readers of his natural history. As a literary genre, such "History" thus juxtaposes two different pictures of reality in the manner that Dorothea von Mücke links to the "seduction of the occult" in European fiction of the later eighteenth century—a juxtaposition that in her supple reading requires the material presence of the print medium to sustain it.[50] Plot's expansive lightning, conveyed through his book, moreover anticipates Benjamin Franklin's later revision of lighting from a pointed and linear communiqué from God to miscreant humanity into a diffuse ambient phenomenon that is mediated into and through only coincidentally linear forms. It is significant, too, that Franklin often demonstrated the conduct of electricity through books: "We electrify, upon wax, in the dark a book that has a double line of gold round the cover."[51] His experiments often begin like that.

So Then

The absorption of fairies and their ilk into the literary chemistry of thin air gives rise to novel circles of belief and experience that are themselves figured in Plot's circles, Marx's rounded "sounds of air," Seneca's coronas, Wilkins's circles, Johnson's "element encompassing the terraqueous globe," and even the "gold round the cover" of Franklin's book. Spheres are literally written into atmosphere: they describe both how it is figured and how it is experienced. Atmospheric knowledge understands these to be the same thing: "If this sentence has a particular atmosphere, how am I to separate it from the sentence itself?" wondered Wittgenstein. "It would never have occurred to me to think the sentence had such an aura. . . . The picture of the special atmosphere forced itself upon me."[52]

Such a picture raises an obvious question about the role that belief plays in our experience of it. One of Johnson's successors, the unlucky lexicographer hired to track the history of the word "atmosphere" in the *Oxford English Dictionary*, found it "curious that the first mention of an *atmosphere* is in connexion with the Moon, now believed to have none."[53] John Wilkins's immediately encompassing element is not just analogous to the written medium that communicates it; it also makes visible the pretentious devices through which belief is at once conveyed and projected. This "orthographic sphere," as Joseph Priestley would call it in a one-off perspective-drawing manual, brings us to the threshold of a distinctively modern realm of literary experience. Apparently ambient, obviously phantasmic, and yet entirely available to the ordinary senses, this experiential mode may be conceptually tenable only through the "willing suspension of disbelief" that Coleridge would make a tenet of "poetical faith."[54] We can now see that an apparently romantic idea that seems to belong to the life story of writing also arises from the air that seems to circle us all as a result of such forms of belief.

Hence Leo Spitzer's justly classic essay on the concept of "ambiance" traced it through words as they involve themselves with experience: "The history of this word cannot be separated from that of *medium*," Spitzer held. Following the work of the Swedish linguist Karl Michaëlsson, he identified the mid-seventeenth century as a high point in this word's long association with "an inexpressible harmony between men, things, and situations."[55] Translated promiscuously as "ambiance" and "milieu," atmosphere's Greek cognate, *periechein*, expresses this harmony by actively marking the interpenetration of Logos and location. Here we find an affinity between medium, perception, and object that remains implicit in modern English words such as ambience and atmosphere. Any such word is what Spitzer evocatively terms a "'warm' abstraction"—an "abstraction," he clarifies, "which is visualized and which has not severed its ties with life, but remains organic and close to the bodily" (Spitzer, "Milieu and Ambiance," 187). Part of atmosphere's warmth, in Spitzer's view, lies in the fact that it is conceived not as deterministic but rather as protective, embracing, and even caressing. One is in the *periechein* as an embryo is in an egg or as a pit is in a peach. Because one is there at once knowingly and unknowingly, the idea, the word, and the thing are experienced as one and the same.

One possible expression, if not prototype, of Spitzer's *periechein* is the ancient amphitheater, itself a fusion of man-made space and natural *chora*. Equally significant, it is in the stage designs of the earlier seventeenth-century theater that Kenneth Olwig discovers the fluent interpenetration of material and ethereal registers.[56] That theater was famously displaced by—or perhaps into—Puritan print culture in the middle of the century. Spitzer accused the enlightened science that followed of chilling his warm abstraction and breaking the harmony it guarded. But if we consider that the literate culture to which Puritan textuality gave rise rather to have absorbed and rearranged the blended medium of the theater, then we can imagine that Spitzer's ambiance was creatively simulated in the material form of English writing throughout the long eighteenth century. This may have occurred most explicitly in what we now classify as works of the imagination. Nonetheless, it is also incipient in the works of natural philosophy charged with parsing the *periechein*. The next chapters put all of these works in one room. They have gone willingly, innocent of the different genres into which we order them now.

2

"Other Air": Boyle's Spring, Milton's Fall, and the Making of Literary Atmosphere

How shall we breathe in other air
Less pure, accustom'd to immortal fruits?
JOHN MILTON, *Paradise Lost* (1667–74)

When Adam and Eve lost Eden, they also lost "pure" air. No longer theirs were either Paradise's "gentle gales" (4.156) or the "smooth air," continuous with the "wide transpicuous air" (8.141) of heaven, that once carried divine voices straight to Adam's ear.[1] Adam and Eve lost a transparent and unified element along with their felicitous complicity with it, the sense they were at home there. In return they gained atmosphere as we know it. That, apparently, is what Milton's Satan brings with him when, "like a black mist low creeping" (9.179), he seeps through unfallen Paradise in search of some body to possess. He finds it in a serpent "with burnisht neck of verdant gold" (9.502). "Well stor'd with subtle wiles" (9.184), this gilt-edged medium seems to speak first to the eye; to the fallen apprehension, it thus resembles nothing so much as a book. Said book seduces woman into the linked beliefs that she can see what she hears, hear what she sees, and must, therefore, be experiencing something that is really there.

If Eve, as has occasionally been supposed, is the first reader, then the genre that she first consumes looks very much like modern realist fiction.[2] Before consuming anything, however, Eve must first let herself be led into an apparent world—one to which she must respond as if it were all about her. As for her modern readers, the cues we follow in order to seem to ourselves to inhabit the world that literature makes are ones that we have come to take for granted. We must read back through them to appreciate a great many features of Milton's novel epic: the sense, for example, in which his Eve, no less than his Eden, is said—and so appears—to have an "air." Before Satan seduces her, Eve's "every air / Of gesture or least action overaw[es]" him (9.458–60), upholding a see-through pretense of integrated and encompassing sensation.

Satan himself seems to experience Eve's "air" in this atmospheric sense: he behaves as if it surrounded him even as he appears but to be projecting it. How naturally Milton's narrator thus likens him to "one who, long in populous city pent, / Where houses thick and sewers annoy the air," now gets to "breath / . . . the smell of grain, or tedded grass, or kine, / Or dairy, each rural sight, each rural sound" (9.449–50).

Like Eve's "air," Satan's apparent ability to breathe "each rural sight, each rural sound" as if they were the same makes unique sense within a literary medium that by definition compounds the material and visual image of the letter with the ephemeral sound to which we have made it correspond. Notwithstanding his insistence that he communicates (and is communicated with) only through voice, Milton's epic narrator—a dictator, after all, and not a bard—depends on this same correspondence. He thus must "fit" his "audience" to maneuver within mixed literary matter, to absorb its synesthetic pretenses. This necessity supplies the most subtle and pervasive evidence that we have of *Paradise Lost*'s documentary presence as a fallen medium. Again and again, it is through its invocation of air—air fallen or unfallen, "other" or "pure"—that Milton's literary medium registers its own presence as both an obstruction and a condition of experience.

Milton treats such reflexivity as fallen at best, a symptom of loss. At worst, it is satanic. When Eve plunges into the sphere of undifferentiated impressions that Satan has projected (through the air), we find that "Nature first gave signs, imprest / On bird, beast, air, air suddenly eclips'd" (9.182–83). All these "signs" certify physical nature's divorce from divinity, its new commitment to matter as manifest to the ordinary senses, its insularity, and its potentially cruel duplicity. But of the three impressed signs (bird, beast, and air) who are themselves impressed on Milton's reader, it is "air, air suddenly eclips'd" that his epic narrator mentions twice. We have to assume that this is deliberate, unless, of course, in the act of dictation Milton merely tripped over his tongue, and his amanuensis, assuming intent, committed both airs to the page.[3] Either way, the rise of "murky air" (9.280) still attends that of the "imprest" medium of a written poem. Milton's narrator authorizes that poem by claiming that it first passed through the air, "brought nightly to my ear" (9.48). As this claim is delivered to the reader's eye, however, the visual signs that deliver it remind us that, henceforth, knowledge of the mediating body of the air will itself be mediated through *them*. Formerly employed as a prose pamphleteer immersed in the print culture of the Puritan commonwealth, Milton had some experience treating the written word as if it were air. His defense of a free press, *Areopagitica* (1644), not only pretends to be air ("a SPEECH") but wants its readers—while knowing they are not—to see "books, with dangerous frontispieces," as if they were the same as "the airs and madrigalls that whisper softnes in chambers."[4]

More than twenty years later, the angels who fill the pages of *Paradise Lost* would brandish swords, not pens, let alone moveable typefaces. But their aeriform message-bearing makes Milton's angels visible as hopeful sublimations of writing, tokens of its potential to transcend its own limitations. No less than letters aspire to do, Milton's angels instantaneously transmute the "spirituall" and the "corporeall."[5] The "plumes" that Raphael "shake[s]" (5:286) as he flies toward unfallen Adam playfully remind us that when Milton's contemporary, the grammatologist John Wilkins, considered "the swiftnesse of information" suddenly possible through widespread writing, he found that "amongst all created substances, there are not any of so swift a motion as Angels or Spirits."[6] Modern print's swiftness and seeming ubiquity renovated the aeriform angels of old as viable figures, at once proleptic and epiphenomenal, for the very characters of the alphabet. As late as 1696, George Hammond's *Angelographia* would still identify graphic signs with angelic messengers, and it was certainly in this spirit that Wilkins cited one cabalist who had "call[ed] each character by the name of spirits," albeit with the unhappy effect that "under this pretence," he managed "also to deliver some Diabolicall Magick."[7]

The defaced characters that compose *Paradise Lost* likewise afford equal "pretence" for demons and for angels. This indifference perversely approximates the unity of Eden's original air. In Milton's book, both fallen *and* unfallen air are mediated through a literary illusion of audition. Both confuse letter and figure, medium and message. The difference is that, after the fall, air becomes, primarily and in effect, a physical thing. As before the fall, it is still at once expressive and expressed. But as a material body foremost, unpredictable and autonomous, air now risks compromising the task of mediation that still falls to it. Ever threatening to close back on itself, it marries "sense" to "experience," commanding each to forsake all others. Milton's postlapsarian narrator could not feel the bind more unpleasantly or acutely. He complains of the "cold climate" that "damp[s] my intended wing / Deprest" (9.44–45) and doubts that his "argument" is "sufficient of itself to raise / [his] name" through that climate's oppressive, self-absorbing murk.

With this complaint, *Paradise Lost* ceases merely to invoke an "Argument" and its shortcomings: it now evokes the physical environment in which this poem was somehow composed. In its haste to claim Milton as a civil war writer, or to hear in his voice the last gasp of humanist poetics, critical tradition often forgets that it was Restoration London that provided the scene of his novel epic's transcription over the first half of the 1660s. Besides plague in 1665 and the Great Fire of 1666, that metropolis suffered what Milton's contemporary John Evelyn diagnosed as the more ordinary "Evill so epidemicall" of air pollution. Its source believed (erroneously, as it turns out) to be smoke from the industrial burning of sea coal, the London smog was a surefire dampener of the intended wing. Evelyn, for his part, aimed to ameliorate this ambient "Evill" and believed that he could best do so by way of the page. Anticipating Milton's Eden, Evelyn's 1661 treatise *Fumifugium; or the Inconveniencie of the AER . . . of London* recommended the "Dissipation" of inconvenient air by way of pastoral "Closures, or invironing Gardens." These Evelyn would have stocked with any flowers whose "ravishing Perfumes" were "aptest to tinge the *Aer* upon every gentle emission at a distance."[8]

To its seventeenth-century citizens, London's "inquinated" air also seemed to have impressed itself on other, less easily distanced vapors, namely, the chill, viscous air of the British Isles.[9] The depressing "cold Climate" of which Milton's epic narrator complains brings this other air into his poem too. It has to: English writers were practically bound to see themselves as encumbered by the yoke of the insular atmosphere, especially in the last decades of the Little Ice Age when, as Alvin Snider has shown, relentless cold defined English "body worlds."[10] A lengthy "Digression on Air" in Robert Burton's portly *Anatomy of Melancholy* (1628) expressed the hereditary sense that the chill, ponderous air of northern Europe—in the insular case further burdened both by its congress with a wet Atlantic wind and by the fogs that rose from indigenous marshland—was a real encumbrance. The distinctively English malady of melancholy was in turn referred directly to the dank density of the air that native English speakers had to breathe. Burton insisted that, if freedom is to be had, "the rectification of [such] air is naturally required" and that "the medium must needs be good, where the air is temperate, serene, quiet, free from bogs, fens, mists, all manner of putrefaction." Air can be "good" only when it frees itself from the "bridle" of melancholy that the British insular climate imposes on it.[11]

Burton's *Anatomy of Melancholy* proposed no path to liberty. But Milton and his fellow Puritans had aimed to "rectify" Burton's gloomy (if symptomatic) determinism through the energy of their own revolt. As they expressed, and actualized, their spiritual and physical autonomy through the dissemination of print, English writing promised better air. Even as it was turned into the physical forms of writing, such air found itself spiritualized as well. Thus in turn was it potentially restored to something of its original freedom and unity. The Puritans' investment in letters is obviously of a piece with Milton's attempt, through reflexive literary language, to explicate air's necessarily oppressive body as a medium of eventual salvation. In *Paradise Lost*, it is only by moving, willingly, into and through "other air" that Adam and Eve can hope to rehabilitate that air and thereby restore the "pure, now purer" (4.153) kind that they have lost.

So when they exit their own "closure," their first assignment is to figure out the "other air" of this world. That other air is visible to us as a protoliterary body, at once invisible and palpable, transparent and obscure.[12] As Karen L. Edwards and Joanna Picciotto have recently shown, reading the book of nature and reading the books of *Paradise Lost* are potentially complementary activities; both are ordered along the open-ended experimental lines through which experience can be parlayed into knowledge.[13] Furthermore, though, Adam and Eve wander into a world of knotty lexical dilemmas that is explicitly presented as a "clime." Not merely analogous to the literary medium but visibly arising from it, this "other air" reminds us both of that medium's limits and of the imaginative and expressive possibilities that arise from them. In the early nineteenth century, Coleridge was to internalize (and express) both when he recorded his own "continual Dream, representing visually & audibly all Milton's Paradise Lost."[14] Suspended between the "visua[1] & audibl[e]," Milton's airs condition new forms of experience in a world mediated through material forms—a world as distant as it is close at hand.

Literary Pneumatics

Milton's "air" marks his participation in an age of literary experiment. And as just such an experiment—as an attempt, anyway, to discover by way of literature what worlds that same literature might discover, convey, and re-create—*Paradise Lost* coheres with countless new experiments in natural philosophy. This coherence has attracted recent scholarly interest, with Edwards perhaps having done the most to establish similarities of both object and methodology between Milton's densely realized landscapes (and the curious humans who want to make sense of them) and emergent natural philosophies (and philosophers). Edwards ties Milton to early botany and zoology and thus to natural philosophies grounded in substances available to the abetted, often even the naked, eye. But what of the natural philosophy of the air? Confronted with the "torrid heat" and "vapor" of their contradictory "new clime," Adam and Eve are in the end (or is it the beginning?) the first pneumatic chemists.[15] This peculiarly fits them for Milton's own time and place. When England's restored monarch Charles II founded the Royal Society in 1660, around the time that Milton began composing *Paradise Lost*, many hoped that the king would make one of the society's first practical objectives the purification of English air. The means to this end, however, turned out to be relatively abstract: the strict analysis of "common air" as a purely material body. It fell to Robert Boyle (1627–91) to show that matter was all there was to it, and to that end he endowed air with what seemed to be strictly physical attributes of weight, "spring," and ultimately "history." He is fabled to have done so by way of a visualizing machine: the air pump.

It has always been easy enough to see what an air pump *is*: diagrams and artifacts abound, even now. It's become almost as easy to say what an air pump can do symbolically: thanks to Boyle's most influential modern critics, Steven Shapin and Simon Schaffer, we are almost painfully aware that the Royal Society enshrined the air pump as an emblem of its own modernity, a token of its linguistic and philosophical commitment to a materialist understanding of the physical world.[16] As late as 1768, Joseph Wright would reduce the descending dove of the Holy Spirit to a mere cockatiel suffocating in the pump's glass receiver. But if Wright's *Experiment on a Bird in the Air Pump* affirms mechanistic materialism as the most prestigious language of nature, the experiment at its center also elicits shock, sorrow, and unmistakable fascination from the new Eves who watch it work, and even as it traps light, the air pump does nothing to dispel the shadows around them.

Wright's painterly equivocations urge us to read the air pump both as an agent of enlightenment and as a symptom of postlapsarian frailty, an instrument of transcendence that presupposes entrapment in the physical world of mediated forms. In much the same spirit, Samuel Johnson's *Dictionary of the English Language* (1755) noted that when this "machine by whose means the air is exhausted out of the proper vessels" was first designed by the German-speaking Otto von Guericke, it "laboured under several defects." Johnson wanted to turn the air pump's story into a redemptive one, so he added that Boyle, heroically and "with the assistance of [Robert] Hooke, removed its several inconveniencies."[17] It is in turn as a triumphant icon conveying English ingenuity, sociability, and practical intelligence that the air pump appears in the frontispiece of Thomas Sprat's *History of the Royal Society* (1667). But what conveys the air pump? When its story surfaces in Johnson's lexicography, as part of the history of an English compound word ("AIR PUMP"), the air pump's explicit medium—English writing—becomes apparent. In this context, "air pump" specifies innovation within written English; it marks *both* the language's gritty historicity and its perfectibility, Johnson's anthem throughout his *Dictionary*, after all. Just so, today the air pump is reproduced, distributed, and best known not as an object, or even strictly as a picture of one, but rather as cover art for the Norton publishing company's canonical anthology of eighteenth-century English literature.

Nor is it out of place there. To most eyes in its own day, too, the air pump appeared primarily in and through the equally clarifying and obfuscating technology of English writing, at the entrance to two works of literature. Both Boyle's *New Experiments, Physico-Mechanical, Touching the Spring of the Air* (1660) and Sprat's *History of the Royal Society* were deliberately published in English rather than Latin and in an English that openly aspired to cure itself of empty beauty and false sublimity. The question these works thus open—what should modern English writing be and do?—returns us to Milton. As an enemy combatant obviously committed to what language cannot directly evidence, that Puritan would not be expected to have aligned himself formally with Boyle's project. As his friend Gilbert Burnet put it, Boyle had "abstained from purposes of marriage"; he was an aristocrat and a royalist who identified himself first with his birth family and second with the notional one of Oxford's Invisible College, the chrysalis of the Royal Society.[18] Milton, by contrast, was a thrice-wedded Cambridge radical determined to treat the air as a potential medium of divine inspiration.

Yet as Desiree Hellegers's recent consideration of these two writers suggests, *Paradise Lost* begs to be read in tandem with Boyle's *New Experiments* and its many sequels. As Hellegers mentions, Milton tutored Boyle's nephew in the 1640s and probably played some role in Milton's survival at the Restoration. Hellegers's Milton is an ingrate, though. His depiction of the "science" of astronomy in *Paradise Lost* rebukes the Royal Society's attempt to "assert an interpretive monopoly over the book of nature" and bids instead to reintegrate elite natural philosophy with "the material conditions of everyday life" (Hellegers, *Handmaid to Divinity*, 131, 138). Aided by poetry's hesitation between different registers of meaning, this Milton holds even a logic of the stars accountable to the vegetable matter of Eden, as tended by Eve.[19]

To link Boyle and Milton through the air, however, yields a different sense of their relationship. As a subject of knowledge, for one thing, air compelled even languages that did not classify themselves as poetic to hesitate. And it was itself equally the property of the exceptional and the everyday. In any case, as contemporaries and inhabitants of a single island (one where the study of air was *in* the air, so to speak), Milton and Boyle by definition breathed the same air, and Milton knew of Boyle's work on air through the letters of their mutual friend Henry Oldenburg.[20] Like Milton's, Boyle's "eyes became so ill" that he dictated some of his work, and England's "cold" air hampered him quite as much as it did Milton's epic narrator (Burnet, "Memorandum," 27). But set analogy aside. It is when we consider *how* Boyle wrote about the air that his differences from Milton reveal themselves as differences *with* him. These are differences conditioned by the common yet potentially transcendent medium that both called "air." And so to Boyle.

Boyle's Air of Composure

When we speak of modern notions of the air, no figure springs more immediately to mind than Boyle. Every student of pneumatic chemistry knows that it was Boyle who first feigned air to be a "physico-mechanical" body, corpuscular, "elastical," weighty, and, for those very reasons, expendable. In the wake of Shapin and Schaffer's groundbreaking analysis of Boyle's pneumatic experiments of the 1650s, 1660s, and 1670s, we are now well trained to see through the pretenses of this claim. It was, after all, only by eliminating air in the air pump's glass receiver that Boyle managed to discern, let alone describe, any of its material properties. Yet it was only because air possessed these properties that Boyle managed to eliminate it in the first place. The ironies that beset this interminable exchange between what's there and what, apparently, is not have been obvious to everyone from Thomas Shadwell—whose 1676 farce, *The Virtuoso*, caricatured Boyle as that irrepressible accountant of nothing, Sir Nicholas Gimcrack—to Shapin and Schaffer themselves, whose Hobbesian interpretation of Boyle's experimental method holds its materialist sleights of hand responsible for the descent of the ideological mist that is modern science.

Strictly speaking, it wasn't Boyle, or his experiments, or even his air pump, that can be said to have created, more or less out of thin air, modern science's impression that air exists as a "whole parcel . . . compose[d]" of "elastical bodies."[21] First credit here must go to his many literary "composures."[22] Shapin and Schaffer were happy to extend such credit. They pointed out that Boyle's published accounts of countless experiments involving the air pump, a few onlookers of good repute, and a great many very unlucky small animals were so incredibly (or rather so credibly) detailed as to convert a glass jar from which absolutely everything had evidently been expressed into both the agent and emblem of an entire, and entirely modern, "literary technology." This "technology" is often described as a sociable, empirical, and circumstantial manner of supposing, trying, and communicating reality. In its singular address to the mind's eye, Boyle's writing is supposed to have pushed all other ways of doing business out of the emporium of knowledge. But other scholars have sprung up to highlight Boyle's lifelong attachments to what look (from a materialist point of view) like airy nothings: alchemy and apparitions, faith healing, and French romance.[23] Clearly the most Christian of virtuosos' famed skepticism was at least as likely to uphold archaic styles of faith in nothing much as it was to support new modes of belief in what at least appears to be something. How might we reconcile this Boyle with the other one?

One way would be to look at what his writing says about itself. For Shapin and Schaffer, Boyle features as an expert manipulator of the peculiar vacuum of the written word. He is well versed in its illicit power to bring to the mind, unbidden, a vivacious and convincing image of certain invisible objects. This take certainly comports with Boyle's exhaustive and well-sealed descriptions of ordinary surroundings—the circles of credible gentlemen, the glass receiver, the quivering linnets, and the cloth, its blue "as deep as that of the sky in a fair day" that in one experiment so poignantly fades in the open air.[24] Yet Boyle's pneumatic "composures" also speak at every turn of their own *dis*composure. They dwell obsessively on the accidents and misfortunes that befall them. They apologize for the pressure of circumstance that has left them riddled with omissions and unsightly turns of phrase, for too rapid transcription and strange absence of thought. *New Experiments* opens almost abjectly, with a guilty eye to the gaps between the "Experiments" to follow; its "Proemical Essay" admits to "transgress[ing] the laws of oratory." Meanwhile, Boyle pleads, a "distemper in my eyes" kept him from writing "so much as one experiment, [or] even to read over myself what I dictated to others" (*New Experiments*, 1:2). Throughout *New Experiments*, its reader is exposed to a nebulous yet pressing climate of missed "opportunity" and distracting "avocation," one that Boyle's words resist with conspicuous difficulty. These surroundings are not described with the same pragmatic buoyancy that the scenes of experiment are. In fact, they are not described at all, but instead announce their presence by oppressing the phrases that strain to account for *them*. So recedes the language of witness and reference that links Boyle's writing with the confident and denotative Dutch realist canvasses of his own day and fingers him as one of the English novel's many possible fathers.[25]

The air of discomposure that Boyle's writing wears can look like a plain-dealing pose assumed so as to stimulate readerly investment in our narrator's moral integrity and thus in the conceptual integrity of his entire project.[26] But might not this same air also be bound to Boyle's inscrutable and incoherent subject matter—to, in a word, *the* air? To speculate thus is to discover in Boyle's sentences the same species of catastrophe that so often beset the experimental spaces to which those sentences presumably refer. There, Boyle often laments, vats, vessels, and bladders were ever apt to break, bowled over by the wind, shattered by plummeting temperatures, or stretched to bursting by bellows applied with excessive zeal. Such disasters, quite as much as his success in the art of suffocation, apparently persuaded Boyle that air can only be discerned in its effects on visible bodies. Why can't materialized words count as such bodies? From this perspective, Boyle's air is not just what his writing is about; rather, air appears to be about—literally around—*it.*

Hence might spring the hybrid "contexture" of his prose. Its layers produce the sense that—akin to the first aristocrats, Adam and Eve, alienated from their native air—Boyle was not entirely at home even in his mother tongue. When in the early eighteenth century the popularizing royal physician, Peter Shaw, set out to condense, abstract, and mechanize Boyle's pneumatic writing, he first rooted out "the quaint, antiquated, or uncouth expressions [Boyle] sometimes uses."[27] As Shaw explained, without apology, "It seemed convenient to make him wholly English, to throw out the harsher words or phrases; and so far to alter the style, that it should not offend a merely philosophical ear" (Shaw, *Philosophical Works,* 1:iii). But the Boyle that Shaw amended suggests that, far from free or pure, even the most common and straightforward English words are by turns impeded and released into shape by their own conditions of utterance. These conditions both include and may be expressed as the air about them.

One whisper to this effect materializes in the first of the forty-three *New Experiments* that Boyle published just after Charles II's restoration in 1660. Here Boyle paints one of his most famous pictures of "our air." Boyle's air "either consists of, or at least abounds with, parts of such a nature, that in case they be bent or compressed by the weight of the incumbent part of the atmosphere, or by any other body, they do endeavour, as much as in them lieth, to free themselves from that pressure, by bearing against the contiguous bodies that keep them bent; and, as soon as those bodies are removed, or reduced to give them way, by presently unbending and stretching out themselves, either quite, or so far forth as the contiguous bodies that resist them will permit, and thereby expanding the whole parcel of air, these elastical bodies compose" (*New Experiments,* 1:11). Though anyone curious about "our air" might want to think of this representation as an image of that "body," it is not eidetic in the least. If anything, Boyle's words achieve immediacy haptically, by drawing us into their shapes and inclinations, their bents and stretches, and their points of compression and resistance. A whole flotilla of Latinate adjectives and adverbs can solicit no picture from the reader's mind, and one feels less pressure to visualize the subject of this elastic sentence than to move with (or against) its sixteen awkwardly contiguous parts.

Similar expressions of air's deeper composure animate the whole ungainly corpus of Boyle's pneumatic writing. Its hapless anatomizer must somehow cut through a revised edition of *New Experiments* (1662) that included a defense of Boyle's "Doctrine" against the apostasy of Thomas Hobbes, a two-part *Continuation of New Experiments* (1669 and 1690), and countless fugitive writings on the relationship between flame and air, on respiration, on the "Subtility and Efficacy . . . of Effluviums," on the atmospheres that surround hard bodies, and much more that were printed in the *Transactions* of the Royal Society and elsewhere throughout the 1670s. Boyle's pneumatic writing culminated—or perhaps simply collapsed under its own weight—in the magisterial yet strangely fragmentary *General History of the Air* that appeared just after his death in 1692. This means that, in all, Boyle churned out hundreds of pages concerning "that thin, fluid, diaphanous, compressible and dilatable Body, in which we breathe and wherein we move" (*General History*, 5:612). The pages themselves are anything but thin. Still, in the internal curvature and reflexive nuance that creates the impression of piled "contexture," Boyle's writings bear traces of the air their titles so often declare them to be "touching." That air in turn presents itself most immediately not as a physico-mechanical body but as an extremely and uncomfortably intimate medium for transporting and, as Boyle put it, "exhibiting certain phaenomena." This medium Boyle's literary one less conjures than expressively conveys, with difficulty (*New Experiments*, 1:6).

Why the difficulty? In a turn that brings his writing into line with the complaints of *Paradise Lost*'s freezing, down-at-the-mouth English narrator, Boyle's singular habits of speech and mind left him with no conception of the air that could be entirely separated from his experience of it as an impressive medium. Its primary identity as such a medium makes Boyle's air, for all its modernity, surprisingly difficult to tell apart from the ether of the ancient and medieval metaphysicians, even as his epistemology, in this matter at least, merged with the open forms of wonderment that we find in the pages of heroic romance.[28] Indeed, in his pneumatic writing it is impossible not to feel air's affinity with the occult (insensibly material) "qualities" that medieval philosophers, including Aquinas, had deemed accessible only to the imagination, because only the imagination was like them in kind. As Keith Hutchison has shown, natural philosophers of Boyle's day—Boyle, in his view, among them—did not so much rule out the possibility of the occult as insist that it can be explained. Other historians of science have gone on to propose that certain instruments—since ruled entertaining as opposed to knowledge-producing—briefly afforded alternative modes of visual, hence seemingly explanatory, access to the occult; Athanasius Kircher's magic lanterns, for example, did so by projecting images in dark rooms lacking in "place or form."[29]

As Kircher evidently was not, Boyle was all about see-through pumps in well-lit rooms. Nevertheless, in his *Continuation of New Experiments* (1662), he confesses that, throughout his pneumatic studies, the "production of a new phaenomenon is a thing to be aimed at."[30] If the "phaenomenon" of Boyle's air was to achieve any of the force of novelty its author plainly hoped it would, it had to appear not just as a "thing" but as the reflexive emanation of the English "composures" that spoke of it, even as air continued to inform them. As it circulates through these possibilities, Boyle's air seems to be of a piece with Milton's. Its very difficulty fits it as the cure of all, a paradoxical source of amelioration and conceptual liberty.

What We Understand by the Air

Just what, according to Boyle, *is* air? As late as 1692—more than three decades after he had undertaken his first investigations into the matter, and a few months before he was to draw his own last breath—he still does not appear to have been disposed to say. The first heading in his posthumously published *General History of the Air* is not "Air" but "What We Understand by the Air" (*General History*, 5:612). Well, then, what do we?

In the pneumatic experiments that he began performing just before the Restoration, Boyle claimed to capture what he called "common air," separating it at once physically, conceptually, and linguistically from the supersubtle aether that his predecessors from Aristotle to Descartes had presumed to surround it and to direct its action.[31] It is crucial to note that these acts of separation are one. For all its incoherence and lack of conclusion, Boyle's *General History* was still designed to "serve to some men as a common place for a history of the air" (*General History*, 5:610). As such, it interlays the "common air we breathe and move in," or "atmospherical air" with "aetherial or celestial substance" (*General History*, 5:613). Unlike that notoriously insubstantial substance, common air is no oxymoron. If veiled from direct observation, it is nonetheless a material body, endowed with a shape and a behavior. These Boyle renders concrete, if not visible, in common air's signature and decidedly nonethereal quality of "spring."

"Spring" here may be seen as something besides the measurable and defining attribute—the characteristic—that it of course remains. Spring is also a way Boyle found to figure common air's difference (even its near-satanic freedom) from the infinite fluency of aether. This was a freedom perversely beholden to what we might call air's historicity, as Joseph Priestley was to recognize a hundred years later when he contrasted the "scene for ingenious speculation" that the ether provided to the fact that "real history resembles experiments by the air pump."[32] On one level, Boyle's elastic, common, knowable air—"that thin, fluid, diaphanous, compressible and dilatable body, in which we breathe, and wherein we move, which envelops the earth on all sides . . . but yet is so different from the aether"—is to his familiar and transparent prose style as that very aether was to the esoteric symbologies Boyle and his fellow professors of the Invisible College so famously sought to dislodge. But Boyle's separation of "common air" from "aetherial or celestial substance" through the "common place" medium of his pneumatic writing also discloses some ironic continuities. One is evident in the extreme difficulty a reader will encounter in telling "thin, fluid, diaphanous" air from the aether it's not. Compressibility and dilatability, as suggested, bear the burden of distinction. Nor do they bear it very well, because aether too (and increasingly) was supposed to stretch and relax.

Paradise Lost exploits these attributes in its chronicle of the angel bodies that, waging war in Heaven, "in the air / Made horrid circles" (6.304), parting only until "th' ethereal substance clos'd" (6.30). As for Boyle, the true difference between ether and common air, apparently, lies in what the speaker leaves out. This includes exactly the identification with the inexplicably occult that Milton embraced and that may be traced back to the pre-Socratics at least. Boyle's "common air" will have nothing to do with that. It is all about *us*. The syntax of the sentences within which Boyle said as much would put the emphasis on the preposition—the air is all *about* us. This order of "aboutness" is what justifies air's study in the first place; as *New Experiments* would have it, Boyle's "choice of subject" is due to "the air being so necessary to human life [that] creatures that breathe, cannot live many minutes without it"—which is to say outside it: "The ambient air" is "that whereto both our bodies, and most of the others we deal with here below, are almost perpetually contiguous" (*New Experiments*, 1:6). Boyle's air is thus about us in another sense. As we both breathe in it and breathe it in (and out), we become, if not one with it, at least one means whereby it moves.

It is thus surprisingly hard to see, let alone say, how the "common air" expressed from Boyle's air pump differs from its linguistic expression. It is because of this expressive unity that Boyle could only call what remained in his evacuated pump "nothing"; he had to say that it was nothing both in order to establish that air is *not* nothing and in order to secure its distinction from that pure and possibly intelligential body that had long been denominated aether. As his antagonist Thomas Hobbes recognized, the vacuum's status as a linguistic placeholder could not be separated from its status as a real thing or indeed as a true notion. And from here it was but a short step to aether's own practical conflation with nothing. Boyle was in effect abandoning, in favor of common (and humanly communicable) air, the frame of "aetherial or celestial substance" long believed to surround—and hence determine—the "common air we breathe and move in." From a literary perspective, Boyle's grounding of common air left his readers enveloped in nothing else, a state as parlous as that of Adam and Eve, just exiled into "other air / Less pure" than that of their native Paradise.

Except for one thing. Boyle evicted his reader from the aether into this new and common air in such a way as to make the comprehension of that air contingent on his own linguistic medium. Hence, in Boyle's *General History*, we find the heading "What We Understand by the Air" as opposed to plain "Air." The heading refuses to say whether its understood subject is air itself or the word "air." This is Boyle's famed scrupulosity speaking, of course, but its effect is less to limit what we understand air to be than to expand the sphere of what we might understand "Air" to *mean*.[33] If Boyle refuses to embed "Air" in the aether by which it had been previously shaped and constrained, the infinitely creative field of human language takes over this contextualizing role. It is from within that prolific medium that we are bound to witness air's peculiar power of "exhibiting certain phenomena" from sound and smell to color and even light.

Of course, we now suppose that Boyle overestimated common air's capacity for mediation and exhibition. It doesn't carry color or light, for example, and ironically enough current electromagnetic theory has resorted again to something very like aether (as opposed to air) to account for light's transmission through space. Still, after Boyle, English investigations of the air "almost perpetually" about us were to pivot on language's own "perpetua[l]" turn into the common air that it would communicate. This turn becomes analyzable and available to common scrutiny only when human conventions of expression become visible, as they do so uniquely within the written medium. And it was precisely in deference to that medium that Boyle refused to extricate "atmospherical air" from the material practice of language. For instance, in order to gratify Oldenburg's pressing "desire to know . . . not what the true nature of air is, but what I guess concerning its chiefest property or attribute," Boyle's *General History of the Air* tenders a tripartite anatomy of that "thin, fluid, diaphanous, compressible and dilatable body." There are, first, those airs that "under the form of vapours [are] in a word . . . made to diffuse themselves in the atmosphere" (*General History*, 5:613). Then there are the "more subtile" corpuscles that the sun and stars emit to "produce what we call light" and that may be "dispersed through or harboured in the intervals of the stabler or grosser corpuscles of the atmosphere." Boyle introduces a third kind of air only "because you expect from me a distinguishing quality, which may put a difference between the parts already named of atmosphere" and "compared with which I have not yet found any, whereto the name of air does so deservedly belong" (*General History*, 5:614).

It's this third "sort of particles" that made Boyle's name, supplying the subject matter of his main pneumatic project. "This sort of particles," Boyle concludes, "are those, which are not only for a while, by manifest outward agents, made elastical, but are permanently so, and on that account may be stiled perennial air" (*General History*, 5:613). Boyle here proposes that perennial (alias common, or atmospherical) air differs from the diffuse and subtle versions of itself formally and conventionally, but perhaps not essentially—hence the almost gratuitous extraction of a different name from it. Aside from this name, perennial air seems to be distinguished only by its perennially "elastical power." This power is not just an objective property, though it may still be that primarily: we certainly do not want to deny that Boyle's primary concern was to deliver the facts. All the same, "spring" also functions in the "understand[ing]" very much in tandem with "stiled" language, bringing a difference forward from a background of apparent similarity. At the very least, Boyle's manner of separating air from aether—of "stil[ing]" it "perennial" in order to "put a difference between the [other] parts already named of the atmosphere"—invites us to see his experiments through the screen of their expressive medium.

Rose-Mary Sargent submits that for Boyle the *act* of "writing was part of the learning process, and thus a crucial part of experimental activity itself."[34] But what of the *fact* of writing? Writing as fact—as alphabetical forms that, it is consensually agreed, appear on a page—turns out to be a key source of Boyle's ostensibly nonliterary technology, the air pump. He adopted that device after reading, "in a late mechanical treatise," of "the pneumatical engine, that I use to call the wind-gun, . . . which being charged by the sole impression of the air, would, by the violence of the contracted Boreas, send forth a leaden bullet."[35] Figural density ("the violence of the contracted Boreas") and rhetorical self-consciousness ("that I use to call the wind-gun") both signal experiment's origin in a literary "impression." Reflection on that impression gives rise to new ways of experiencing the air—of inhabiting it conceptually.

Consider how Boyle, in his honorable way, negotiated the conflict that had long dominated Western European air studies. That "noble controversy" pitted plenist against vacuist. Boyle refused to take sides, holding that there is in fact "no need to have recourse to a *fuga vacui*" to explain what remains where air is not, for the spring of that air "seems to depend clearly upon the weight of the atmosphere . . . which I take to be the grand immediate agent in these phenomena" (*Continuation of New Experiments*, 3:198). In "tak[ing]" the atmosphere itself—or at least its weight—as "the grand immediate agent" of actions once attributed to aether, Boyle was redirecting attention from the relatively crude question of whether nature really does abhor a vacuum to a very different one: what is the condition under which a vacuum might or might not appear to be? As Boyle conceives of it, atmosphere, no less than the word for it, manifests this possible condition. This grand immediate agent's indifference to the truth status of the proposition "vacuum" legitimates the chemist's claim that his "visible experiments" with "atmospherical air" were intended (or simply tended) neither to "convince unbelievers" nor "to prevent or remove a scruple" but rather, and merely, to produce "a new way of making an estimate" (*Continuation of New Experiments*, 3:184). Something thus was added to a simple representation of the physical world. Boyle had found a way to release that world's motives and meanings through a literary action.[36]

The Spring of the Air

Before Boyle, air, locked in aether's grip, was supposed to twist and turn, to push and pull, to flow and even fall. But to spring? Boyle's air seems novel in part because of its elasticity—a quality that, we've seen, his reader can experience more immediately than she can picture elusive air itself. Indeed, when Boyle's pneumatic writing resists the quasi-referential protocols of virtual (and virtuous) witnessing, it is most often in the interest of making spring apparent. So is sealed an animate affinity between the spring of the air and the figures of speech.[37] This affinity in turn supports the homology in which spring is to air as figuration is to language. Spring and figuration make the presence of, respectively, air and language known. So are opened new realms to be comprehended, even as their boundaries are marked.

"Spring" is most obviously and basically a "physico-mechanical" property. It equals resistance to pressure and a body's tendency, once that pressure is removed, to restore itself to its original shape and size, "unbending and stretching . . . so far as [any] contiguous bodies . . . permit." But Boyle's experiments tended to emphasize less spring's autonomy than the limits within which that autonomy expresses itself. In fact, spring needs a wall to hit if it is to happen in the first place, and Boyle's experiments brim with instances in which he discovers this truth, the most notable being his demonstration that suffocation is the effect not of air's withdrawal from the hapless body that once contained it but of air's violent recoil *within* the body that has been deprived of air *without.* At the same time, the "contiguous bodies" that squash and then release corpuscles of air, thus bidding them to spring, cannot be said to be more than springy air themselves.

Perhaps to escape this circle—or at least to distract himself from it—Boyle early on abandoned the question of spring's value in helping aerial bodies regain their true figures. But he ratcheted up its value in helping human understanding figure air's elasticity to itself. It's best, he thus decides in *New Experiments,* to "speak of" air "under the notion of springs." This "notion," he supposes,

> may perhaps be somewhat further explained, by conceiving the air near the earth to be such a heap of little bodies, lying one upon another, as may be resembled to a fleece of wool. For this (to omit other likenesses betwixt them) consists of many slender and flexible hairs; each of which may indeed, like a little spring, be easily bent or rolled up; but will also, like a spring, be still endeavouring to stretch itself out again. For though both these hairs, and the aërial corpuscles to which we liken them, do easily yield to external pressures; yet, each of them (by virtue of its structure) is endowed with a power or principle of self-dilation; by virtue whereof, though the hairs may by a man's hand be bend and crouded [*sic*] closer together, and into a narrower room than suits best with the nature of the body; yet, whilst the compression lasts, there is in the fleece they compose an endeavour outwards, whereby it continually thrusts against the hand that opposes its expansion. And upon the removal of the external pressure, by opening the hand more or less, the compressed wool doth, as it were, spontaneously expand or display itself towards the recovery of its former more loose and free condition, till the fleece hath either regained its former dimensions, or at least approached them as near as the compressing hand (perchance not quite opened) will permit. (*New Experiments,* 1:15)

Boyle appears to "compose" this analogy in such a way that his reader may picture aërial corpuscles as she would the "many slender and flexible hairs" in which a "fleece of wool" consists. Like any one of those hairs, an aërial corpuscle can be imagined to stretch itself out again once that coercive hand relents and releases it from constraint. And Boyle's "likeness" does. of course. work on this level. The image of curled "hairs" seems to let us see what air must be like, while at the same time showing us if not air itself then at least its defining motion.

This motion is all we actually *do* see of the air. It substantiates both air and hair to become itself the only true substantive in this passage. Hence, the word "spring" actually appears as a noun rather than a verb. At the very least, air's shape—its figure—can be neither told nor perceived apart from what it apparently does. As if to seal this bond, Boyle deliberately "omit[s] other likenesses" between hair and air. In compressing the range of likeness, Boyle again throws the conceiving mind back on its own devices, allowing it not to see spring as an object but to experience it, synesthetically, as a dynamic figure. Here we find form without fixed content. Any form in turn is defined as much in relation to other, equally contingent forms (including notional ones) as to whatever it might happen to contain and carry.

This dynamic, so wonderfully enacted in the spring of the air, proves consonant with Boyle's formal thinking about figuration. It is especially consonant with his thinking about air's redemptive power for those bound to the material world. That thinking is characterized by—perhaps it even springs from—deep skepticism concerning the virtue and efficacy of strict analogy. In a late essay on method, *The Christian Virtuoso* (1690), Boyle specified that analogies ought to give "strength" rather than "light." They are, he insisted, "not always bare pictures and resemblances, but a kind of arguments; being oftentimes, if I may so call them, analogous instances which declare the nature or way of operation, of the thing they relate to, and by that means do in sort prove, that, as it is possible, so is it not improbable, that the thing may be such as it is represented."[38] Boyle's "if," "as," and "not improbable" create a hybrid condition in which the figure of spring may animate the hair-air analogy, even as spring remains a real and defining attribute of both hair and air. Here Boyle also echoes Milton's archangel Raphael, who teased a prelapsarian Adam that "lik'ning spiritual to corporal forms, / As may express them best" may not foreclose the possibility that the "forms" compared may be "each to other like more than . . . is thought" (5.573–76). Raphael too hints that analogy may fold in on itself, then spring back as an active, indeed autonomous, expression of its own subject matter. In Boyle's postlapsarian case, the so-called analogous instance opens up something further still: the peculiar notional space of things "not improbable." But this is something it can *only* do when the strict equivalences of analogy are so constrained that they stop pretending to give light.

Boyle's analogous instance of airy hair cannot help but bring Milton's Eve to mind. Even before she is seduced by fiction, Eve is immersed in a wilderness of mediating forms. Watery mirrors, wedded vines, the caresses with which Adam conveys all higher knowledge: this is all she knows. Eve's own most knowable feature seems to be her hair, those "golden tresses [that she] wore / Dishevell'd, but in wanton ringlets wav'd" (4.303–4). Milton's snarled (unto Boylean) syntax links the "wav[e]" of Eve's hair—not, notice, the hair itself—to "the Vine [that] curls her tendrils, which impli'd / Subjection, but requir'd with gentle sway" (4.306–7). As a straight analogy, this figure would appear to identify Eve with Eden's visible vegetable substance. But, of course, it is not a straight analogy at all: it is the "wave," not the hair itself, that's identified with its environment of mediating forms. In turn, the pressure that Milton puts on this word links Eve to the medium that communicates her, not to any physical object to which that medium refers. Hence—as Satan later realizes when he seems to breathe Eve's "air"—what Eve's hair instances, in Boyle's sense, is her lived identity with paradisal air, air suspended after all in "th' eternal Spring" (4.268). The wave of Eve's hair is an elastic figure that already turns away from a trope transmitted through antique myth toward the spring of Boyle's air. Eve's springy hair scarcely requires the "gentle sway" of the demonstrative hand because it already "impli[es]" it as a real pressure within the material linguistic medium that conveys her.[39]

For Boyle's reader, meanwhile, the spring of the air is bound up with the spring of analogous instance. Boyle uses analogous instance not just to show us what the spring of the air *is* but also to involve us in what it *does* within a verbal and conceptual framework. It's worth observing that one thing spring does is identify "hair" and "air" with one another, revealing them as aspects or appearances of the same encompassing whole. In speech, it is air itself that would do this work, because "hair" sounds all but the same as "air." On the conspicuously silent page, however, it is the spring of analogous instance that achieves this end. In the process, expression is integrated with interpretation, and the text of experiment is turned into the ground of potentially infinite speculation. So while "spring" remains a matter of fact about common air, and a dynamic within expository discourse, it also marks out the styles of expressive thought within which air may be construed and communicated. In Boyle's analogous instance, these aspects of spring are naturally coordinated by the "compressing hand (perchance not quite opened)." In literary space, meanwhile, spring is naturally communicated through the hand's work of writing—work that Boyle's own writing naturally discloses.

"Stuttring Habitude"

Our involvement with air is, by definition, lifelong. When Boyle reflected directly on his, he did so not in the waxing language of modern science but rather through the allegedly waning one of romance. His fragmentary autobiography displaces his life story onto the adventures of one "Philaretus." Motherless Philaretus spends his early days at Eton absorbing, simultaneously, "good Aire" and the prose romances that seem to "divert his Melancholy."[40] The downside, Boyle recalls, was that these "Fabulous and wandring Storys" also "prejudic'd him, by unsettling his Thoughts, then they could have advantag'd him." To make matters worse, "they accustom'd his Thoughts to such a Habitude of Roving, that he scarce has ever been their quiet master since, but they would take all occasions to flinch away, & go agadding to Objects then unseasonable and impertinent" ("Philaretus," 8).

One such unseasonable object was apparently the "Aire," which, like Philaretus's own mind, seemed wont to rove. Before long, a "Prosperous Puffe of Wind" blows the budding scholar to France ("Philaretus," 13). Later, in Geneva, a storm puts him "in a Fright with such loud Claps of Thunder (which are often times terrific in those hot Climates and Seasons) that he thought the Earth would owe an ague to the Aire" ("Philaretus," 16). In that "dismall Tempest, the Winds were so loud as almost drown'd the noise of the very thunder." And yet this "happy Storm" performs "an operation upon [him] resembling that it had upon the Ground." It induces a spiritual conversion in which the young Philaretus feels "such kind & geniall showres, as water'd his parch't and almost wither'd graces" ("Philaretus," 16). Here "Aire" materializes as a figure of speech that unifies thought and experience in a single, resilient unit of expression. Similarly, a wall collapses while Philaretus is in bed, raising clouds of dust "so thicke that he might there have stifled, had he not remembred to wrap his head in the sheet; which serv'd him as a strainer, through which none but the purer Aire could find a passage" ("Philaretus," 7). One might say that "purer Aire" finds passage as much through the reflexive expression of a subject as through a posited object.

Adventitious, opaque—if also animate and potentially redemptive—the "Aire" of juvenile romance shadows the physico-mechanical body made visible in *New Experiments* and its sequels. To it, the "Life of Philaretus" adds another, more bodily experience of the air. By and large, Boyle's physical body goes unmentioned in the "Life"; the only physical detail on which he dwells is the "stuttring Habitude" that he retained into adulthood ("Philaretus," 2). At that later juncture, Boyle's physical body appeared to his contemporaries (or was at least represented by them) as an image of his inward moral stature, a conveyance through which his virtue could be viewed. In the flesh, Boyle was, as his eighteenth-century biographer and editor Thomas Birch put it, "tall of stature, but slender, and his countenance pale and emaciated" (Birch, *Life*, 1:cxxxvi). Portraits often caught the ethereal beauty that was to become a trope for the transparency of his scientific exposition.[41] As Evelyn put it, the distinctive "contexture of his body . . . appeared to me so delicate, that I have often compar'd him to a Chrystal or Venice-Glasse, which tho wrought never so thinn & finely, carefully set up, would outlast the harder Metal of daily use: and he was withall, as cleare and candid; not a bleb, or spot to tarnish his Reputation."[42] Like his air pump's vitreous bell, Boyle's physical "contexture" resembled see-through prose. It marked his character, and was it.

Our modern paradigm of character, derived from the Greek natural philosopher Theophrastus, prioritizes this kind of visual impression.[43] But that paradigm masks another sense, at once deeper and more superficial, of human beings as themselves impressed by what lies outside them. An analogy would be Boyle's air under pressure from a hand "perchance not quite opened." In a series of reflections heavily influenced by the tragedy of Boyle's French contemporary Jean Racine, the modern-drama theorist Jean-Louis Barrault reminded himself that "all around us is the outside world, which presses upon us and makes proposals to us." Indeed, "just as the earth is surrounded by an atmosphere, the living human being is surrounded by a magnetic aura which makes contact with the external objects without any concrete contact with the human body." Barrault makes this "aura, or atmosphere" the foundation of an alternative theory of character in action, for any actor "must first of all be aware of this boundless contact with things." Such contact renders the performance of character a kind of expressive suffocation, for "there is no insulating layer of air between man and the outside world. Any man who moves about causes ripples in the ambient world. . . . The external and the internal world weigh upon each other in the body of man." To act, in turn, is more than the projection of a character, the imposition of an impression on the face or indeed on the witnessing eye; it is also to convey an individual's "ambient world," the one that weighs upon him.[44]

Barrault's exemplary actor is the "tragic mime," exiled from ordinary speech but for this very reason preternaturally expressive of the "ambient world." In their reviews of Boyle's performance as a social actor, his contemporaries seem to have taken his character in this sense—as he did himself. The "Life of Philaretus" begins with Philaretus's "Acquaintance with some Children of his owne Age, whose stuttring Habitude he so long Counterfeited, that he at last contracted it." Boyle moralizes this "stuttring Habitude" as "possibly a just Judgment upon his Derision," one that "turn[ed] the Effects of God's Anger into the Matter of his Sport" ("Philaretus," 2–3). His habit of impeded speech, though it began as a counterfeit, is here also a realization, an effect that expresses the informing contexts both of God's invisible wrath and of social milieu. With respect to his "stuttring Habitude," Boyle further reports, "divers Experiments, believ'd the probablest meanes of Cure, were try'd with as much successlesnesse as Diligence" ("Philaretus," 4). These early, diverse, nameless, and finally failed "Experiments" could be counted as Boyle's first with air. No less than the "Aire" of romance, they subtly informed his later "experiments, physico-mechanical," especially to the extent that these too were experiments with "Effects."[45]

The Anglican divine Gilbert Burnet interpreted Boyle's stuttering as a "visible" marker of supernatural presence, and in his eulogy for the chemist declared that Boyle never spoke God's name "without a visible stop in his Discourse."[46] Boyle himself also characterized his "stuttring" as a "Habitude"—a symptom of heteronomy acquired socially *after* he had first acquired a fluency akin to the prompt (and promptly lost) eloquence with which an unfallen Adam speaks in Milton's *Paradise Lost.* For Boyle to have presented his "stuttring Habitude" as an epiphenomenon of language, not an obstacle to its acquisition, is to cast every word as a conveyance itself conveyed and thus as an exhibition of its own conditions. These become apparent only when language's work as a simple medium is frustrated, its spring out of neutral mediation and into interested—not to mention interesting—figuration laid bare.[47]

Boyle's "stuttring Habitude" also seemed to reveal the invisible powers that compelled it. Evelyn, for his part, specified that Boyle's speech impediment involved something besides the relatively common and uncontrollable repetition of assonants and alliteratives. Rather, "in his first Addresses, being to speake, or Answer; he did sometimes a little hesitate, rather than stammer, or repeate the same word" (Evelyn to Wotton, 89). By contrast, the Italian virtuoso Lorenzo Magalotti noted "some impediment in his speech, which is often interrupted by a sort of stammering, as if he were constrained by an internal force to swallow his words again and with the words also his breath, so that he seems so near to bursting that it excites compassion in the hearer."[48] But no matter who recorded it, Boyle's manner of speaking (or not) leaves the impression that language, with its producer, is ultimately contained, even compelled, by what it seems simply to contain, whether that be Magalotti's "breath," Burnet's name of God, or the vacant tracts that, Evelyn suggests, stretched between Boyle's every word.

Speech pathology forces language to register its own invisible conditions. This recognition hovers over otherwise disparate representations of Boyle's impeded speech and casts new light on the transparency aimed at in his scientific work on air. In Magalotti's extraordinarily vivid evocation, for example, Boyle's "sort of stammering, as if he were constrained by an internal force," sponsored a scene not unlike those that unfolded inside Boyle's own air pump, where air likewise springs back with the lifting of constraint. Just so, the "compassion" that Boyle's "sort of stammering" evidently excited redoubles the affective responses to his own experiments with suffocating birds that he himself records—"the pity," for instance, "of some fair ladies," which on one memorable occasion "made me hastily let in some air at the stopcock" until the gasping animal was presently recovered, and in a condition to enjoy the benefit of the ladies compassion" (*New Experiments*, 1:106–7). It's customary to read the settings of Boyle's experiments as mises-en-scènes framed by the cool gaze of a few good gentlemen and then turned into virtual experience by Boyle's scenic prose. But Boyle's overdetermined manner of speech brings us to a very different conception of what and how experimental space represents.

The forms of knowledge that spring up in such obviously mediated spaces differ from the materialist and analogy-bound ones that the natural philosophy of the period is supposed to have sponsored. Their exemplary practitioner, moreover, is not Milton's percipient unfallen Adam but his Eve. Even before the fall, Eve is embedded in mediating forms that she is obliged to know, and register, both synesthetically and as an inevitable environment. At the very least, the community of "compassion" and "pity" that surrounds Boyle's "gasping animal" (or philosopher, just as the case may be) is a feminine one, very much as it would be in Joseph Wright's pictorial *Experiment* of 1767–68. Boyle himself appears to have imagined the English language itself as effeminately penetrated (and thereby rendered awkwardly pictorial) by the conditions of its own utterance. By his own account, he "us'd to liken Writers in English Verse to Ladys that have their Pictures drawne with the Clothes now worn which tho at present never so rich & never so much in Fashion, within a few yeares hence will make them looke like Antickes" ("Philaretus," 10).

Boyle's pneumatic experiments are not versified, but they *are* in English, a written English that Boyle presented as isomorphic with his own thwarted speech. The narrator of *New Experiments* thus confesses himself "wont to speak rather doubtfully, or hesitantly, than resolvedly, concerning matters wherein I apprehend some difficulty" (*New Experiments*, 1:2). Boyle's avowed model here is none other than Aristotle, the substance of whose pneumatic theory Boyle of course rejected. Now, though, Aristotle springs back into favor by virtue of the fact that he "somewhere notes, that to seem to know all things certainly, and to speak positively of them, is a trick of bold and young fellows; whereas those, that are indeed intelligent and considerate, are wont to employ more wary or diffident expressions" (*New Experiments*, 1:2). Invoking Aristotle as a rhetorical authority, if not a scientific one, allowed Boyle to correlate stammering with the epistemological and ethical styles of diffidence and skepticism. As Milton's Eve and Boyle's Philaretus both discover, diffident expressions are also figures of speech's imperfection. As such, they indicate the ever-shifting grounds on which we decide what words might mean, the incoherence, malleability, and transience of the worlds they make.

Recess

Boyle's most immediate heirs in the consolidating discipline of pneumatic chemistry used the spring of the air to designate Boyle the heroic enemy of all supernatural forms. Boyle's first comprehensive editor, Thomas Birch, trumpeted nothing more loudly than the fact that he had "discovered and demonstrated the elastic power or spring of the air, and by this means exploded the notion of a *Fuga Vacua*, and shewed, that the strange effects, which were before ascriped [*sic*] to that imaginary cause, arise merely from the native self-expansion of the air." For Birch, Boyle seems to have struck a blow for enlightened freedom from "imaginary cause." But his use of the native self-expansion of literary figure to do the deed merely transfers the credit for "strange effects" to the media of supposed enlightenment themselves. Precisely because of its potential to be held in common, the page of philosophical transaction is, of course, one of the most vital of these media. Its visibly mechanical composition makes it the perfect medium of a mechanical universe, one where, as Birch puts it, "the most admirable things, which have been taken for the effects of substantial Forms . . . may be the result of the mere texture and position of parts" (Birch, *Life*, 1:lxi).

Why was it so hard to see that Boyle's quantifiable "parts" of air can also harbor more mysterious, and liberating, "Forms"? Paradoxically, Boyle's indefatigable Newtonizer Peter Shaw seems to have seen more than Birch did, noting that in Boyle's experiments the elements "are scrutiniz'd and tortured by experiments, to confess their natures, offices, uses, and the wisdom and design of their creation [and] 'tis by this means that he shews us the very various and surprizing properties and effects of that familiar fluid, the air, or atmosphere" (Shaw, *Philosophical Works*, 1:vi). But in the end Shaw, too, scurried to uncouple Boyle's writing from the air it renders. Boyle, as we have seen, had written that his reader would "easily suppose, that the notion I speak of is, that there is a spring, or elastical power in the air we live in" (*New Experiments*, 1:11). He went on to enact in literary space all the ways in which "our air either consists of, or at least abounds with, parts of such a nature, that in case they be bent or compressed by the weight of the incumbent part of the atmosphere." Shaw distilled this elastic performance into one terse declaration: "To render our experiments more intelligible, we must premise, that the air abounds in elastic particles, which being pressed together by their own weight constantly endeavour to expand and free themselves from that force" (Shaw, *Philosophical Works*, 2:426).

Shaw's description of the spring of the air is nothing but a description, in the most limited sense of that word.[49] It flat out refuses to spring. Because they do not exert any kind of syntactic pressure on its own constitutive elements, his sentences appear to exorcise Boyle's air of its encumbering reflexivity. But if one response to Boyle was to express all of air's imaginative potential from it, his air's affinities with the romantic and spiritual airs of the past, other readers were determined to restore it. Drawing insight as much from Boyle's fugitive *Notes . . . About the Atmospheres of Consistent Bodies* and his thoughts on the *Efficacy and Subtility of Effluviums* as from his physico-mechanical model of the air, they populate the following chapters. Having begun in Milton's Eden, though, this one will stop in the Royal Gardens at Richmond.

In one of those gardens stood a pair of architectural conceits, both commissioned by Caroline of Anspach, the German-speaking queen of George II. One was a small library known as Merlin's Cave. Opposite this indigenous "Fairyland" could be found the so-called Grotto, a structure (in the words of Edmund Curll's 1736 *Rarities of Richmond*) "very Gothique, being a Heap of Stones thrown into a very artful Disorder, and curiously embellished with Moss and Shrubs, to represent *rude Nature.*" In this degenerated Eden, "as soon as you enter you perceive in the Front a very curious Busto of the Celebrated and Honourable Robert Boyle, Esq.: Illuminated behind with Rays of Gold."[50] Four other worthies were also featured in Caroline's grotto: Newton, Locke, and the natural theologians Samuel Clarke and William Wollaston. All the same, if the *Gentleman's Magazine* of 1733 is to be believed, "the Bust of Mr. *Boyle* st[ood] higher than these, on the pedestal, in the inmost, and, as it were, the most sacred Recess of the Place; behind his head, a large Golden Sun, darting his wide-spreading Beams all about, and towards the others, to whom his Aspect is directed."[51]

This Boyle's informing context is the shady enclave of the grotto—a recess structurally and materially the same Merlin's Cave, and not so different from the one in Paradise where Milton's Eve "sat retir'd in sight" (8.41), bound in a world of mediated forms. In the manner of that same recess, Caroline's grotto was soon transposed into the modern mediating form of print, the only place it survives today. The *Gentleman's Magazine* of 1761 catches sight of it for the last time, "a grotesque building, which seems as if it had stood many hundred years." By now the grotto is "adorned with a kind of ruinous angular pediment; the stones of the whole edifice appear as if rudely laid together, and the venerable look of the whole is improved by the thickness of the solemn grove behind." But if Boyle's grotto has become "gothique"—more spuriously ancient and inward than radiant or aspiring, more atmospheric, really—one thing remains the same. "In an alcove is the truly honourable Mr. Robert Boyle."[52]

3

"Discontented Air"; or, *The Rape of the Lock*

Airs, languid airs are breathed around; the atmosphere is perfumed with affectation.
WILLIAM HAZLITT, *Lectures on the English Poets* (1818)

And that, as far as Hazlitt was concerned, was as close as Alexander Pope ever came to the "poet of nature." The fabled garden at Twickenham notwithstanding, the natural world in its ostensibly unmediated forms does not appear to have been one of his long suits. Thus, while Pope's writings and genius are almost always scrutinized in light of their contexts and milieus, these informing environments are usually taken to have been social or cultural, political or aesthetic—to have been, in short, formal themselves.[1] "He could describe the faultless whole-length mirror that reflected his own person better than the smooth surface of a lake," was Hazlitt's verdict. And it's true; if ever we are so lucky as to come across a body of water in Pope's poetry, the surface is all we will find.[2] Behold "the musing Shepherd" of *Windsor-Forest* (1713): he peeks into the "Glass" of the river Loddon and, perceiving nary a drop of water, contents himself with images of "headlong Mountains," "downward Skies," "absent Trees," and "floating Forests [that] paint the Waves with Green" (211–16).[3] In Pope's writing we are never outside formal mediation, and consequently we never seem to be outside.

So it is interesting to read *The Rape of the Lock* (1712–14). Apart from the early appearance of its heroine, Belinda, "on the Bosom of the Silver *Thames*" (2.4) and her severed tress's spectacular "Progress thro' the Skies" (5.133) at the end of the poem, what little happens in Pope's mock epic does so behind closed doors, in a world shaped by human hands; at the end of the day, the skies and the Thames have come to look like stage sets themselves. And yet this poem would seem to be as fraught with weather as any gothic novel. Breezes blow through every canto, sometimes mounting into veritable gales; mists and vapors rise one minute, only for lightning to flash the next. "Pungent Grains of titillating Dust" (5.84) are flung. Some of these renegade phenomena are easily contained as social metaphor, clearly designed to press home the satiric moral that we are adrift in a world where pretenses—the "airs" that human beings affect—matter far more than anything that, well, matters. "Airs" here carry the acculturated sense of studied attitude, a sense compatible with the more presumptuous figures of speech, such as allegory and personification, that carry *them* (at one juncture, "Affectation" herself "faints into Airs" [4.30]). But Pope also brought plenty of "grosser Air" (2.83) into his poem, and that air matter-of-factly denotes the material body to which Boyle recently had assigned the physico-mechanical properties of spring and weight. Unencumbered with significance, this literally "discontented Air" (2.79) appears to be at odds with Pope's other air of studied formality.

Appearances, needless to say, can be deceiving. For the rival airs of *The Rape of the Lock* make peace as they bind this poem to the branch of contemporary natural philosophy devoted to what was beginning to be called the atmosphere. As we will see, English aerial philosophers writing in Boyle's wake sought a language that could adequately convey both the form of the air about us and the story of its actions and assorted contents. Such a language proved very difficult to come by. For all the lift that Boyle's pneumatic physico-mechanics might have lent it, the study of the atmosphere would not assume a recognizably modern shape, with any settled predictive power or broad cultural authority, until the early nineteenth century, well behind physics, chemistry, biology, and the like. Meteorology still lacks the prestige of those disciplines, even if it remains the only one with its own satellite channel.

In retrospect, such humiliation looks like something we can explain. Often more giddily speculative than strictly lucid, early atmospheric writing seldom drew hard lines between formal contrivance and forces not humanly compelled. Edmund Bohun's 1671 *Discourse concerning the Origine and Properties of Wind* concludes that there is "little more to enquire concerning the *Naturall*; then may be advantageously Explicated from the *Artificiall* Winds" that he had recently whipped up in his wind-making machine, the so-called aeoliphile. As a matter of fact, ventured Bohun, "we imagine the Atmosphere to be as one immense *Aeoliphile*."[4] More than sixty years later, Pope's crony John Arbuthnot, in his *Essay Concerning the Effects of Air on Human Bodies* (1733), was still distinguishing that "wonderful Fluid" by its facile passage between what we shape, and see, and what we do not. "The Air" is not only all about us, all the time, Arbuthnot stressed, but is also "one of those *ingesta*, or things taken inwardly." Ingestion gives "things" a different, less legible complexion when they reappear.[5]

Its suspense between gross and refined, literal and figurative, inward and outward registers (and techniques) of meaning did make the air we breathe a player in early social science. In a work that anticipates that discipline—the witty, weighty *Characteristicks of Men, Manners, Opinions, Times* (1711)—Arbuthnot's Whig contemporary the Earl of Shaftesbury complained that, long ago, "every one took the air and look which was natural to him," only in modernity to assume "one single peculiar air, to which it was necessary all people should conform."[6] Not unlike such modern sociologists as Georg Simmel and Erving Goffman, Shaftesbury was committed to what he influentially termed the sensus communis—a sphere of shared impressions that Shaftesbury imagined to be upheld, if not constituted, by just such "airs" as *Characteristicks* gauged. As for natural philosophy—in matters pneumatic seldom quite distinct from social philosophy—the apparent fact that people both inhabit and make a habit of the air had always to be taken into account. Not that this dispelled the supernatural aura once spotted by Hippocrates. "Air is what he means by one of the Powers of the Universe," Arbuthnot explained, "which he says Nature cannot overcome" (Arbuthnot, *Essay*, viii).

Planets can have atmospheres. But if "Air," always implicated in the human, belongs neither outside "Nature" nor altogether within it, might not a poem have one too? Pope invites us to think so. The invalid's anxious sensitivity to extremes of temperature pervades his letters, moving many a biographer to mention that he was "extremely sensible to cold, so that he wore a kind of fur doublet, under a shirt of a coarse warm linen." A bodice and waistcoat went on top.[7] Come the 1930s, Edith Sitwell was quick to attribute Pope's notoriously "bad temper" to his "extreme sensibility to cold." By literalizing it, Sitwell turned Pope's "temper" into both a term of knowledge and a condition of that most sociable and yet visceral of emotions, empathy. She positively shuddered at the iron brace that held the poet's twisted body "as the cold may hold together the bones of the dead."[8]

We might expect a cult of "sensibility" to come out of "the cold." What does instead is a diffuse yet sensuously embodied aesthetic response, one held together by the terms in which it is communicated. Thus Dame Edith mused that "there is nothing cold about [Pope's] poetry" (Sitwell, *Alexander Pope*, 14) and coolly praised his failure to please the "cold, damp mossiness" of Victorian tastes, as typified in Matthew Arnold's "chilblained, mittened musings" on the moral shortcomings of his verse. In Sitwell's view, its reader can only shiver with delight on perceiving that "the words 'airs' [and] 'hairs' that ripple through *The Rape of the Lock* have a little waving sound, like the slight breath of a summer air," the whole poem every bit "as fresh as the summer air blowing down the dew" (Sitwell, *Alexander Pope*, 275). A generation later, Cleanth Brooks's classic study of *The Rape of the Lock* would frolic far less wantonly on the surface of the senses. Yet Brooks too distinguished Pope's mock epic by reference to the quality of its apparent air: "Swift penetrated the secrets of the Lady's Dressing Room with what results we know," the new critic reported. "Belinda's dressing table, of course, is bathed in a very different atmosphere."[9]

Of course. But just what *is* the "atmosphere" that Brooks assumed his own reader would recognize? What does it mean to say, as a matter "of course," that a literary work "is bathed in" one? What is doing the bathing, for example? If a piece of writing may be said to be bathed in an atmosphere, and if we all somehow "know" what that means, does this make social convention and literary experience legitimate foundations of natural knowledge? Such questions have long hovered just above the history of *The Rape of the Lock*'s reception, where readers often sound like meteorologists, ever assessing Pope's "air."

"Air" is a slight syllable, and therefore, as Sitwell appreciated, it is appropriate to the insignificant matter at the heart of Pope's best-loved satire. That would be the coquette Arabella Fermor's "hair," which the poet subtilizes still further into his iconic Belinda's objectively nonexistent lock. Nonetheless, over what became the five cantos of her hair's ironic chronicle, the word "air" sustains more than one meaning or manner of making it. That word's currency within an interpretive community now three centuries old suggests that these meanings and manners may be reconciled only when all are subjected to a complex aesthetic register, one uniquely accountable to the visible medium that makes it possible. As it develops just such a register, *The Rape of the Lock* draws inspiration from the atmospheric writing that once experimented in Pope's own alphabetical medium with much the same matter that he did.

Fields of Air

It is nothing new to read Pope in relation to his era's emergent languages of nature. But the pioneer of that approach, Marjorie Hope Nicolson, saw a disparity between Pope's literary practice and those languages. She gave priority to the latter. Nowadays we're more likely to perceive imaginative writing and natural philosophy, especially in the early decades of the Royal Society, as coextensive, embedded in the same body of social practices, employing many of the same epistemological and figural procedures, and differing most significantly at the level of how and where the knowledge that each produced was valued.[10] Perhaps an even more considerable difference is one that Pope's poem makes apparent in its engagement with the air studies of his day: that of how each was believed and of how each conditioned its consumers to believe, and thus to inhabit, it. When it came to air, the differences between these nascent faith communities could be vanishingly small. "Hear and believe!" Ariel enjoins Belinda as he whispers his version of natural philosophy, most of it a fluent calculus of the "thousand bright Inhabitants of Air" who surround her (1.35, 28). Here belief is not just *about* what is about one; it is itself about—around—the aptly vapid believer, and ultimately it conditions what she can and cannot see.

Belinda's prototype, Arabella Fermor, is still notorious for what she could not see when she read *The Rape of the Lock*—namely, all of its satire. But was she—is Belinda—any more vapid than the sociable virtuosi of her day? As Boyle discovered when he tried to discriminate among the many "corpuscles" that evidently composed it, what we attempt to unify by calling it "the" atmosphere is by definition disparate and unstable; because it is all-encompassing, this same atmosphere refuses to be isolated as a truly separate object of inquiry. Nor did religion and superstition gracefully (if ever) surrender their interpretive authority when it came to the air about us, so long conceived as the medium of ethereal being, if not of ether itself . . . whatever that was. And despite new or improved instruments of detection and quantification—the hygrometer, the thermometer, the barometer, the air pump, and the aforementioned aeoliphile—this same air, atmosphere's sine qua non, stayed stubbornly out of sight. It was perceptible only indirectly, through its effects on the things it surrounded and by which it could, indeed must, be absorbed. "Air is not visible," was the first thing that Pope's friend Arbuthnot would find to say about it (Arbuthnot, *Essay*, 1).

Of course, Newton's gravity is even less in sight, and we have seen that enlightened natural philosophy never put occult qualities outside its ostensibly ocular purview.[11] But air's invisibility is unique in that its desultory contents (dust, clouds, raindrops, and the like) so powerfully create the impression that it is somehow seen.[12] As Arbuthnot put it in his study of air's many "Effects," "what we see in the Stream of Light let in by a small Aperture into a room is not Air, but Dust and other Bodies floating in the Air" (Arbuthnot, *Essay*, 1). As if to point up the limits of Lockean "Understanding," any plausible conception of "atmosphere," let alone "Air," thus had to be projected explicitly from within a cloud of secondary impressions; certainly, it could not be traced to sense experience *stricto sensu*. This is much the situation into which Pope's Belinda awakens at the beginning of *The Rape of the Lock*. She is immediately caught in a cloud of miscellaneous airborne impressions: the thump of a slipper, the tick of a clock, the whisper of a sylph. Just as immediately Belinda is cajoled to "believe" in this sphere of pure effects as if they were the "things" that produce them.

Contemporary aspirants to knowledge of "the" air found themselves in similar circumstances, compelled to believe something about ambient mediating forms before they were in a position to see anything at all. It is in this regard that Pope's rippling, reflexive "Fields of Air" resemble those we find in his day's natural philosophies of the atmosphere. Pope's air cites Boyle's depiction of "common air" as a material body governed by the physico-mechanical law of "spring"—a pivotal word in Pope's own invocation to his two designated readers, the "approv[ing]" John Caryll and the "inspir[ing]" Arabella Fermor (1.6). We have seen that Boyle's spring was both a matter of and a manner of conveying fact. Just so, when Ariel's "Airy Band" of sylphs "their careful Plumes displa[y]" (3.115), or when their "Colours change where'er they wave their Wings" (2.68), Pope's poem registers on-the-ground attempts to organize the diffuse aerial phenomena we call weather into a stable object of knowledge.

This had been the ambition of the English air pump's other fabricator, Robert Hooke. Even more than Boyle's *New Experiments*, Hooke's air studies of the 1660s approached their elusive subject matter not just via literary figuration but also through certain graphic innovations, particularly as these interacted with parts of speech. Reproduced as a showpiece in Thomas Sprat's 1667 *History of the Royal Society*, Hooke's "Method for Making a History of the Weather" (fig. 5) thus contrived a glossary of animating adjectives from "*Cleer*" and "*Checker'd*" to "*Hazy*" and "*Water'd.*" Arranged in columns that could be read "at one View" either up and down or sideways, Hooke's graph of secondary characteristics promised to bind written words to the transient appearances of the sky in such a way that the former might in itself yield experience (hence communicable knowledge) of the latter.[13] The adjectives that Hooke laid out ("clear blew," "a little checker'd," "red and hazy") seem to correspond to some independent object. But because that object is by definition evanescent, more pressing are these words' glancing interactions within a field of alphabetical impressions. To move among those impressions, however visually or notionally or indeed superficially, is in effect to practice moving among changing atmospheric circumstances. In turn, the reader who has learned to do so is likely to be the one who moves easiest among the sylphs who surround Pope's Belinda. No less than Hooke's "clear blew but yellowish" appearance of "the faces or visible appearances of the Sky," those "lucid Squadrons . . . dipt in the richest Tincture of the Skies" (2.56, 65) convey the air that conveys *them* through purely secondary characteristics such as color

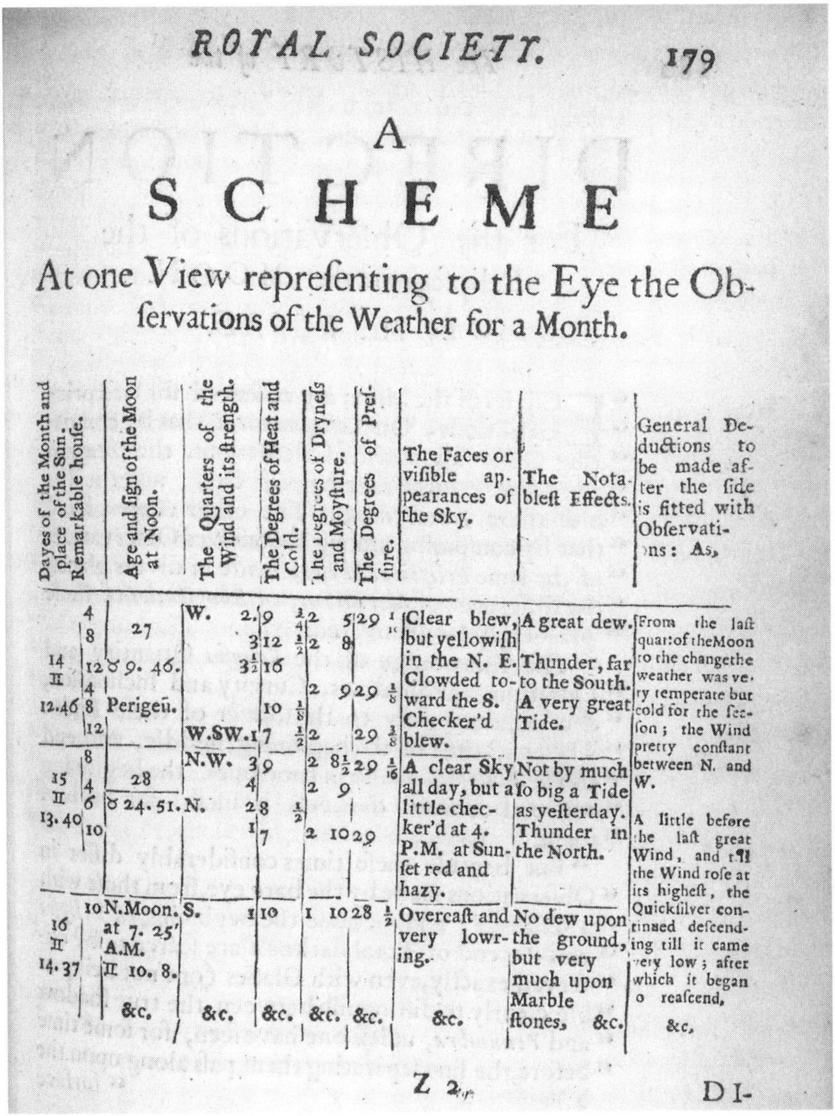

ROYAL SOCIETY. 179

A
SCHEME
At one View repreſenting to the Eye the Obſervations of the Weather for a Month.

Dayes of the Month and place of the Sun. Remarkable houſe.	Age and ſign of the Moon at Noon.	The Quarters of the Wind and its ſtrength.	The Degrees of Heat and Cold.	The Degrees of Drynefs and Moyſture.	The Degrees of Preſſure.	The Faces or viſible appearances of the Sky.	The Notableſt Effects.	General Deductions to be made after the ſide is fitted with Obſervations: As,
4 8 14 ♊ 12 12.46 4 8 12	27 ♉ 9. 46. Perigeü.	W. 2. 3 3½ W.SW.1	9 ¾ 12 ½ 16 10 ⅛ 7 ½	2 5 2 8 2 9 2	29 1/16 29 ⅛ 29 ⅜	Clear blew, but yellowiſh in the N. E. Clowded toward the S. Checker'd blew.	A great dew. Thunder, far to the South. A very great Tide.	From the laſt quar:of theMoon to the changethe weather was very temperate but cold for the ſeaſon; the Wind pretty conſtant between N. and W.
8 15 4 ♊ 6 13.40 10	28 ♉ 24. 51.	N.W. 3 4 N. 2 1	9 8 ½ 7	2 8½ 2 9 2 10	29 1/16 29	A clear Sky all day, but a little checker'd at 4. P.M. at Sunſet red and hazy.	Not by much ſo big a Tide as yeſterday. Thunder in the North.	A little before the laſt great Wind, and till the Wind roſe at its higheſt, the Quickſilver continued deſcending till it came very low; after which it began to reaſcend.
10 16 ♊ 14.37	N.Moon. at 7. 25' A.M. ♊ 10. 8.	S. 1	10	1 10	28 ½	Overcaſt and very lowring.	No dew upon the ground, but very much upon Marble ſtones,	
	&c.	&c.	&c.	&c.	&c.	&c.	&c.	&c.

Z 2 DI-

FIGURE 5. Robert Hooke, "Scheme of . . . the Weather," from Thomas Sprat, *The History of the Royal Society* (1667). Photograph: The William Andrews Clark Memorial Library, University of California, Los Angeles.

and motion. The sylphs' "purple Pinions" (2.70) and "azure Wand[s]" (2.72), their "glitt'ring Textures of the filmy Dew" (2:64) all report simultaneously on the medium that gives rise to them and on the world that that medium conveys.

No less continuous with contemporary aerography was Pope's rhetoric of "inspir[ation]" (1.6). This he deployed ironically in his opening lines and then went on to literalize as the breathing in of such befuddling substances as snuff or coffee fumes. Such figures of inspiration particularized the contemporary conviction that—especially in England, and even more especially in the places where English folks now congregated to make culture—the air had gone bad. If now swept clear of the angels and demons about which nobody seemed to agree, it was all the fuller of fumes. Everyone could probably believe in those. So there was something to be gained by saying that a person could practically see the air these days. Apparently, you could trace it through the impressions it left in the pathologies of respiration—asthma and allergies, "vapours," colds, and catarrhs—that seemed to multiply in an ever more urbane modernity. No wonder Arbuthnot was obsessed with "the Effects of the Air on Human Bodies." These "Effects" could be as social and performative as they were bodily, matters at least as much of consensually verified figure as of sensible fact.

Arbuthnot, for his part, claimed that "air operates sensibly in forming the Constitutions of Mankind, the specialties of Features, Complexion, Temper, and consequently the manners of Mankind" (Arbuthnot, *Essay*, 147). The notion that the air informs those who habitually breathe it was as old as Hippocrates, of course; in England it had been in the air at least since Burton's 1628 *Anatomy of Melancholy*. Burton's long "Digression on Air" holds that air "is a cause of great moment in producing [melancholy] or any other disease, being that it is still taken into our bodies by inspiration, and our more inner parts."[14] But when they spoke (or wrote) of air's operations, Arbuthnot and his contemporaries (including ideological opponents such as Shaftesbury) seem to have balanced those operations' apparent power to "cause" internal bodily states with pleased admiration of their shared "Disposition"—a common position, posture, or pose struck as much by the air as by the "Inhabitants" who after all exhaled as often as they took it in "by inspiration."[15]

As Arbuthnot saw (or at least put) it, there appeared to be an especially "great Conformity between the constitution of the Air and that of Diseases" (Arbuthnot, *Essay*, 125). The physician George Cheyne had made the "Distempers, Spleen, Vapours, and Lowness of Spirits [that] are in Derision called the ENGLISH MALADY" his business. And these, he surmised, were not simply the result of "the Moisture of our Air [and] the Variableness of our Weather." They could also be understood as indirect expressions of that moisture and variability.[16] Richard Blackmore's influential *Treatise of the Spleen and Vapours* (1729) would likewise spurn the notion of mechanical cause to propose that the "Natives of this Island" are merely *disposed* toward those ailments "from the peculiar Constitution of the Air they breathe."[17] The ancients were convinced that the vapors "consisted in the elevation of dark Fumes and Exhalations from the Matrice which ris[e] up in fruitful, but unwholesome Clouds" (Blackmore, *Treatise*, 100). But Blackmore proposed a mutually expressive transaction between the vaporous body and its environments. The Cave of Spleen where Pope's Belinda so sulkily sojourns captures the purely superficial—if reciprocal—sense of atmosphere that Blackmore's vapors project. "Sheltred close from Air" (4.21), certainly located neither around Belinda nor inside her, the cave seethes with phantasmic "Mists" and "various Tempers" (4.61). Such condensations both are and express their surroundings. Pope's neighbor Mary Wortley Montagu would make the satiric most of this double capacity when she pretended to see "perpetual fogs enclose" his celebrated grotto at Twickenham.[18]

As for Belinda's Cave of Spleen, "no cheerful Breeze this sullen Region knows, / The dreadful East is all the Wind that Blows" (3.19–20). Pope's famously slippery set piece might look like exactly that—a set piece. Nonetheless, it actively binds his cartoon of the cave to high-minded medical writing about "the *Hysteric* . . . fit" (4.60). Blackmore pitied the hysteric racked by "intestine Tempests" and "flatulent Effluvia," the internal "vapours and conflicting Winds" that could so easily "arise to a raging Storm" (Blackmore, *Treatise*, 19, 144). As it is in Belinda's cave, this inward climate might well be related to the outer one causally or analogically; we still think about moods this way. But regardless of the nature of their supposed confluence, what held fast were the superficial figures that conventionally mediated between internal states and outward appearances. Claude Du Bosc's atmospheric "sculpture" for canto 4 (fig. 6) harnesses the order of experience that arises when such figures gain primacy and expository authority. He thus manages to draw his viewer (Pope's reader) into the cave's spherical form. Here the clouds not only create depth but *are* it as they boil above and below, in foreground and background. The image creates its peculiarly mixed impression of ironized interiority through the clouds' delicate suspense between individual projection and objective phenomenon. Their outlines trace Pope's "Goddess with a discontented Air" (4.79), even as she appears to be tracing the conditions of her own appearance in graphic space.

Thought characteristic of insular Britain, a baleful atmosphere was easily implicated in disorders other than those of mood. Boyle had been fascinated by the "poisonous steams" that seemed to rise from everything from spiders to carnations and Spanish gloves, provoking allergic reactions in the human bodies bound to receive them through the very air they filled.[19] Fast on Charles II's restoration, John Evelyn's twice-printed *Fumifugium* (1661), as we saw, vividly realized what its subtitle depicted as the "Inconveniencie

FIGURE 6. Louis Guernier and Claude Du Bosc, *The Cave of Spleen*. From Alexander Pope, *The Rape of the Lock* (1714). Photograph: The William Andrews Clark Memorial Library, University of California, Los Angeles.

of the AER and Smoak of London." And as his 1707 *Treatise of the Asthma* chronicled a modern epidemic of shortness of breath, John Floyer found not just that "the Changes in the Spirits of Asthmatick Persons . . . succeed on the Changes of Weather" but that the "Spirits of the Asthmatic" cannot "bear the Air of *London*, whose Smoak, like Fire itself, irritates [them] into an Expansion."[20] Such "an Expansion" both interprets and expresses that humanly impressed air.

To infer the character of atmosphere from its effects on the formal structures that in turn affect *it* was to withdraw from the language of immediacy that was beginning to dominate the languages of nature. It was to seek refuge in a sphere of vapors—of strange shapes and uncanny conveyances whose autonomy and materiality were so bound up with their immateriality that they might be most faithfully accounted for in the compromised language of art. Even in Bernard Annely's drily titled *Theory of the Winds* (1729), for example, "air is a yielding Body [that] may be impressed by innumerable Modes and Actions."[21] So while Arbuthnot paid lip service to Boyle's "elastic" corpuscles of pure air, he devoted most of his attention to the affectations they effect. His "Atmosphere" molds all of its inhabitants into their characteristic forms: "No Vegetable or Animal," Arbuthnot insisted, "can expand its Fibres, in their Natural Figure, but in a Fluid that resists equally the Elongation of its Fibres. The Pressure of the Atmosphere keeps both Vegetables and animal Fibres within certain Limits of Accretion." This will "give a human Creature the due Figure which Nature affects" (Arbuthnot, *Essay*, 23).

Arbuthnot's "Atmosphere," however, develops shape and depth because it is always half-created by "the Steams, Effluvia, and all the Abrasions of Bodies on the Surface of the Earth" (Arbuthnot, *Essay*, 11). This closed circuit of mutual "Fermentations," interchangeable with affectations, renders any experience of the atmosphere an encounter, at once social and intimate, with spectral projections on which consensual meanings are ever only temporarily conferred (Arbuthnot, *Essay*, 15). Arbuthnot, like Evelyn, insisted that air's "occult Qualities" had no place in his treatise (Arbuthnot, *Essay*, 42). But his attempts to graph the many strange and delicate transactions between "human Bodies" and the air about them brought him into deeper contact with what remained unknowable: "It is impossible to conceive the Result of all [air's] Operations in a heterogeneous Mixture," Arbuthnot wrote. "Mankind may be sensible of their Effects, but can never know their Nature" (Arbuthnot, *Essay*, 16).

Because it shows both imaginative speculation and social manner to be conditions of individual sense experience, Arbuthnot's claustrophobic yet contingent representation of "Atmosphere" parodies claims of empirical certainty, leaving its reader room to suppose that such claims might be nothing more than positions temporarily adopted within the vaporous sphere of mediating forms.[22] Even if "there are Congelations of the redundant Water Precipitations, and many other Operations unknown to Art, producing a great variety of effects," the air is still "the instrument of all those Operations when perform'd by Art; and those heterogeneous Bodies act upon one another in the Air itself, by many unconceivable ways" (Arbuthnot, *Essay*, 16).

Arbuthnot's atmosphere was the "unconceivable" effect of artful air, a way of downplaying heterogeneity while yet displaying it. As such, airy "Congelations" were implicated in human pathologies of inspiration not as a cause but rather as an explicit condition of things "becom[ing] visible" (Arbuthnot, *Essay*, 6). Now even celebrated tools of lucid modernity such as the weather glass cease to appear as instruments of enlightenment. They betray their close kinship with consensual modes of image making, and their work is to apprehend outsides that cannot be told from what lies within. Thus could Bohun "imagine the *Atmosphere* to be as one Immense *Aeoliphile*." Charting "the wonderfull effects of Winds," in which "*invisible Agents* are diffus'd every where," he found he could "believe much of that *Naturall Magic*, by which distant Bodys act upon each other (though we usually ascribe it to *Occult Qualities*) . . . do proceed from no other cause than the secret operations of Winds: For they are the *Carriers* of the Universe, and transport from one place to another, odors, diseases, [etc.] of which we can give no account how they came there; as vegetables that spring on the tops of houses" (Bohun, *Discourse*, 207).

What made it possible to "believe" such things? Perhaps it is the manifest affinity between Bohun's lettered medium and his aeolian "*Carriers* of the Universe," at once everywhere and nowhere, many and one. Contemporary writing about bad air often took shape within the conspicuously literary traditions of satire and pastoral. Long before Tobias Smollett's 1771 novel *Humphry Clinker* would caricature the "thousand frowzy steams" that compose "the fragrant aether we breathe in the polite assemblies of Bath," Arbuthnot put a Juvenalian twist in the "sulphureous Steams" of "the Air in which People of Fashion pass a great deal of their time" (Arbuthnot, *Essay*, 102).[23] He may well have been inspired by Evelyn's indignation at contemporary pollution. Anxious to awaken London readers to the truth about the "AER" about them, Evelyn had rejected air as a "sublime Subject." As would Arbuthnot, he aimed to concentrate on "the less pure, more turbulent and dense, which for the most part we live and breath in," that taken in by "common Respiration."[24]

Literal though his mind aspired to be, however, Evelyn still detected a dark, coercive power at work in the "AER." Like Bohun's windy "*Carriers* of the Universe," Evelyn's bad air is an omnipresent, and, "as it were, universal *Medium*" that "suffers . . . infinite mutations, both from superiour and inferiour causes, so as its accidental effects become almost innumerable" (Evelyn, *Fumifugium*, 14). These effects include new mutations, themselves "affected, . . . which render [the air] noxious to the *Inhabitants* who derive and make use of it for life" (Evelyn, *Fumifugium*, 13). Bad air's most insidious "effects," though, were unaccountably visible. They included "congestion so misshapen" that it can never be penetrated, streets "narrow and incommodious" thanks to an "ill and uneasie form of pavement," and even a "troublesome and malicious . . . Disposure" of spouts and gutters. Air pollution was an "Evill so epidemicall" that it had "rendered" London itself "a *Labyrinth* in its principal passages," right down to the "Deformity" of its wharves (Evelyn, *Fumifugium*, 8).

Here "AER" doubles as a perverse writing function, a condition of significant form. Through its invisible agency, visible marks appear automatically, taking too infinite a variety of twists and turns, the devil always in the endlessly prolific details. To ward off the intimate—and infinite—possibility that lurked in such details, Evelyn projected a counterart of edenic "Closures, or invironing Gardens." Such containing forms attempt to realize a fantasy of distanced communication: the flowers and shrubs that Evelyn imagines for his restorative "inviron[ments]" are the ones "aptest to tinge the *Aer* upon every gentle emission at a great distance." Evelyn's own writing, of course, could not literally convey their "ravishing Perfumes" (Evelyn, *Fumifugium*, 47). Nor could it transfer meaning through the air to a king praised as "the very Breath in our Nostrils." (Evelyn, *Fumifugium*, 4). Such limitations give Evelyn's writing an uncanny resemblance to the misshapen "Disposure[s]" that register man-made air's active presence. Hence even Evelyn's pet remedy is imagined through literary contrivance, as a "Catalogue of native *Plants*, and such as are familiar to our country and *Clime*, whose redolent and agreeable Emissions would even ravish our senses" (Evelyn, *Fumifugium*, 47).

At once abstracted and immediate, the artful remedies of pastoral form were finally of a piece with bad AER's own horrid arts of imposition and "disposure." Both arose from the page, in that act diverting it from its charge of virtuous persuasion toward autonomous art. Indeed, to readers of our own day, it often appears that Evelyn's work "is genuinely literary" and even deserves to be regarded as if it were "a painting or a statue."[25] Evelyn himself seems not to have minded that *Fumifugium* had no special effect on royal air policy; it was enough that its sovereign reader had declared himself "much pleased with it"—"was," Evelyn boasted in his diary, "pleased to discourse to me about my Book inveighing against the Smoke of Lond" [*sic*].[26] And quite as Evelyn may have desired Charles's pleasure more than anything else, so do many accounts of *Fumifugium* offer mainly graceful synopses of its "plot." One even locates this work's cultural significance in its novelistic power to "remin[d] us that the fascination of history lies in the discovery of situations similar to ours and yet distinctively different."[27]

As it conspired with literary art, Evelyn's "Smoak" may be said to have given rise to a social and moral conception of atmosphere, one that both puts it outside us and affords us a place inside it. Other iterations of bad air included what Pope, in his later *Essay on Man* (1733–34), would christen "quick effluvia."[28] Effluvia were invisible particles, themselves neither quick nor dead, that Boyle and others supposed might travel through the air from one body to another, bearing odors, spreading illness, and in turn guaranteeing indirect influence. As its chronicler Gordon Chalmers has shown, "effluvium" was a novel word in its own right, born into English print only thirty years before Pope was born. But from the moment of its conception, it transmitted ancient theories of emanation. These not only united actions in the physical world with those of divinity and human sympathy; they also fluctuated incessantly between literal and metaphorical poles of reference.[29]

Natural philosophers such as Boyle, Bohun, and Kenelm Digby are easily seen merely to have arrested this fluctuation, materializing and mechanizing the "effluviums" in question. Fixing the word in writing gave such naturalizing philosophers a way to picture, jointly if indirectly, the possible insubstantiality of solids, the solidity of insubstantials. As Bohun put it, citing Epicurus, "Continuall effluviums do issue from all material concretes." Consequently, there is "scarce any Body so solid, which pays them not some tribute as they passe" (Bohun, *Discourse*, 135). Such reciprocity made the word "effluviums" itself a dynamic figure for the modes of literary experience through which it was encountered. As Sophie Gee has recently shown through a reading of Pope's own *Dunciad* (1728–42), the imagery of effluvia thus seems to have been custom-built at once to regulate and to revile the bad literary matter that seemed to be everywhere in early eighteenth-century England.[30]

Boyle's own 1673 *Essays of the Strange Subtility, Great Efficacy, and Determinate Nature of Effluviums* finds strangeness and subtlety perfectly compatible with efficacy and determination. This duality is for Boyle inseparable from the dynamically mixed conditions of the communicative act to which writing gave material form. He "hop[es] the reader will not think it strange, not to meet with in the following papers, a more close and uniform contexture of the passages that make them up." In order to ameliorate this "strange[ness]," he "seasonably inform[s]" that reader, one Pyrophilus, "of the rise and occasion of penning them," presenting his atmospherical "Papers" as, in effect, emanations from his earlier writings on niter (Boyle, *Strange Subtility*, 3:659). Here, effluvia's peculiar qualities realize the figures of speech that conveyed those qualities. Just so, in order to transmit the idea of effluvia to his reader, Boyle cast about for some small object that, because it "can be managed with our hands, . . . may help us conceive [the] wonderful minuteness" that is effluvia's signature. He finally hit on a "gilt wire as slender as a hair" for this very purpose (Boyle, *Strange Subtility*, 3:663). In something of the way that Gee's research into eighteenth-century effluvia coincided with *The Scandal of the Season* (2007), her atmospheric fictional interpretation of *The Rape of the Lock*, Boyle's effluvial analogies tempt us back to Pope's poem. Without exactly grounding Pope's satiric proposition that Belinda's hair is a trifle light as air, Boyle's "hair" stimulates an interior awareness of the figural layering through which propositional language achieves its most sensible effects.

"Effluviums" can be apparent only in literary environments where signs are, paradoxically, both abstracted and available to at least two of the ordinary senses. It was in such an environment that Boyle could hope *both* "to manifest [the] truth" *and* to "help your imagination to conceive, how great a space may be impregnated with the effluxions of a body," as in the case of the emanations that rise from "the substance left upon the grass, or ground, by the transient tread of a partridge" (Boyle, *Strange Subtility*, 3:674). In turn, the "free atmosphere" that rises from the "exhaling matter we are speaking of" may grow effectual and ponderable only by means of the sensibly insensate literary imagination. At the very least, Boyle deemed it "not improbable" that "no sense can immediately perceive" effluent particles (Boyle, *Strange Subtility*, 3.675). In written form, "effluviums," in turn, make nothing so apparent as the mediated nature of what one believes to be true. As they visibly compromise any possible boundary between literal and figural meanings, "effluviums" foreground the role that imagination, projection, and provisional belief play in the getting—and conveying—of collective knowledge. Indeed, Boyle predicts that those who confine themselves to "the negative informations of their senses" will automatically deny "the power and efficacy of effluviums" (Boyle, *Strange Subtility*, 3:678).

What about "effluviums" in action? Only in Pope's *Essay on Man* does anyone actually "die of a rose in aromatic Pain" (1.199), but Boyle had plenty of allergy anecdotes to retail. His pages teem with high-strung poets and histrionic ladies-in-waiting collapsing under the weight of the otherwise invisible "effluviums" supposedly emitted by damask roses and other deceptively pretty offenders. These lively verbal sketches coupled indirection with emphatic detail to make the invisible properties of the atmosphere seem visible to Boyle's own readers. But to believe that there are such things as "effluviums," you need to believe that "one body [can] communicate to another a solitary and naked quality, unaccompanied by any thing corporeal to support and convey it" (Boyle, *Strange Subtility*, 3:671). If Boyle here seemed merely to be describing "Effluviums," he also revealed the "strange subtility" through which written words work the rooms of collective belief.

Impressed by the human figures that it impresses, deludes, and sickens, the smoke- and snuff- and scent- and steam-infested air in *The Rape of the Lock* shows Pope to have absorbed his era's copious writing of the atmosphere.

As a satirist, Pope was no doubt enticed by the fact that it was often bad air that made atmosphere seem most visible. Still, he brought the literature of pneumatic fact into his poem not skeptically or parodically, as Swift would have done, but with the intention—or at least the effect—of producing an atmospheric reading of his own.[31] If the atmosphere in question is the perfumed indoor one of pure affectation that Hazlitt registered, Pope's moody *rapprochement* with it differed little from the prolific figures that others were using at once to read and to represent what John Wilkins had depicted as the "Sphere of vaporous Air which encompasses our Earth."[32] Such a body can, evidently, be perceived only through mediating forms. Of these, the human form is one, and human forms of speech are another. But writing would appear to be the most revealing one of all.

Pathologies of Inspiration

Pope, like Arbuthnot, suffered regular attacks of asthma, and it so happens that asthma is one of those pathologies of respiration that make it hard to say where the person ends and the air begins. Poised to make it easier to do so, however, were shallow, charming figures like that of what asthma's first modern historian, John Floyer, called "Effervescence of Humours." Indeed, in an anticipation of Coleridge's psychosoma, Floyer also maintained that "the Asthma" expressed itself simultaneously in "the Fancy" and in "real Distempers" (Floyer, *Treatise of the Asthma*, 11).[33] Pope's early biographer, Owen Ruffhead, reported that at death Pope "yielded his breath so imperceptibly, that the people [who did] most constantly attend him, could not tell when he expired."[34] Pope's life had been consumed with an ever more epic battle against that same air. This battle the poet appears to have waged partly through written correspondence with a circle of fellow sufferers that included Arbuthnot. "I have these 3 months or more, been advancing to an Asthmatic Complaint, (from one Cold to another, as I believe)," Pope wrote. "It is now at such a height, that I can scarce walk, or go up a pair of Stairs, or move much in my body, without quite losing breath."[35] It is unclear whether the poet felt the air to be closing in on him or depriving him of its ministry; his tendency to impute his many ailments to the wind suggests more the former than the latter. Still, air's touch was precious too: "I am grown so tender as not to be able to feel the air," Pope once confided to Hugh Bethel, adding a few weeks later that only a good bloodletting saved him when he "stopt . . . almost from breathing," and later still complained of "shortness, & sometimes almost Stoppage of breath" (*Correspondence*, 4:461, 517).

"Shortness" and "Stoppage," of course, are two of the metrical signatures of Pope's heroic couplet, even as he drops air's name there over and over again. He even prefaces his 1717 translation of Homer's *Iliad* by commending the "*Aeolic*" dialect, "which often rejects its aspirate."[36] In Pope's aspiring version of that epic, a "vital Air" (Pope, *Iliad*, 1.112) springs everywhere. Breathed by Achilles and Hector alike, this air is made by the inspiring gods to harbor all manner of thing from "vapours of a noxious kind" (Pope, *Iliad*, 8.141) to "living lightnings" (Pope, *Iliad*, 8.267) and "driving snow" (Pope, *Iliad*, 19.381). The pastorals that Pope published at the tender age of seventeen were borne aloft by the same "vital air," which in turn wings Belinda's lock to its ethereal apotheosis. Who is to say that this is not also the breeze that wafts its way through Pope's early "Ode on Solitude," its speaker so "content to breathe his native air, / In his own ground" (3–4)? Like his friend John Gay, Pope composed more than a few lines "on a FAN," and the reader of *Windsor-Forest* sails through "Airy Wastes" where "high in Air *Britannia*'s Standard flies" (111) and shot lapwings plummet, "leav[ing] their little lives in Air" (130–34). Pope sent a verse epistle, borne on the "Wings of Winds," up to a rising expert on the air, John Arbuthnot himself. Even at the bottom of that fetid swamp of perverted inspiration otherwise known as *The Dunciad*, a lonely "child of Heat and Air," the butterfly, valiantly "swims along the fluid atmosphere" (4.422–28).

This particular butterfly manages to transcend the satiric matter of hot air and noxious wind, an ethereal figure of unanticipated beauty. Its flight makes apparent the contours of the "atmosphere" it plies. That atmosphere seems at least as much to rise from the body of Pope's poetry as to stand apart from it as an independent point of reference. Its delicate realization of its own environments can make the "subtile body" of Pope's poetry seem a uniquely convincing form of natural philosophy, convincing because its most basic concern is with the conditions of the experience that it provides. Ruffhead certainly took Pope that way when he declared that the poet had "penetrated into the general ground and reasons of speech." In Ruffhead's view, Pope had even "reduced his natural talent for poetry to a science," thereby delivering "proof in all the sublime of poetic thought and expression" (Ruffhead, *Life*, 20, 51).

Proof of what we are not told, but even Pope's most irritable contemporary reader, John Dennis, likened the sylphs of *The Rape of the Lock* to "Eels in Vinegar" that "are only to be seen thro *Microscopes*."[37] At the same time, the air in Pope's poems limns states and orders of belief undreamed of in the standard transactions of the Royal Society and left implicit even in the popular atmospherics of an Arbuthnot or Evelyn. As David Fairer has beautifully marked out, Pope's clouds and vapors are traditional figures for the centripetal labors of the imagination.[38] His "air" mediates between consensual and freely speculative versions of experience, not simply equalizing them but endlessly framing and reframing them, the one within—then around, then again within—the other. So, in exchange for the supposedly reassuring grounds of probability, do we gain the sense of provisional conviction. As Pope ambivalently realized, this sense materializes, if ever temporarily, through the printed pages of his work.

How else to account for the "lenient air" with which our speaker longs to "extend a Mother's breath" (Pope, *An Epistle to Arbuthnot,* 410) in Pope's epistle to that splendid composer of "the Effects of Air," John Arbuthnot? As is well known, Pope's own mother was dead by the time he composed these pious lines, and so what looks like ethical obeisance toward the living actually secures the frame of expressive convention as the *Epistle*'s absolute horizon. More exactly, lenient—softening, permissive—air here mediates between ethical, experiential, and aesthetic propositions to project an environment consisting only in the mixed media of proposition. Because it is nonpropositional from the get-go, such an environment can certify nonsensory events as legitimate versions of experience. Meanwhile, in the first three of Pope's four early *Pastorals* (1709), air appears to be little more than a figure commissioned to transmit Virgilian tradition. The "vital Air" (73) of *Spring,* the "cool Gales that fan the Glade" (73) in *Summer,* and the "Winds" that in *Autumn* "appear to breathe, the waving Woods to move" (41) are strictly imitative, their transitive power "to move" but the effect of grammatical misdirection. But in Pope's fourth pastoral, *Winter,* the wind picks up, gaining sudden and lively immediacy via the shepherd's elegiac lament for one most fortuitously named Mrs. Tempest, which no doubt impressive person happens really to have died in England's Great Storm of 1703–4.

As Pope's note to *Winter* so modestly puts it, "Mrs. *Tempest*'s Death gave a Propriety to this Eclogue, which in its general turn alludes to it" (135). In his fourth pastoral, that is, our poet hit on a new direction for the wind to take and, by means of a pun, harnessed its now evident power to "turn allu[sion]" into virtual identity. The discovery that unadulterated convention could accommodate matter of fact with no damage to its figural integrity fanned the flames of epic ambition precisely to the extent that that same ambition was convenient to ironic self-reference. In a few short years Pope's sardonic "Receit to Make an *Epick* Poem" would list "Tempest[s]" as ingredients vital to this literary form as he conceived of it. The inspired poet is advised to add the mythic forms of "*Eurus, Zephyr, Auster* and *Boreas*" to the natural ones of "Rain, Lightning, and of Thunder (the loudest you can)" and to "mix your Clouds and Billows well together till they foam." Above all, you should "brew your Tempest well in your Head, before you set it a blowing."[39] A good tempest, it appears, might well tear things apart, but it also coordinates expressions and impressions with apparently random phenomena. Made visible, if facetious, this integrative activity may in turn give rise to a uniquely reflexive order of experience—one that the poet, quite unfacetiously admiring the work of "warm" Homer, made out to be "an ampler *Sphere* to expatiate in," a "new and boundless Walk for [the] Imagination" (Pope, *Iliad*, 5). This "walk," or "*Sphere*," becomes apparent through the figures of the very "imagination" that it contains, in much the manner that our atmosphere only becomes apparent to us through the transient shape of a cloud.

Even in his personal correspondence, Pope appears to have delighted in the way that, when we actually see them, "atmosphericall" words can seem to spring open into their heterogeneous referent. "Sir," begins a representative letter to the Bath physician William Oliver, "Sir,—Since the Congelation of all my Faculties by this Frost, you will not wonder that I am not un-thawed before Eleven in a morning, or I had been with you at ten this day. I find myself less encumberd by my [own] Planetary Crust, than [by] this Atmosphericall Shirt over it" (*Correspondence*, 4:218).[40] As Pope turns it inside out, reversing the natural crust and the man-made shirt, the "Atmosphericall" presents itself as the condition of both sense experience and figurative expression. In this manner it guarantees the aesthetic experience that it elucidates.

Vital Air and Paper-Durance

Pope's physical fragility, it seems, had everything to do with his ability to conceive of atmosphere as he did. Here we might summon the ghost of another eminent invalid, Virginia Woolf. Woolf's 1926 reflections on literary atmosphere took shape in the context of her famous lament that "English . . . has no words for the shiver and the headache," and she maintained that it is those stranded in that intimate microclimate of the shiver and the headache who win permission "to look . . . at the sky."[41] Because their relationship to that sky is hypermediated (by fever, say, or chills), her invalids "discover [it] to be something [rather] different from" what it is ordinarily supposed to be. Indeed, it is not a "what" at all but rather an "incessant making up of shapes and casting them down," a no less "incessant ringing up and down of curtains of light and shade," and, as we saw, an "interminable experiment with gold shafts and blue shadows" (Woolf, *On Being Ill*, 6). As we also saw, Woolf's atmosphere is an "endless activity" not distinguishable from those of hallucination or art. No object is revealed to the convalescent transfixed by the physical atmosphere, nor is a setting for any other thing. She, or he, is left with a fleeting impression contained—and contained by—the perpetually suspended participial phrase that expresses it.

Woolf reckons that literary "atmosphere" also hangs suspended between impression and expression. Flipping listlessly through a biography by the now-forgotten Augustus Hare, she endures an opening "plethora of uncles and aunts" in which "we flounder and threaten to sink" by reminding herself that "there is such a thing as atmosphere, that the masters themselves often keep us waiting intolerably while they prepare our minds for whatever may be—the surprise or lack of surprise" (Woolf, *On Being Ill*, 24). In its literary sense, Woolf's "atmosphere" is a condition of ambiguous precipitation, a subjectless stranding among effects that may or may not also be causes themselves. In this, it differs little from the "incessant making up of shapes and casting them down" that goes on all around us, all the time.

Such making up and casting down might with justice be called precipitation. And the problem of precipitation bothers Pope's mock epic narrator at the beginning of *The Rape of the Lock*. The first thing he wants to know is "what dire Offense from amorous Causes springs" (1.1). Of course, this isn't phrased as a question at all: what we have, rather, is simply a springing from amorous causes—a reflexive movement that precipitates a narrative that is in reality quite as empty of truly dire offense as it is of genuinely amorous cause. It's only appropriate that this narrative should draw near its close with Belinda's defender, the truly trivial Sir Plume, brandishing his "*clouded Cane*" (4.124) while her ravisher, the baron, invokes the "vital Air" (4.137). "Empty" and sylph-infested, "gros[s]" and "ethereal," air has by canto 4 contained (if barely) a veritable plethora of stupid events that have no truly meaningful precipitants and that mean nothing—to say the least.

And yet, for all its vapidity, Pope's discontented air remains more "vital" to this poem than to any other that he wrote. Why? We could point to *The Rape of the Lock*'s satiric proposition that Belinda's trifling world is as light as air and, like it, amounts to nothing. Of course, we have seen that the air actually felt pretty heavy to Pope and his contemporaries, and anyway there is the inconvenient truth that Pope's poetry holds the fixed analogy that such propositions entail in some contempt, preferring the more dynamic figural machinery of metonymy and metamorphosis.[42] From this point of view, it isn't that Belinda and her world are as insubstantial *as* air so much as that air is common to both her world and the real one it references. If air does anything, it precipitates these worlds' perpetual conversion into and out of one another. This would be how Pope's poem manages to dismiss the bubble-bright world of the historical Arabella Fermor while also (as is so often observed) participating in that world.

What if we were to go further and see Pope's chronically discontented air as inspired by (and from) contemporary expressions of the atmosphere as they had taken shape within the emergent "sphere" of modern letters, an intrinsically social element common to poets and natural philosophers alike? To place Pope in this milieu would be to confront the possibility that his poem's propositional features (Belinda is like air; her airy world refers to Arabella Fermor's) are contained and conditioned by the very "sphere," at once tangible and transient, that so many of his contemporaries had specified and that his future readers would recognize as "atmosphere."

One place where this "sphere" makes itself apparent is at the acoustic and metrical level of Pope's verse. Readers quickly notice Pope's quasi-musical reliance on sense-echoing sound, and so, literally, on the fluctuation of air currents. It's an effect well captured in the "little waving sound, like the slight breath of a summer air" that Sitwell claimed to hear whenever she read *The Rape of the Lock* (Sitwell, *Alexander Pope*, 275). And yet, as Joseph Warton pointed out, sound's grounding in sense is almost purely literary; it is, like Belinda's airy hair, "in Paper-Durance bound" (4.105). Pope mastered the art of prosody, according to Warton, by "copying [Dryden's] harmonious versification." In turn, his poetry springs from a fiction of aerial undulation, a virtual experience whose medium is the airtight and thence most convenient one of English letters.[43] Samuel Johnson too noted the necessarily visible conditions of what appears to be purely acoustic. "Dwell[ing] a little on the celebrated paragraph [in Pope's *Essay on Criticism*], in which it is directed that 'the sound must seem an echo to the sense,'" Johnson noticed that to the extent that this "precept" is even observable, it may be traced to two conditions: to the merely conventional supposition that "sounds are estimated by their meaning" and to the decidedly nontonal effect of "motion," a rate of progress across the page. Here the sense *that* sound echoes sense is merely an effect of the strictly literary phenomena of "delay" and "rapidity." In Johnson's book, Pope's plain-scouring swift Camilla, so often cited as a case in point of sense-echoing sound, is nothing but an "experiment upon *sound* and *time*." With this "experiment," fleeting sound has little—but material sound effects much—to do.[44]

It may then be that air is so "vital" (and yet potentially so toxic) to *The Rape of the Lock* precisely *because* it can't be told apart from such effects—effects of whose social orchestration Pope was so agonizingly aware. From this point of view, the atmosphere that Cleanth Brooks found to "bath[e]" Pope's poem (and Hazlitt to "perfum[e]" it) presents *The Rape of the Lock* as a hazy fiction. As in all fiction, supposedly mediating forms are in fact all that is there. But at least with respect to the "Paper-Durance" of the printed page, they *are* there, and they are there more definitely than anything that that page might posit. Indeed, such fictive forms are the one dimension of literary experience that is *not* posited, remaining for all its cloudiness the enveloping and determinate condition of any single position that a literary work may be said—or indeed perceived—to take.

His commitment to this barely endurable condition explains why Pope's epistolary address to his designated reader presents the coming poem first and foremost as a fiction. Here the poet loftily reminds the fittingly christened Arabella Fermor that in the pages before her "the Human Persons are as Fictitious as the Airy ones" and identifies those pages with "a French Book . . . which both in its Title and Size is so like a *Novel*, that many of the Fair Sex have read it for one by Mistake" (p. 217). Especially when personified in the sylphs that Pope seems to have lifted from that "French Book call'd *Le Comte de Gabalis*," it is his poem's air and its nebulous contents that most conspicuously mark it off as a literary fiction, "made of nothing," as Hazlitt put it. Pope himself confirmed as much. He insisted that his poem's contents were first "communicated with the Air of a Secret" and let even a relatively casual acquaintance know that "I see people so very busy about nothing, that I fancy, I am no improper historian to write their actions. It would be filling the paper as they live their lives."[45] Between seen "people . . . busy about nothing" and the nobodies of Pope's poem we have the practical condition of "filling the paper." This condition is an activity performed in the manner of weather by an "it" rather than an "I." "It" mediates between two pure forms and thereby turns them, in effect, into content. In this manner, "it" temporarily assuages the worry about writing's material body, at once too ephemeral and too substantial, that would plague Pope his life through.

The Rape of the Lock gives rise to—even as it makes apparent—the environment in which it may be perceived and at last (if always provisionally) interpreted. Even the poem's least lenient contemporary critic, Dennis, thus rightly traced one source of its generative power to "the *Puns* which are spread throughout it." These, Dennis judged, "bear the same Proportion to *Thought* that *Bubbles* hold to *Bodies*, and may justly be compared to those gaudy Bladders which Children make with Soap; which, tho' they please their weak Capacities with a momentary Glittering, yet are but just beheld, and vanish into Air" (Dennis, *Remarks*, 47.) Although it might seem that Dennis was merely scoffing at Pope's appeal to "weak Capacities," part of what seems to have irked him was the fiction maker's evident contempt for any distinction between figures of speech and the world they mediate. Dennis very accurately perceived *The Rape of the Lock* to "vanish into Air" at the point of each of its puns. Just as accurately, he took these vanishing points for evidence of the constant oscillation between figure and ground that keeps us from locking *The Rape* within a single and unified interpretive frame, even as it permits us to enter this poem *as if* it were a world where that is possible.

As it is, the most convincing readings of *The Rape of the Lock* express precious few convictions about this poem, preferring instead to take it in light of its "effects." In the space of but two pages, F. R. Leavis's classic essay thus speaks of the poem's "total effect," its "kindred effect," its "exaggeration of effects," the "success of [its] effects," and its constitutive "blends of other effects."[46] For Leavis, these were all "effects" of immediacy that make Pope's poem seem more like the lifelike "Love Song of J. Alfred Prufrock" than, say, "The Lady's Dressing Room." William Wimsatt likewise noted the "phonetic effect [that] serves to give thickness or texture to the meaning." In his view, the "total effect" of Pope's poem "is a concreteness which makes a difference between poetry and science."[47]

By "science" Wimsatt can't have meant the atmospheric science of Pope's day, from whose often ironic concreteness the significant "difference" of Pope's own language is hard to make out. Here Dennis's complaint about what "Proportion . . . *Bubbles* hold to *Bodies*" is perversely instructive. Dennis's intended slight looks slight indeed if we grant air the definite, impressive, and singularly elastic body that it assumes in contemporary atmospheric writing. And it's all the more so because we've seen that writing encounters atmosphere as both an illusion and a condition of experience—as a sphere, that is, of conditional belief. Within this (atmo)sphere, it's only natural that no action in Pope's poem should be performed more often than that of breathing, or give rise to more of those offensively realizing puns, as when, for instance, "incens'd *Belinda*" flings "a charge of Snuff" at the baron, "just where the Breath of life his Nostrils drew" (5.80).

The Airy Band

The main device by which Pope "drew" attention to his poem's atmosphere is, of course, the sylphs. Ariel's "Airy Band" seems literally to have risen from the 1712 version of *The Rape of the Lock*. There, as is well known, Belinda's aeriform guardians never did appear. But to the 1714 version they are, no less than "Air" itself, "vital." Pope's reader first encounters the sylphs outside his poem's frame, in the letter to Arabella Fermor that airily introduces them as subtle emanations from the expiring coquette—that altogether fascinating creature who just happens to be his reader too. So if the sylphs are, as Pope introduces them, mere "Machinery, *Madam*," they are of the same make and model as Bohun's aeoliphile. Both generator and explicator of many an "*Arti-*

ficiall win[d]," that machine was of a piece with the air it made apparent. To read Bohun on wind—which he wonderfully characterizes as diverse "elasticall *Reciprocations* of the Atmosphere"—is to grant said wind little more than an air of probability, if that (Bohun, *Discourse*, 22).

Here is where Pope went Bohun and the other artful atmospherists one better, for in *The Rape of the Lock* we have *only* the air of probability, and it is an air so ostentatiously assumed that it wafts nothing but *im*probability our way. As Pope's own editor, William Warburton, complained, his sylphs have nothing to do with any "System in . . . nature." They traffic instead with entire kingdoms of groundless belief—"the Legendary stories of the Guardian Angels, and the Nursery Tales of the Fairies."[48] The sylphs certainly can't be believed to be true "heavenly Appearances." But in particularizing and making *seem* visible the obviously *in*visible air of *The Rape of the Lock*, the sylphs at once animate and elucidate the second nature of modern fiction. This nature, we are made to understand, is neither divorced from the phenomenal world nor just a simulacrum of it. Instead, the two possibilities are compounded, and realized, through their common air.[49]

Pope's sylphs appeared at a transitional point in Western literature's long history of representing supernatural and mythic figures—a point at which, in William Melmoth's words, such figures had ceased to "make their appearance in person and as agents" and now could manage to "enter only in simile or allusion."[50] Nobody who was anybody seemed to believe in them as independent entities anymore. As Earl Wasserman put it, "No longer sanctified by popular belief," such "piece[s] of machinery" had lost their "air of probability."[51] Or at least they had unless we are talking about a world whose constitution—whose disposition, anyway—is believed to be fundamentally figural. In this case, to "enter . . . in simile or allusion" must be to enter more or less in reality. In the same spirit, Johnson's "Life of Pope" observes that the sylphs are not protected by that air of probability. In Johnson's words, "The employment of allegorical persons always excites conviction of its own absurdity: they may produce effects but cannot conduct actions; when the phantom is put in motion, it dissolves." But it is precisely out of probability's evaporation that Johnson credits Pope with having "brought into view a new race of Beings, with powers and passions proportionate to their operations." Gamely taking in the novel atmosphere that arises, Johnson finds that "in this work are exhibited, in a very high degree, the two most engaging powers of an author: new things are made familiar, and familiar things are made new. A race of aerial people, never heard of before is presented to us in a manner so clear and easy, that the reader seeks for no further information, but immediately mingles with his new acquaintance, adopts their interests, and attends their pursuits, loves a sylph, and detests a gnome" (Johnson, "Life of Pope," 4:71).

Apparently, it is through the sylphs that Pope's reader seems to herself to move about in his work. Courtesy of them, she experiences that work more as an environment than as one of those pricey and beguiling *objets* to which it is so often likened. Yet this experience cannot be one of naive immersion, for it is also the sylphs who in their flagrant artificiality make apparent *that* this same reader absorbs and thus shapes, even composes, the world in which she appears to be immersed. We will not encounter this dynamic again until the heroines of Ann Radcliffe's gothic fiction form such eerie intimacies with their misty surroundings. Johnson appreciates the sylphs as "phantoms" that *don't* dissolve, even as to his eyes Pope's "race of aerial people" uncannily transposes the "familiar" and the unfamiliar, or "new."

Here Pope's sylphs chart the sphere of fiction—a sphere whose only true claim to probability is, precisely, its possession of the air of it. Pope's novel sylphs simply make this air apparent, and with it an irreducible fact about the kind of formal social life that ultimately has nobody in it. Dennis was surely right to find that the sylphs have "no Manner of Influence upon what [Pope] calls his *Poem*, not in the least promoting, or preventing, or retarding the Action of it" (Dennis, *Remarks*, 3). But if the "*Poem*" is approached in terms of actualization, as opposed to "Action"—in terms, that is, of its dynamic projection of its own qualities and conditions through language—the sylphs promote a great deal. John Sitter finds that they "show the show," David Fairer that their main task is to unbind words from their moral meanings so that certain expressions (err, loose, wild), are "demoralized" into terms of pure movement (Fairer, *Pope's Imagination*, 11).[52] Both readers see the sylphs both to be and to make atmosphere in just the sense in which Pope and his contemporaries apparently understood it. In the contexts that their wonderfully nuanced readings create, the word "air" does not just signify a world of appearances but makes it. So are revealed the apparitional, possibly arbitrary conditions of experience itself.[53]

So Pope naturally pronounces his sylphs "the best-condition'd Creatures imaginable" and promises Arabella Fermor that "mortals may enjoy the most intimate Familiarities with [them] upon a condition very easy to all true Adepts." If at last that "condition" just seems to be the ironically impossible one of the "inviolate Preservation of Chastity," in a larger sense the condition that Pope makes familiar is the mixed and reflexive belief state that his poem requires if it is to be perceived at all. Even then, Pope's "sylph machinery" shows us what we see *through* far more convincingly, and directly, than it gives us anything to see (p. 218). So did one Victorian peg that machinery as "an *undoubted* work of fancy: . . . 'A Midsummer Night's Dream' seen through chocolate fumes."[54] Out of such an aperçu too rose Sitwell's exultation in Pope's poem as "a miracle of summer air, airy and glittering"—"a poem so airy it might have been woven by . . . sylphs . . . floating like a little wind" (Sitwell, *Alexander Pope*, 275). As they appreciate the way Pope's sylphs make the compounding of medium and message visible, Sitwell's words echo those of Pope's contemporaries when they tried to look at the air but could only see what they were seeing it through. Was not, in his effusive *Description of Ventilators* (1743), the pneumatic chemist Stephen Hales about to celebrate "that genuine Cordial of Life: . . . that wonderful Fluid the Air"? Air "produces [its] surprizing Effects," Hales marveled, only by "infinite Combinations with natural Bodies."[55] Pope's sylphs are nothing if not the "surprizing Effects" of such unlikely "Combinations."

The Art of Atmosphere

And Gotthold Lessing surely came across them. In *Laocoön*, his influential 1766 essay on the contingent limits of painting and poetry, the German poet reflected on the artistic problem of how to make invisible actions visible. This is obviously less of a problem for the poet than it is for the painter, who must find some visible instrument for suggesting to those outside a painting the presence of what cannot be seen by those within it. A device that achieves this end would thus signify not just the presence of the invisible but that of the artistic medium itself: it would perfectly compound form with content, medium with message. According to Lessing, "The means which painting uses to convey to us that this or that object must be thought of as invisible [within the frame of art] is a thin cloud veiling the side of the object that is turned toward the other persons in the picture."[56]

Lessing is registering atmosphere. And in a pictorial context, atmosphere's work is twofold. First, it tells us that something isn't seen even though it is there. Then, because this is a little different from telling us that something seen is not there, atmosphere also marks a difference between those who are inside a work of art and those who are outside it. In this way, atmosphere allows a representation to surpass its own limits, as happens in a different way when someone in a painting is shown to hide himself behind a cloud. "In this case," Lessing writes, "the cloud is a true hieroglyphic, a mere symbol, which does not render the rescued hero invisible, but says to the spectators: you must imagine to yourselves that he is invisible" (Lessing, *Laocoön*, 68–69). In either case, atmosphere doesn't simply mark the presence of mediation. Precisely in so doing, it compels an experience of immediacy. Directly addressed at the level of the representation, the viewer disappears into the world of appearances that is, after all, a world just as much as it is an appearance.

Lessing tells us that artists borrow this technique from language, and in particular from that of Homer, whose *Iliad* has its gods wrap heroes in clouds when they wish to render them invisible. Painting "appropriat[es] the Homeric mist" to "every instance where the spectator is supposed to see something in the painting which the characters themselves, or some of them, cannot see." So is poetic mist transmuted into "an arbitrary sign," one notably short on even the stability that many arbitrary signs manage to secure, because "it is used both to render the visible invisible and the invisible visible" (Lessing, *Laocoön*, 69).

David Wellbery notes that Lessing's aesthetics presuppose "an alphabetized culture for which language is not one medium among others, but, as it were, the meta-medium." In Wellbery's view, "This aesthetic loses its evidentiary force at that historical juncture when aesthetic media become pluralized."[57] This was, of course, Pope's "juncture"—one of which, if ambivalently, he often made the most, for example by translating Homer's *Iliad* profitably enough to finance a grotto of his own. In that work's mocking predecessor, Pope was to bring to epic the sense (which eluded even Milton) that the medium of letters may be a determining climate in its own right. In the sylphs, he takes on the problem of mist that wasn't one for Homer by pretending to tackle it as a painter might. The question of how to make the invisible visible to those *outside* the frame, even as it remains invisible to those *inside* it, is registered as a pictorial dilemma. But in treating it, if disingenuously, as such a dilemma, Pope both reinforces the hyperconventionality of Belinda's social world *and* realizes that world for his reader. Thus when Ariel "appears" beside Belinda's bed at the beginning of canto 1 and "seem'd to her Ear his winning Lips to lay, / And . . . in Whispers said, or seem'd to say" (1.24–25) what he does say, the cloud of "seem[ing]" surrounds him in such a way as to keep him invisible to Belinda even as he becomes visible to us. What Pope makes us "see," however, is not actually Ariel but rather the "Machinery" by which this happens. He emulates the artist who would drop a veil of mist to hide a god from the mortal she or he addresses, thereby making the artistic medium, not the god, *the* vital condition of experience itself: in effect, the god.

Pope's painterly take on the atmospheric potential of literary language within an "alphabetized culture" may help to account for his sylphs' visibility in *The Rape of the Lock*'s vibrant iconographic afterlife. Their ascent parallels English art's mounting ambition to make air both visible and inconceivable apart from the medium that conveys it, an ambition, of course, most completely realized in J. M. W. Turner's cloud-ridden canvases. Overseen by Pope and engraved by Louis Du Guernier and Claude Du Bosc, the earliest illustrated edition of *The Rape of the Lock* tendered sylphs as chubby *putti*, improbably suspended in midair by inadequate small wings. Matters were little different in Simon Gribelin's engravings for the 1717 edition of Pope's *Works*, or even in Anthony Walker's single illustration for Warburton's nine-volume edition of 1751, where Walker's image of Belinda at tea and the baron about to snip the lock even integrated the sylphs into a decorative frame above a mantelpiece, assimilating them to the environment of static art in which Pope's human characters seem to move. But when Hazlitt's contemporary Thomas Stothard watercolored the sylphs, he decided to render Ariel at least as a creature of nature. After William Blake suggested that Stothard model his sylphs on butterflies, the artist (as had Pope when he put poor old Mrs. Tempest into his winter pastoral) resolved that he would "paint the wing from the butterfly itself" and for that purpose went straight out to trap one.[58] As Stothard supernaturalized Ariel, he turned the sylphs from baroque cherubic decor toward native faerie lore, foregrounding their affinity with Hippocrates's supernatural air, that "Powe[r] . . . that Nature cannot overcome."[59]

Henry Fuseli was to take matters a step further. In his phantasmagorical *Dream of Belinda* (1789–90), a topless classicized Belinda dozes through a palpably erotic dream, the sprite Umbriel bolting peevishly from her and a garlanded head sylph descending from above, miniature toilette table in hand.[60] On Fuseli's canvas, Ariel's form dissolves into the air around him to convey the sense of an atmosphere equal parts ethereal and material, one that blends into solid cloth on one side and darkness on the other. Even Umbriel is drawn into its luminous, transmissive folds. Fuseli saw, and showed, that it is the sylphs who both figure and secure the atmosphere of Pope's poem only because they so conspicuously appear to rise from its material particulars. So do they reveal the merely apparent nature of the so-called natural world. The gothic atmosphere of Belinda's *Dream*—like that of the work that Fuseli's witty and uncannily similar *The Nightmare* inspired, Mary Shelley's *Frankenstein* (1818)—is in this sense actually the "Machinery, *Madam*" of enlightenment.

It may be because it experiments with the emergent second nature of modern literary experience that *The Rape of the Lock* became a vital frame of reference for many women novelists later in the eighteenth century. Charlotte Lennox, Frances Burney, Maria Edgeworth, and the young Jane Austen were all interested in female misconceptions of the social environment. They tested that environment's relationship to fictional form through the semitransparent media of heroines with such inherently air-born names as Belinda,

Arabella, Evelina, and even Ariella, as Burney's "elastic" heroine Camilla was originally called. These writers' experiments with (and within) the airscapes that condition their nonexistent heroines' reality carry forward Pope's visibly unseen labor in *The Rape of the Lock.*

Pope's sylphs themselves show up everywhere in eighteenth-century literature vexed with the question of how women read fiction and fascinated by the nature of their belief in it. In *The Fan* (1749), a "heroi-comical Poem in Three Cantos," a symbol of fashionable languor (etymologically derived from the Latin *ventus*) is also a tool of airborne art. Vain Lydia's vapors are precipitated not just by "the loss of [her] Fan at a late assembly" (which catastrophe compromises her "Modish airs") but ultimately by *The Rape of the Lock*, whose author is praised for having "freed himself from the shackles of the old poetic faith, and given us . . . a creation of new beings from the Rosicrucian scheme." These "beings" mark out the credal penumbra that novelistic fiction had begun to project: Lydia's fan bears a scene from Fielding's novel *Tom Jones*, the "painting" of which, "tho' not real, shews what a rich field for the pencil, there is in the scenes of Tom Jones."[61]

In Marmontel's "The Sylph-Husband"—one of twenty moral tales translated into English in 1764—the fair Elisa rejects her husband's advances, having found only "in fiction something to engage, to move, to melt her. The fable of the sylphs was in vogue." Indeed, "these brilliant chimeras had in her eyes all the charms of truth. In short, Elisa believed in sylphs." When her desperate husband impersonates one of that airy tribe, Elisa falls in love with the realized "pictures of imagination." In this way, she "pass[es] by an inconceivable prodigy from illusion to reality." Marmontel's "inconceivable prodigy" transcends social satire as it displays the lines along which distinctions between "illusion" and "reality" are made.[62] Similar lines are drawn in "The Sylph" (1790) by the English poet Anne Francis. Here a woman is admiring a pot of ginger when "a radiant Sylph in dazzling white / Then burst upon my wondring sight." The creature's charge is

> to lure the mind [and] lead the roving eye
> Through nature's fair variety;
> Or with the quainter turns of art—
> To greet the sight, and chear the heart.[63]

And in the anonymous "The Hobgoblin, or the Amorous Sylph" (1746), a French girl is wooed by an apparent sylph who plants *Le Comte de Gabalis* in her room, leaves love poems and pictures inscribed to "Belinda," and even stocks her closet with butterflies. He eventually materializes in the form of those creatures "whose wings were variously diversified, with various colours, that the ablest painter in the world, would have been very much put to it, to have surpassed these masterpieces of nature."[64]

In all these manifestations, a sylph reveals an order of nature whose foundation is formal representation and then nurtures the acts of mind that can knowingly inhabit that nature. In "The Lady and the Sylph: A Visionary Tale," appended to the Marquise de Lezay-Mamezia's *Letters from Julia . . . to Ovid* (1753), the lady in question reflects that "there is something in the air of Truth which commands Belief" and naturally allows her own thoughts to be "fixed in the sylphs, whom I had never looked on in a chimerical light—and whom I had always wished to be as real as I now found they were."[65] Even the popular Covent Garden entertainment *The Sylphs* (whose "Airs and Chorusses" [*sic*] materialized in print in 1774) employed the atmospheric figure of the sylph to show the conditions under which illusion is maintained and navigated in social life as well as on the stage.

The conventions that shape those overlapping spheres are once more on display in Georgiana Cavendish's 1779 epistolary novel *The Sylph*. Cavendish's book went into three editions in the year of its publication, and it took its epigraph from *The Rape of the Lock*, drawing further inspiration from Germain François Poullain de Saint-Foix's one-act play *The Sylph* (1771), Saint-Foix himself having adapted Crébillon Fils's 1730 story of the same title. Cavendish's Julia, a Welsh ingénue, is unhappily married to a London libertine. A childhood admirer appoints himself her secret guardian and guide and, in the persona of a sylph, writes her letters meant to guide her through the maze of metropolitan high life. This is to teach Julia how to maneuver through its pretenses without being touched by them. The sylph himself materializes only within the double fiction of a masquerade, "the most elegant creature that imagination can form." The effect is sealed when "on each shoulder was placed a transparent wing of painted gauze."[66] After Julia's husband shoots himself, she makes her way through this layered pretense and finds romance. As *The Sylph*'s recent editor, Jonathan Gross, points out, Cavendish thus absorbs (and perpetuates) Rosicrucian spiritualism as a legitimate erotic language. But Julia's story also grounds female experience in a sphere of airs that are only apparent. Survival depends less on seeing through these airs than on learning to see *with* them.

Cavendish was distantly related by marriage to the pneumatic chemist Henry Cavendish, and she contributed financially to the establishment of England's Pneumatic Medical Institution in 1799. We will visit that less than venerable institution in chapter 9, but to the extent that it was to treat the experience of air more in social and aesthetic terms than in therapeutic ones, we may cite it as further evidence that Pope's sylphs did anything but vanish into thin air. As for *The Rape of the Lock*, Pope's "Airy Band" (3.113) appears one last time, just as Belinda's ravished curl makes its dazzling ascent from the lower atmosphere to the ethereal sphere above, "the Heav'ns spangling with dishevel'd Light" (5.130). Or they appear to appear: before this lucent display, the "aerial Kind" are not the objects of perception so much as they are its vibrant medium. Invisible, because assimilated, readers, they are naturalized citizens of the literary sphere. As such, their remaining charge is only to "behold [Belinda's hair] kindling as it flies, / And pleas'd pursue its progress thro' the skies" (5.131–33).[67] As they conduct her to this vision, Pope's living reader is conditioned to stop making sense of his aeriform poem from outside its self-created premises. It must be from within them that she will be most "pleas'd" to see Belinda's airy lock achieve its brief—thus lasting—sublimity, even as it is lost in darkness and distance.

4

Novel Atmographies: Eighteenth-Century Weather Writing and the Atmospheres of *Robinson Crusoe*

This the Book, [which] is the same as the Sky.

THOMAS APPLETREE AND AUGUST KLEINZAHLER, "The Exquisite Atmography of Thomas Appletree" (1703/2010)

If we are to be, we must have a where.

DANIEL DEFOE, *Serious Reflections on Robinson Crusoe* (1720)

Exact contemporaries though they were, Daniel Defoe and Alexander Pope today are seldom spoken of in the same breath. It is difficult to imagine anything further flung from Pope's shapely couplets than Defoe's turbulent prose, and few politics were more likely to ruffle the feathers of the Roman Catholic Tory satirist than those of a Whig dissenter still often seen as the last of the Puritans. Pope's sylph Ariel might have sprung, at least nominally, from the enchanted islands of Shakespearean romance, but he would surely have withered in the capricious air of Crusoe's desert one. And if both Pope and Defoe felt compelled to write about the Great Storm of 1703, Pope did so only within the stylized confines of his fourth pastoral, *Winter*—a work that bears little resemblance to *The Storm* (1704), Defoe's ramshackle *Collection of . . . Disasters* published the same year.

And yet, buried in the latter, it is possible to find "A PASTORAL, Occasion'd by the Late Violent Storm."[1] At the very least, shared subject matter registers if not exactly a common air, then certainly a preoccupation with the kind that is uncommon. In Defoe's case, even more than Pope's, this preoccupation mediates between two notable dimensions of his fiction: its commitment to matters of material fact as these are parlayed into formal realism and an abiding fascination with the supernatural.[2] Echoing Walter Scott's perception that Defoe's "air of writing with all the plausibility of truth" in fact involved the magical manipulation of appearances, John Bender has recently formulated the maxim that "the apparitional is deeply woven into the very fibre of realist narration."[3] And, indeed, in much the way that Pope's "supernatural Machinery" tinkers with the "one System in all nature which was to his purpose, the *Rosicrucian Philosophy*," Defoe's interest in what he termed "the History and Reality of Apparitions" was but the other side of his interest in those natural signs known as weather.[4] As these two interests converged, both Defoe's most influential novel, *Robinson Crusoe* (1719), and his least classifiable one, *A Journal of the Plague Year* (1722), turned the page of British literary fiction into a living register of its own environments. This chapter considers *Crusoe*'s atmospheric dimension in light of contemporary English writing about the weather—a connection that has remained virtually invisible in criticism of the novel.[5] The next chapter will turn to air's supernatural properties as Defoe wove them into the lexical fabric of his *Journal*.

Novel Tempests

Depending on the reader, *The Life and Strange Surprizing Adventures of Robinson Crusoe* can seem to have come from almost any "where," from spiritual autobiography and travel writing to legal casuistry and popular economics.[6] But it is hard to find a more immediate family than English writing about the weather. Preceding the systematic study of "meteors"—notable weather events—as it consolidated over the course of the eighteenth century, this popular and necessarily homegrown genre was one in which Defoe dabbled more than once. *The Storm* aside, he went on to contribute vivid entries to both the *Weekly Journal* and *Mercurius Politicus*, documenting the widespread "Terror and Amazement" that attended a northern European hurricane in 1717.[7] The storms that come up throughout *Crusoe* to terrify and amaze its protagonist thus make the novel look like a sequel to Defoe's earlier treatments of singular weather events. Conversely, the lucrative novelty of what remains the most copious of those treatments, *The Storm*, was bound up with that of what could almost be seen as its own eponymous main character, the catastrophic wind that ravaged Defoe's own island in the winter of 1703. "Such a Tempest never happen'd before," *The Storm* avers, "as that which is the Subject of these Sheets" (*Storm*, 25).

As recent studies stress, its subject matter rendered the English weather writing of Defoe's day as generically incoherent as the nascent novel itself.[8] Robert Markley attributes that writing's formal heterogeneity, like the inconsistencies within and between its truth claims, to the unrepresentability of natural disasters, on the one hand, and, on the other, to a systematic "ecocultural" understanding that the English landscape comprises many "complex sites of climatological, moral, and economic interaction."[9] Yet as it virtually parodies the haphazard and fragmentary nature of contemporary weather writing, *The Storm* suggests that natural disasters are altogether *too* representable and that Defoe would include aesthetic and imaginative elements among the complex interactions that extreme weather events stage.

Such elements potentially introduce exactly the distance that natural disaster otherwise, by definition, lacks. That distance in turn introduces a new order of proximity: an uncommon sense that is at once linked to matters of fact and free to experiment within—not merely with—them. "Here," Defoe writes, "we see Effects but cannot reach their causes" (*Storm*, 11). *The Storm* teems with eyewitness accounts of the devastation that "the Subject of [its] Sheets" has wrought. These blow up against pastoral poems, citations from the ancients, moral reflections on the late "Tempest," barometric readings arranged into tables, and regular evocations of holy scripture as a symbolic and conceptual frame—one that turns "the storm" (and *The Storm*) into an occasion for thinking about the forms in which we make sense of things. "We read in the Scripture of Two Great Storms," Defoe reminds his reader, "one past, and the other to come. Whether the last be not Allegorical rather than Prophetical, I shall not busie my self to determine" (*Storm*, 17).

Leaving aside the fact that we would expect Defoe to treat it foremost as historical, the question of whether the storm at issue was "Allegorical" or "Prophetical" intrigues for many reasons. For one thing, insofar as Defoe wryly leaves this question unanswered, he replaces both interpretive models with a third one based on exactly the kind of suspense that neither will tolerate. Second, he so perfectly conflates his "Storms" with "the Scripture of" them that it is impossible to tell whether he is speaking of real meteorological events or merely of their literary redactions. Most interesting of all, Defoe's remark about the "Scripture" of storms turns out to have been itself "Prophetical." People were also to wonder how to read Robinson Crusoe's "Surprizing Adventures," and in his "Serious Reflections" on it (1720), Defoe would "affirm" that that "story, though allegorical, is also historical."[10]

His equivocation puts this founding work of English fiction in the same indeterminate category as the "Scripture" of storms. Like other works in a tradition traceable to the book of Job, *Crusoe* seems to be two very different things: a primarily religious text indebted to allegedly residual forms of Puritan thought and experience and a "historical" narrative that anticipates emergent and open-ended forms of realist narration.[11] In both readings, *Crusoe* proves to be what Leslie Stephen called "a kind of allegory." Both, allegorically, set the novel at a fixed distance from its supposed referent.[12] Defoe, however, seems to have reconciled these two symbolic possibilities by invoking a third, one that simply abolishes the interpretive distance on which the others depend. "The story," he added, is also "the beautiful representation of a life of unexampled misfortunes, and of a variety not to be met with in the world" (*Serious Reflections*, v). His words raise the possibility that *Crusoe* is better engaged in terms of aesthesis than in those of historical or religious allegoresis—in terms, that is, of how imaginative writing figures forth the conditions of our experience of it and registers its potential effects on us.[13]

Though we don't often think of him this way, Defoe apparently admired the world-creating "Method of great Men in the East in the Ages of Hieroglyphicks, when things were more accurately Describ'd by Emblems and Figures than Words."[14] To the extent that "Describ'd" might mean "encompassed," we're here able to see how and why his contemporary Joseph Browne could have charged him with "always raising Apparitions in the Air."[15] It is when we appreciate Defoe the artist and conjuror—the describer, so to speak, of certain figures in the air—that we can finally move outside the question of what "kind of allegory" *Robinson Crusoe* might be. And yet, ironically enough, to do so is to return to Defoe's treatment of "the Air" itself, for in contemporary writing about the states of that air as expressed in weather, "Words" often aspired to the condition of "Emblems and Figures." They did so, moreover, in the interest not just of representing or interpreting a strange and surprising world but of making one.

"It"

Just how does weather figure in *Robinson Crusoe*? Its protagonist's moral life as the interpreter (thus molder) of his own experience commences with the fear he feels the first time a "Wind beg[ins] to blow" his ship off course.[16] As for plot, the same wind may be held responsible for most of it, even as Crusoe's survival on his island depends in large part on what sort of relationship he can form to a seemingly unceasing barrage of thunderstorms, dry spells, earthquakes, rains, and gales. Crusoe's obsession with such bewildering atmospheric events, however, far outstrips his human need for shelter from sun and storm, his agrarian need to master the cycle of the seasons, and even his mortal need to know what God might be trying to tell him before it is too late. It seems vital that he write the weather down—so vital, in fact, that Crusoe apparently starts keeping a journal for that very purpose. Throughout its pages, meteors matter even more than Christmas itself, which Crusoe marks as follows:

> Dec. 24. Much Rain all Night and all Day; no stirring out.
> Dec. 25. Rain all Day.
> Dec. 26. No Rain, and the Earth much cooler than before, and pleasanter. (*Crusoe*, 116)

As David Marshall elaborates in his artful reading of Crusoe's journal as autobiography, the journal's status within the larger body of his narrative is questionable, if not downright bizarre.[17] Crusoe, of course, claims to have set down his impressions of island life in writing he later "copied" into the larger narrative of his "Life." Yet the relationship between the journal and the text that surrounds and supersedes it is virtually impossible to chart. Its commencement is boldly posted with the words "THE JOURNAL" to establish a simple figure-ground ratio between that document and its narrative environment. But Crusoe warns his reader that "THE JOURNAL" is bound to peter out, for in time, "having no more Ink[,] I was forc'd to leave it off" (*Crusoe*, 52). No *X* marks the spot where the narrative background reclaims the foreground. Imperceptibly and obscurely, we find ourselves at some point outside the journal again. Who knows when or how we got there?

Here it is doubly significant that most of the dated entries that compose Crusoe's journal begin with weather notations: "*Nov.* 7. Now it began to be settled fair Weather. . . . *Nov.* 13. This Day it rained . . . but it was accompany'd with terrible Thunder and Lightning" (*Crusoe*, 54). Such notations suggest that the journal's relationship to its narrative environment may not be that of figure to ground. We might do better to describe it in terms of the atmospheric connection between the rain whose fall here makes such an impression and the "it" that rains. At any rate, it is unclear which narrative frame is in authority at several of the most impressive points in Crusoe's account as a whole—which is indeed figure and which (if there is any) is ground. For this reason, Crusoe's writing begins to look less and less like the grounded account it seems to be, and more and more like an impression in its own right.

Crusoe's embedded "journal of every Day's Employment" (*Crusoe*, 51) has been identified with the residual genres of spiritual autobiography and the emergent ones of how-to manual, mariner's log, and middle-class memoir. But in its impressionistic form, as well as in its content, it closely resembles the weather journals that numerous literate Englishmen kept in Defoe's day. Intimately involved with the local weather that they recorded, from rain and shine to hurricanes and drought, such journals hovered outside the printed accounts of meteorological events that, like *The Storm*, were centered in London and intended for general consumption. Largely unpublished, they have not been part of our picture either of the period's literary activity or of its knowledge culture, for all that they participated indirectly in both. Recently, however, Jan Golinski has retrieved and very subtly appreciated the "speciall Language" that English weather journals could employ as they blended Renaissance symbologies of signature and correspondence with up-to-date physico-mechanical narratives of causation.

Golinski's deft analysis of one such journal—Thomas Appletree's Worcestershire weather diary of 1703—highlights both its heavy cargo of figurative language (the impending sky is "a tall Fresco ceiling," the "Atmosphere" itself "loaded & varnished with Bulging, dull swelling BasReleive clouds") and its author's nagging, at times overwhelming, consciousness of linguistic inadequacy to the phenomena at hand. "Our Language is exceeding scanty & barren of words to use & express ye various notions I have of Weather &c," complains Appletree, "& I cannot invent a Language commensurate to [the] vast & infinite Properties discoverable in meteorology."[18] Appletree's journal has resurfaced in our own day in purely lyric form, as witness August Kleinzahler's "The Exquisite Atmography of Thomas Appletree" (2010). A poem composed wholly of fragments from Appletree's journal (as transcribed by—mediated through—Golinski), "Exquisite Atmography" ends with Appletree's internal gesture toward "this the Book, . . . which is the same as the Sky over Edgcock."[19]

What is most striking about Appletree's kind of writing is that the more it foregrounds the act and fact of writing ("this the Book"), the more immediately it renders an impression of "the Sky" and all that happens there. In the eighteenth century, similar "atmographies" sometimes found their way into popular print culture indirectly via such works as Defoe's *The Storm*. Their relationship to those works mirrors the relationship that Crusoe's weather journal has to the novel *Robinson Crusoe* itself. This allows us to expand Kleinzahler's evocative term, "atmography," so as to encompass all the diverse and visibly contingent forms that weather writing assumed between the middle of the seventeenth century and that of the eighteenth.

Early English meteorology has been broadly interpreted as itself an interpretation of the English atmosphere and thus as a venue for ideological conflict between Christianity and natural philosophy, as well as between competing regional and class interests.[20] But to think of it as atmography is to foreground the graphic properties through which this kind of writing supplied English readers with a visual, if noneidetic, image of their own atmosphere. The word also captures the reflexive and analytic potential built into those same properties. "Atmography"—as opposed to the anachronistic term "meteorology"—may be seen to have framed the English atmosphere as an "it" so obscure that it could only by mediated through "impressions," "appearances," and—Defoe's favorite word throughout *The Storm*—"Effects." Indeed, these were contemporary terms of art for beautiful (the rainbow, the aurora borealis) and sublime (hurricanes, hail) weather events. Even as such "impressions" briefly illuminated the negative spaces around them, they were increasingly experienced as part of the system of graphic impressions—the writing—that presented them to an even more literate English public as if for the first time.

We've already seen that even Aristotle's *Meteorologica* began by confronting air's involvement with our linguistic impressions of it. In Defoe's own British Isles, Aristotle's meteorology had been Christianized into a public place of assignation between the visible and invisible worlds. The notion that atmospheric aberrations (shooting stars, lightning bolts, rainbows, and the like) are both terrestrial media and celestial messages had hit a high note during the English Civil War, when Puritan redactions of the skies—angelic battles in the clouds, portentous comets, and the like—vied with Church of England representations of the same meteors to prognosticate (or posit) Britain's political future.

That internecine conflict, however, was eventually subsumed under the one that scholars such as Lorraine Daston and Michael McKeon have lately charted. Unusual or startling weather events that had been represented by both religious factions as intentional and divinely instituted signs now buckled under an emergent view of the same events as inherently insignificant facts. Meteors erstwhile viewed as supernaturally caused events were now often cast as the effects of natural forces—forces themselves subject to physical laws that could not be understood as the effects of anything other than themselves. In the wake of such debates, weather signs could be consistently legible only in relation to one another *as* signs. This novel sense of the skies as a system of cross-referencing signs synchronized itself with the experimental literary forms that had begun to supplement, if not supplant, oral and traditional ones.[21]

And what of the dynamic background against which—the environment within which—these same signs appear? "It," in this instance, would be accessible only as the negative space of the meteors suspended in it or, at a second remove, through formal readings of such novel weather-reading devices as the thermometer, the barometer, and the hygrometer. The leap was not even necessarily all that long: hygrometers, for example, measured humidity in the atmosphere by exposing stacks of paper to the air and registering their degree of saturation. In any event, the represented readings of all such devices—the writing in English of "the" English weather—would then achieve, in the wholly mediated environment of the printed page, something akin to the lyric conflation of "Book" and "Sky" that we find in private atmographies such as Appletree's. Above all, written readings of the air would perforce register the states that make knowledge possible, the shifting frames and contexts within which particular events are meaningfully perceived, experienced as stable forms, and at last inhabited conceptually.

Exceptional weather events such as the Great Storm of 1703 or the aurora borealis of 1716 precipitated pamphlet wars that exemplify both the period's figurally mediated approach to atmospheric states and the uniquely reflexive forms of knowledge that could arise from that approach. As Vladimir Janković has shown, these debates pitted a religious discourse of sublunary revelation against the apparently secular insistence that the meteors in question were purely physical events. Janković, however, takes controversy over what the northern lights *were* to be the same as controversy over what they *meant*; he thus interprets them in political terms as a point of collision between the superstitious laboring classes and the arbiters of a repressive modern state. The same controversy is also highly legible, as for Daston and McKeon, as a point where rival epistemological habits clashed. Yet disputants on both sides seem to have entertained a view of this splendid and enigmatic "fulguration" as a "theater of light" that defied "perfect Descriptions" even as it could be viewed as continuous with the "Glory wherewith our painters . . . surround the Holy Name of God" (Janković, *Reading the Skies*, 69).

When printed depictions of English weather events are seen as representations of perceived representations, they may also be seen to have shared a single sphere of appearances. A reader may experience this sphere both immediately *and* with an ironic awareness of the mediating frames within which sense is made of—and communicated about—its apparent contents. This possibility already lurks in Aristotle's *Meteorologica*, which notes that, in the air, "everything happens . . . with a regularity less than" it happens anywhere else. The result is an ethereal parody of the way "opinions recur in rotation among men."[22] Here atmosphere may only be seen when it is imagined as an unrestricted and dynamic field of reflexive figuration. This field is conceptually unified, and so rendered psychologically habitable, only by the observer-dependent laws of novelty and irony.

In English letters, those laws are made visible, and even given a new twist, by the Anglican bishop Thomas Burnet in his enormously influential *Sacred Theory of the Earth* (1681). Burnet surmised that our planet originally hung untormented by climatic variation; meteors, he maintained, are an epiphenomenon of the flood, and the first of them—the rainbow, of course—instantiated an entirely novel signifying system. Previously, meteorological signs, "like words, signifie any thing by institution, or may be appli'd to any thing by the will of the Imposer and are signs of something present, and . . . signifie only by use and repeated experience." But now, Burnet stressed, "we are speaking of signs of another nature, given in confirmation of a promise . . . and given with design to cure our unbelief, or to excite and beget Faith in . . . the Promiser." In scanning the ancient lexicon of divine covenant, Burnet's *Sacred Theory* nonetheless assimilates these signs to its own vocabulary ("we are now speaking . . .") as if they were entirely new. Instead of being arbitrarily imposed by divine authority, weather signs appear to create their own systems of credit. Consequently, Burnet supposes, such signs "must be some new appearance, and must thereby induce us to believe the effect, or more to believe it, than if there had been no sign, but only the affirmation of the Promiser; for otherwise the pretended sign is a meer Cypher and superfluity." In fact, a weather sign "must indispensably be something new, otherwise it cannot have the nature, vertue, and influence of a sign."[23]

When Burnet wrote of atmospheric events as if they were unprecedented and autonomous symbolic actions, he adapted them to novel representational forms. These include not just novels themselves, such as Defoe's *Robinson Crusoe*, but also the privately scribbled weather journals that joined sacred theory, natural philosophy, and ideologically slanted reportage to supply both *Crusoe*'s heteroglossia and its immediate literary environment. All these atmographies were implicitly at liberty to devise their own idiosyncratic and reflexive grammars for containing, and conveying, weather events, though perhaps free-form private weather journals such as Appletree's make this liberty most apparent.

The atmographies of the late seventeenth and early eighteenth centuries were ultimately absorbed into "Syllab[i] of the air"—printed charts that both mapped and predicted the weather of the whole of the British Isles and made predictions from within increasingly fixed and coordinated frameworks.[24] Thus when English demographer Thomas Short set out to devise his own "Short Syllabus of the Air, Weather and Seasons," he naturally first consulted with the "many ingenious Gentlemen in different and distant Places [who] . . . have kept Journals of the Weather, Air and it's [*sic*] Temperature." First sight of these journals turned Short into another Crusoe, distraught that "we are left as ignorant of our own Seasons as Foreigners; for we find no Proportion as assigned between our wet and dry Weather."[25] Nor was the history of the English air—including questions such as that of "whether our Air is healthier now or formerly"—easily ascertained (Short, *Comparative History*, 123). Confronted with a plethora of new representations of the air, Short was at first daunted by the "seeming Inequality and Uncertainty of our Weather and Seasons," lamenting that "the great Agreement of Winds, Weather and Glasses in this Island seldom holds good in a fair Trial" (Short, *Comparative History*, 126).

Yet Short came to see "Winds, Weather and Glasses" as coextensive with their graphic elaboration in literary space. This disposed them to give rise visibly to their own interpretive environments. Such a perspective permitted Smart to devise a second-order representation in the form of "Thermometrical Tables" and "Thermometrical Journals." Here the frame within which weather *happens* coincides with the one in which the aerial phenomena we loosely call weather simply *appear*, thereby precipitating their own story and its meanings. So whereas previously "sundry uncommon Phases of the Moon" with winds have been thought to portend different "Sorts of Weather," Short was now able to locate truly reliable "portent[s]" and "presage[s]" in the "Journals of the Weather" that reflexively presented fogs, mists and winds.[26] Once weather signs were seen to share the same space that surrounded the literary signs that mediated them—and vice versa—it became possible to achieve consensus about their meaning. The result was less a set of objective facts about which there could be no debate—an anachronistic ideal in any case—than an agreement to believe something about things that happen in the air.[27] It's this agreement that can give rise to novel orders of reality, and it is a key ingredient of what we mean when we now speak of "atmosphere."

What kind of signs are meteors in this context? Consider the wind. In Defoe's *The Storm*, "Wind" at first simply baffles. Is it a natural phenomenon unencumbered by any penumbra of meaning or the work of a supernatural hand? Is it immediate or the medium of something else again? Is anything behind the "thin Veil of Natural Obscurity" (*Storm*, 15) that less cloaks than *is* a weather sign? If there is, that would justify the Longinian sublime, the deductive ecstasy of Jobean metaphysics, or even those inductive forms of modern natural philosophy that "being at the Consequence, . . . from thence *hunt Counter*, as we may call it, to find out the Cause" (*Storm*, 13). But there may not be anything behind that "thin Veil." In this case, which Defoe certainly entertains without committing himself to it entirely, we would be left only with "those Inscrutables of Nature" of which "the Winds are some" (*Storm*, 15). In the face of such inscrutability, "the Philosopher's Business is not to look through Nature, and come to the vast open Field of Infinite Power; his Business is in the Wood; there grows the Plant he looks for" (*Storm*, 14). His business is the novel.

Speaking truly objectively—beginning, that is, with the relatively intransigent fact that *The Storm* exists only as a physical collection of pages—there is simply no moving outside the wind event, only a complex set of reflexive figural "Emotions" that share the form and dynamic of their apparent referent. Yet it is on these grounds Defoe meets the late seventeenth century's most prolific cartographer of the winds, Edmund Bohun. Cited in the last chapter as a context for Pope's sylphs, Bohun's *Discourse concerning the Origine and Properties of Wind* (1671) presumably stands a step closer to its subject matter than Defoe's heteroglossic *Storm*. Defoe quotes Bohun to the effect that "the Winds . . . are generated in the Intermediate Space between the Earth and the Clouds, either by Rarefaction or Repletion, and sometimes haply by pressure of Clouds, Elastical Virtue of the Air, &c" (*Storm*, 15). As Defoe wryly notes, "Though no Man is more capable of the Enquiry than this Gentleman, yet to the Demonstration of the thing, amounts to no more than what we had before, and still leaves it as Abstruse and Cloudy to our understanding as ever" (*Storm*, 16). Its interaction with Bohun's weather writing renders *The Storm* a satire on knowledge. From an interpretive point of view, it may be daunted by the catastrophe at its heart. Yet thence arises the possibility of reflexive commentary on the business of gaining knowledge itself.

Perhaps most important for Defoe's literary practice, it is by directly engaging the ways the wind *can't* be engaged directly that *The Storm* delivers its own reader to a new order of experience. Defoe himself is explicitly concerned with the "Conduct of the Story" that his book has to tell, and as far as he is concerned, his storm is the "Subject of these Sheets" (*Storm*, 25). "These Sheets," in turn, are what Defoe's reader actually sees, in something of the way that they are also what we say we see in a heavy rain. Hence, just as Bohun's nonexplanation of the winds turns our attention to "the Intermediate Space" that contains and conveys them, so can we read between the lines of what Defoe says *about* the nondemonstrable "thing" to what he says, in effect, *with* it. One thing he says is that the subject of his sheets is "Cloudy to our understanding." This is just a figure of speech, of course. And yet a literal feature of "the Storm"—its cloudiness—is here absolutely equivalent to its figural quality of, again, cloudiness. Here the "Elastical Virtue of the Air" applies straight to the lexical figures that describe it. Those figures sustain not only description or analysis but experience itself. It was this novel possibility that Defoe would test fifteen years later when Robinson Crusoe would find himself under the weather.

About the Umbrella

Weather journals may be understood as a point of transference between weather events themselves and the writing about them that, via print, soon began to spread over the whole of the British Isles. Like both kinds of atmography, Crusoe's "journal of every Day's Employment" begins with the "dreadful Storm" that brought him to his present state, and its first entries are battened down with mention of the "Wind," "Rain," and "Intervals of fair Weather" that framed his first days as a castaway (*Crusoe*, 57). In this manner, he less proffers his reader an immediate experience of the island air than takes up an intermediate position between linguistically experimental weather journal*izers* and published weather journal*ists* like Defoe himself. The former seeded each sentence with "reflections" on the relationship between writing and event. The latter conveyed similar doubts structurally and indirectly, through the juxtapositions of various pieces of atmography.

This genealogy can explain why, for all that it goes on to mark most days in terms of what "the Weather [is] being" at the time, Crusoe's journal never really is about that weather. True, it mentions an earthquake at one point. There is plenty of thunder and lightning, of course, and there is always the wind. It rains about half the time, as Crusoe complains when he devotes some pages of his journal to a calendar of half-months grouped "not into *Summer* and *Winter*, as in *Europe*, but into the Rainy Seasons and the Dry Seasons" (*Crusoe*, 78). Yet we scarcely notice these many splendored atmospheric events in and of themselves. Should a "sudden Flash of Lighting happe[n], and after that a great Clap of Thunder, as is naturally the Effect of it," Crusoe will report himself "not so much surpris'd with the Lightning as I was with a Thought which darted into my mind as swift as the Lighting it self" (*Crusoe*, 45). Indeed, this irresistible "Thought"—worry for the gunpowder the lightning might ignite—continues to "surpris[e]" him whenever "it lighten'd and thunder'd" (*Crusoe*, 47). Particularly when fixed in the graphic space of Crusoe's journal, what this particular meteor brings to mind is not the eidetic image of a stupendous and appalling celestial spectacle, or even that of the threatened gunpowder (to which Defoe's contemporaries believed lightning to be related chemically), but rather the mind itself. And yet that mind is less itself than a figure of transcription through which a reader can "see" nothing so clearly as the mentality that links events with meanings, and images with significances. Just so, and in great part because we talk that way ourselves, we pass over Crusoe's casual reference to the "it" that "lighten'd and thunder'd." But insofar as "it" means the mind as much as it means the world, that world is shown to be not so much perceived as conceived.[28]

In Crusoe the winds can also bring this "it" to mind. About to build a raft to sail to his wrecked ship, Crusoe writes, "I found the Sky overcast, and the Wind began to rise, and in a Quarter of an Hour it blew a fresh Gale from the Shore." Whereupon, we read, "it presently occur'd to me that it was in vain to pretend to make a Raft with the Wind off Shore," and Crusoe shortly abandons the project at hand (*Crusoe*, 43). While we want, I think, to interpret this as the report of a chain of motives and actions—the wind rose; I wisely decided not to make a raft—what Crusoe's words actually register (as do we, if peripherally) is the uncanny migration of "it" from what "blew" to what "occur'd" to the mind. This migration is uncanny because it leads us to believe that "it" is not a "what" at all and that occurrences *in* the world are only with violence told from occurrences *to* the mind.

That amphibious "it" may on occasion assume the force of a spiritual directive. Taking in the "dismal Prospect of my Condition" at one point, Crusoe notes that "as I was not cast away upon that Island without being driven, as is said, by a violent Storm quite out of the Course of our intended Voyage, . . . I had great reason to consider it as a Determination of Heaven, that in this desolate Place and in this desolate Manner I should end my Life" (*Crusoe*, 46–47). Here "it" appears to be first Crusoe's "Condition," then the "violent Storm" that seems to have created it, and then again a "Determination of Heaven." Most significant, these three possibilities find meaningful organization not under the umbrella of religious conviction or under that of a quasi-historical storm or other unfavorable "Condition" but, finally, within a figure of speech, for Crusoe presents himself as "driven, as is said" to his present state. So the medium of presentation (a colloquial English saying) is foregrounded in such a way as to suspend forever the question of whether "it" was a storm or a determination of heaven that drove Crusoe to his present "Condition." Most likely, "it" is Crusoe's "Condition" itself. If so, though, "it" is a condition best grasped as a metatextual state, a briefly realized figure of speech located somewhere between the reader's mind and the printed page, which is to say—literally—in midair.

While our instinct is to take them as objects of interpretation, then, the aerial events such as lightning, wind, and rain that Crusoe mentions really, as he himself puts it, just "happe[n]." In turn, their function is less to generate something we could call *Crusoe*'s ambience than to gesture toward where that ambience already is and the nature of its activity. Indeed, the fact that it *is* active brings to mind *ambience*'s root in the Latin and later French verb *ambire*, whose active meaning of "to move around" complements its now far more dominant but nevertheless passive sense of ambience as something that simply hangs around.[29] And when conspicuously written—then consciously read—these aberrations of the air make briefly but significantly visible the background that surrounds Crusoe and against which everything that he himself does, thinks, and is takes shape. They denote the frame within which he becomes figural and hence transmissible in and through a British insular imagination, which is to say an imagination in rapport with its medium, English.

This process is not visible—nor would it have been possible—without the example of contemporary atmography. "Here" in the storm of 1703, Defoe had written, "is room for abundance of Romance, because the Subject may be safer extended than in any other case, no story being capable to be crouded with such Circumstances, but Infinite Power, which is all along concern'd with us in every Relation, is suppos'd capable of making true" (*Crusoe*, 6). *Crusoe*'s supposed "Editor" appropriates this same "Infinite Power . . . capable of making true" in his prefatory claim that, with respect to Crusoe's story, he "believe[s] the Thing to be a just History of Fact" with no "Appearance of Fiction in it" (*Crusoe*, 3). His irony strikes us all the harder when we come at it from this direction, because unlike Defoe's meteorographic "Romance" of 1703, *Crusoe* only pretends to copy real weather writing (or, at a further remove, to describe real air). It is, in Scott's quotable phrase, concerned merely to construct an "air of truth," not to tell, as *The Storm* had set out to do, the truth about the air (Scott, "Daniel De Foe," 2:274).[30] Still, there is great value in this pretense, for in making it, *Crusoe* may be said to have made the mediating frames that English atmography assumed visible as, themselves, appearances. The foregrounding of Crusoe's weather journal doubles this emphasis. In any case, to apprehend Defoe's novel as a literary experiment of this sort is to apprehend early English fiction's unique and potentially "Infinite Power" to make apparent how and that things rise from the mind—how, as the saying goes, they appear out of thin air.

Crusoe pursues its own point through the impressions it makes, both figuratively—on the reader's mind—and, literally, on the page. Meteors were often called "impressions," and we have already seen how Crusoe's manner of speech uses these impressions to make us aware of the immediate acts of mind that surround and uphold them. *Crusoe* as a whole works the same way, as Scott perceived. The Scottish novelist noted that in its peculiar fusion of "lucky hits and accidents" with "providential occurrences," Defoe's fiction produces "the same impression . . . which the *Book of Martyrs* would on a young devotee or the *Newgate Calendar* upon an acolyte of Bridewell" (Scott, "Daniel De Foe," 2:278). This suggests that the historical and the spiritual converge in Crusoe not just as alternative models of interpretation but also as aspects (conditions, situations, even apparent meanings) of a single (written) "impression." Furthermore, it was this impression that Scott held responsible for the "air of truth about the delightful history of *Robinson Crusoe*"—a history that to his mind had earned "the merit . . . of that species of accurate painting which can be looked at again and again with new pleasure" (Scott, "Daniel De Foe," 2:279).

Scott's remark is telling, and not just because it identifies Defoe's style with the visual one peculiar to Dutch realist painting.[31] That painting is of a famously indoorsy "species," but what is striking about the impression that *Crusoe* in particular made on its English readers is that it seems to have been mediated through an image of the outdoors. We may take as an example George Borrow's literary memory of *Crusoe* as, primarily, a weather picture. Recalling an early boyhood encounter with an illustrated edition of Defoe's novel, the Victorian Bible peddler saw in his mind's eye "a low sandy beach on which the furious sea was breaking in mountain-like billows; cloud and rack deformed the firmament, which wore a dull and leaden-like hue. . . . Reader, is it necessary to name the book which now stood open in my hand, and whose very prints, feeble expounders of its wondrous lines, had produced within me emotions strange and novel? Scarcely, for it was a book which has exerted over the minds of Englishmen an influence certainly greater than any other of modern times."[32] Crusoe's "strange and novel" influence over the English-speaking mind is of a piece with the graphic impressions that English air necessarily made in Defoe's own era.

All of which returns us to Crusoe's curious journal, for as it compulsively registers the state of the air at any given moment, that journal makes this general pattern of impression specific and in some sense visible. Needless to say, this isn't how Crusoe's version of writing to the moment is usually interpreted; more often "THE JOURNAL" seems to work the same way as the other implements of human culture that Crusoe contrives and describes. Like Crusoe's clothes, his hat, his tent, and, indeed, the umbrella that was to inspire a whole product line of "Robinsons" in the nineteenth century, Defoe's writing may always be seen as a form of defense against the elements. From this perspective, its first, last, and best use is to convert an alien sky into a simulacrum of the one above Crusoe's own famously drenched (then merely damp, then drenched again) England. Leslie Stephen thus supposed that Crusoe "is the typical Englishman of his time. He is the broad-shouldered, beef-eating John Bull, who has been shouldering his way through the world ever since. Drop him in a desert island, and he is just as sturdy and self-composed as if he were in Cheapside. . . . He does not accommodate himself to his surroundings; they have got to accommodate themselves to him" (Stephen, *Hours*, 176). And yet it might be more accurate to say that Crusoe uses his journal to make his own inward accommodations with these same "surroundings," thereby visibly producing a binding impression of accommodation.

Here "THE JOURNAL" becomes something that Scott's atmospherically attuned reader can virtually see, a graphic inscription of the horizon where seeing meets believing, and a world presented becomes its presentation to the mind. Having exchanged use value for its aesthetic equivalent, we might also trade the historico-allegorical question of what kind of figure is writing *about* the air (John Bull? A latter-day pilgrim?) for the question of what kind of figure is being written and perceived *within* it. "It," in turn, would be best described not just as an obsessive grammatical feature of Crusoe's writing but also as an impression made by the background that really frames that writing. Entries from Crusoe's first autumn as a castaway do constantly manifest "its" actions: on October 25, "it rain'd all Night and all Day," on November 14, evidently, "it did not rain," and on November 7 "it began to be settled fair Weather" . . . until 13 November, anyway, on which "Day it rained, which refresh'd me exceedingly, and cooled the Earth, but it was accompany'd with terrible Thunder and Lighting, which frighted me dreadfully" (*Crusoe*, 54). In all these instances, "it," if hard to pin down, is certainly active, at least as much so as Crusoe himself, whose own actions "it" may be said to motivate. But what exactly are these actions? And *where* in the world do they take place? Crusoe pretends to tell us: "Much Rain . . . in these Days," he writes, "tho' with some Intervals of fair Weather; but, it seems, this was the rainy Season" (*Crusoe*, 52). "It," in other words, "seems" as much as "it" rains or settles or blows. In the chiasmus of that "it," we catch the novel's language in the act of creating its own environment, graphing through English the sphere of "seem[ing]" in which the novel can at least appear to take place.

Under conditions as favorable as these, certain outlines naturally begin to blur. One is the line of demarcation between a character's inside and his outside. Just so, on June 18, Crusoe remarks: "Rain'd all day, and I stay'd within. I thought at this Time the Rain felt Cold, and it was something chilly" (*Crusoe*, 64). Then the next day: "Very ill, and shivering, as if the Wind had been cold" (*Crusoe*, 61). And on June 23: "Very bad again, cold shivering." Remarkably, indeed, there is no real way to tell what is cold (Crusoe? The rain?). There is only the cold. Strictly speaking, there is only cold. We perceive, of course, that Crusoe himself has *a* cold, a very bad one, and his "Distemper" (*Crusoe*, 66), as he later refers to it, resurrects this unjustly dead metaphor, turning a mere figure of speech back into a warm abstraction. This is the perfect climate for the imagination to take action, and it is significant that while he is in the throes of his cold Crusoe seems to see "a Man descend from a great black Cloud," until "all the Air looked, to my Apprehension, as if it had been fill'd with Flashes of Fire" (*Crusoe*, 75). From here it is but a short step across the room to a salvaged Bible whose written words commandeer this ethereal vision to make an "Impression on my Thoughts" (*Crusoe*, 69).

Crusoe's distemper, in short, creates the immediate impression of a truly novel experience, one in which figure and ground are suddenly, radically reversed. In a later and much-analyzed incident, a crop of barley seems to spring up out of nowhere. Crusoe eventually decides that he himself must have unconsciously scattered the seed some months before, but what unifies the twenty notoriously indecisive pages he spends figuring this out is not ultimately the question of where the corn came from, or that of what to do with it, but the atmospheric conditions that allow it to grow. From a cognitive point of view, we can describe these as conditions of suspended disbelief. And as it happens, the coiner of that phrase, Coleridge himself, turns out to have taken a special interest in this scene, recognizing it as "by far the ablest vindication of miracles that I have met with. It is," he declared, "the true ground, the proper purpose and intention of a miracle."[33] Here the romantic poet converts the literal "ground" in which Crusoe had unconsciously scattered his kernels of corn into the figural "ground . . . of a miracle." It is this conversion—not of things seen but of the ground against which they *are* seen—that outlasts, indeed underlies, both Crusoe's spiritual (or figural) conversion and his agricultural (literal) conversion of virgin soil into crop-bearing field. As for us, we are left with a frame of purely fictional experience, one that generates, visibly, its own limits of meaning, its own present, and its own effectual reality—one that opens, we might say, its very own umbrella.

This, inevitably, brings us to Crusoe's real umbrella. Although it cannot be said to figure prominently in his narrative, that curious accessory appears often enough to create the impression that it is always there. Crusoe's journal has much in common with the umbrella, though not because, as we expect both to, the journal shelters and homogenizes. Their kinship runs more along the lines we appreciate when we note the umbrella's similar ubiquity in French impressionist painting. Umbrellas are umbrellas, after all: overarching frames and visible screens of otherwise disorganized and disorganizing phenomena. Well might Freud have wondered why we forget them all the time.

In Defoe's day, umbrellas were too novel to be easily forgotten: they were recent émigrés from China and more immediately France, with the word "umbrella" itself a *parvenu* to the English language (1611). Because it was such a new thing, Crusoe naturally records the "great Deal of Time and Pains [it took] to make me an Umbrella. I was indeed in great want of one," he confesses, "and had a great Mind to make one; I had seen them made in the *Brasils*, where they are very useful in the great Heats which are there; and I felt the Heats every jot as great here, and greater too, being nearer the Equinox; besides, as I was oblig'd to be much abroad, it was a most useful Thing to me, as well for the Rains as the Heats" (*Crusoe*, 99). Crusoe goes on to map the "world of Pains" he took to fashion his umbrella until at last "I had made one . . . and covered it with Skins, the Hair upwards, so that it cast off the Rains like a Penthouse, and kept off the Sun so effectually that I could walk out in the hottest of the Weather with greater Advantage than I could before in the coolest" (*Crusoe*, 146).

While Crusoe almost never lets us see his own body, the umbrella both figures and lies next to it. It thus not only replicates Crusoe's own hide of skin and hair but also usurps his body's role as the organizing register of alternating heat and coolness. The effect is both to reinforce the "air of truth" that hangs about Crusoe's story and to undermine it, leaving us suspended between conviction of its reality and certainty that it has been made up. Thus the umbrella itself can only have sprung from the "great Mind" Crusoe has to make it. For all the circumstantial description we receive of it, it can only be seen in its likeness to what it isn't, either—on the one hand, to those "made in the *Brasils*" and, on the other, to "a Penthouse."

If Crusoe's rather grudging picture of his umbrella might now strike us as a definitive shift from historical reference to pure figuration, it's worth remembering that the English were slow to suppose that umbrellas were good for much besides decoration. In 1664 John Evelyn got a look at some and called them "fans, like those our ladies use, but much larger and with long handles, strangely carved and filled with Chinese characters." In this sense, what Crusoe has so triumphantly invented is a true hieroglyph, or, as Randle Cotgrave's 1611 *Dictionary of the French and English Tongues* had put it, "a round and broad frame, wherewith the Indians . . . preserve themselves from the heat of a scorching sunne, and hence any little shadow, fanne, or thing, wherewith women hide their faces from the sunne."[34] Unlike the umbrellas about whose effeminate charms his contemporaries harbored so many misgivings, Crusoe's "great clumsy ugly goat-skin Umbrella" is anything but beautiful. But it does appear to keep *him* so, and so it naturally becomes "the most necessary Thing I had about me, next to my Gun" because it preserves his "Figure," keeping the "Colour" of his "Face . . . not so moletta-like as one might expect from a Man . . . living within nineteen Degrees of the Equinox" (*Crusoe*, 146). Crusoe's umbrella makes him to stand out, but only to the extent that, ladylike, he blends in with our impression of him.

Meanwhile, our castaway's own language of impression organizes itself around the famous "Print of a Man's naked Foot" that leaves him standing "like one thunder-struck, or as if I had seen an Apparition." One of the footprint's effects is to transform Crusoe's world into a protogothic one in which extreme cognitive events ("fluttering Thoughts," "frighted Imagination," "Terr[or] in the last Degree") cannot be told from the objects of perception (*Crusoe*, 112). If Defoe were Ann Radcliffe, it might well emerge that this "apparition" is nothing but the work of windy weather. And so, in a manner of speaking, it is. Certainly, Crusoe's meteoric response suggests as much: he is "thunderstruck." This figure of speech supplies, as Coleridge would say, a "true ground" for all the frantic leaps and pirouettes of Crusoe's panicked mind. His being thunderstruck also justifies his failure "to describe how many various Shapes affrighted Imagination represented Things to me in"—a failure that further merges inner and outer mise en scènes (*Crusoe*, 122). We scarcely need description to mediate between imaginative "shapes" and "represented things," for they are one. So are the literal and the figural—like the ground and the miracle in the barleycorn episode—reversed in the novel space of being thunderstruck. Small wonder that when Crusoe manages to convince himself that "there was really nothing in it but my own Imagination" and begins "to peep about again," a second encounter with that literal pied-à-terre gives him "the Vapours again to the highest Degree, so that I shook with cold" (*Crusoe*, 115).

Here we can't be far from the kind of response we say we have to literary atmospheres of all kinds. These ways of speaking about—and gathering—disparate impressions under the umbrella of a cliché are foreshadowed in many of the places where Crusoe at least seems to find himself: beside the footprint, among the barleycorn, under the umbrella, amid the pouring, blustering, and settling of "it." Such textual "wheres" might even leave Defoe's own readers thunderstruck. But in so doing, they mark out the formal space that we share with the fictional form of Defoe's writing. The air literally fills the space between the mind and the page and actively mediates between them. So does it sustain the experience of fiction itself—the true and only atmosphere of *Robinson Crusoe.*

5

Spectral Currencies and the Air of Reality in *A Journal of the Plague Year*

> Many persons in the Time of this Visitation never perceiv'd that they were infected, till they found to their unspeakable Surprize, the Tokens come out upon them, after which they seldom liv'd six Hours; for those Spots they call'd the Tokens were really gangreen Spots, or mortified Flesh in small Knobs as broad as a little silver Peny.
>
> DANIEL DEFOE, *A Journal of the Plague Year* (1722)

Three years after he published *Robinson Crusoe*, Defoe turned his attention to English air, and specifically to that of Restoration London during the plague of 1665. Both in its own time and in Defoe's adulthood, the etiology of the plague was as much a mystery as that of any climatic condition, though both were linked in obscure and complicated ways to the air.[1] To both, thus, the language of "appearance" was very naturally applied. And indeed, the London that Defoe's *Journal* depicts is a protogothic necropolis overrun with "Apparitions" of every stripe. Here, clouds arrange themselves into menacing "Shapes and Figures, Representations and Appearances," while the streets are set with "mournful Scene[s]" in which every "speaking sight . . . has a Voice with it, and a loud one to call us all to Repentance."[2] Their differences from one another notwithstanding, these specters never reduce themselves to mere figures of material, psychological, or moral reality. Their sole purpose seems to be to leave what Walter Scott called a deep "impression of reality" itself, and it is their fidelity to this task that, in Scott's words, makes "the hair bristle and the skin creep."[3] Whenever we open this strange book, such expressions certify the *Journal* as an atmospheric fiction in the modern sense. But the apparent fidelity to sense experience that underlies them helps to explain why, of all Defoe's fictions, the *Journal* is the least comfortably classified as such. We see too why some of the most fruitful and provocative criticism of this manifest piece of ghostwriting turns on its claims to be counted as history, which is to say as a sign of the real.

As he ties *Robinson Crusoe* to the supernatural fantasias *Frankenstein* and *Dracula*, John Bender attributes the (in fact) "gossamer" quality of the factual plain style to that style's "capacity to disappear as writing." While I share Bender's sense that, when it comes to Defoe's realism, all that is solid has always already melted into air, *A Journal of the Plague Year* leads us to the opposite conclusion: what Bender calls the realist "fibre"—or at least a reader's impression of being woven into it—depends on writing's visibility as a mediating frame.[4] As much becomes apparent when we place Defoe's most factual fiction in the context of "real" apparitional narratives.

Close cousins, indeed, of the meteorological "Appearances" that structure *Crusoe*, the specters that jam the airwaves in *A Journal of the Plague Year* are, like them, artifacts of the material print culture from which this bewildering work arose. Here, it seems, writing about apparitions ("things" in the air) was as ubiquitous as writing about the air itself. What I will be calling apparitional writing counts as an inscription of the air because apparitions themselves appeared in the air (or, as we now say, out of thin air) and because information about apparitions was so often represented as having first passed through the invisible currents of oral culture.[5] Although English writers, in defense of their own substantive medium, were more and more inclined to present those currents as unreliable, and even treacherous, apparitions themselves were not necessarily unbelievable.[6] Indeed, five years after finishing *A Journal of the Plague Year*, Defoe himself popularized six prolific decades of English apparitional writing in his cornucopian *History and Reality of Apparitions*; as the lone topographer of his interest in the supernatural has shown, that work consummates a lifelong entertainment of the possibilities that apparitions might be a "Reality."[7] Defoe's investment in these vital traditions is closely bound up both with his literary practice and with the questions of belief that apparitions—or at least the possibility of them—inevitably raised.[8] Because they so explicitly identified textual praxis with problems of belief, we can look to contemporary writing about the "reality" of apparitions for information of our own: information about the realizing strategies that developed alongside the referential and factualist pretenses that characterize the fiction of the period. These strategies build the sense, neither purely spectral nor wholly embodied, that literary fictions have atmospheres.

This delicate suspense is nowhere more apparent than in the strange history of the *Journal*'s reception. As Robert Mayer has shown, that history has always pivoted on the unresolved question of whether the *Journal* itself counts as history: Do the graphic tokens that compose it add up to fact or fiction?[9] In one notorious example of the embarrassment this quandary can create, the eminent twentieth-century epidemiologist L. Fabian Hirst once diagnosed the *Journal*'s subject, the 1665 plague, as—at least in its initial stages—the pneumatic as opposed to the bubonic strain of the disease. A recent literary critic gleefully points out that the symptoms that prompted Hirst to the pneumatic hypothesis were actually presented to him by the *Journal*, fraught as it is with what Defoe and his contemporaries termed the "Tokens" of subcutaneous hemorrhage. Some of these tokens, the *Journal* reports, became visible only at the time of death, a distinguishing feature of pneumatic plague. Awkwardly enough, though, the *Journal* was composed almost sixty years after the plague in question, and its narrator, Defoe's fabrication, in point of fact lacked eyes to lay upon the mostly invisible tokens in question. Barring some primitive memory (Defoe did turn five during the plague year), the most we can say is that, in the words of the aforementioned critic, Defoe "saw these 'Tokens' in his imagination." His narrator, strictly speaking, cannot be said to have seen them at all.[10] And yet, for the modern historian of contagious disease, their ideal or imaginary or phantasmal glow seemed to have been produced by their practical currency in the world of actual events. It seemed in turn possible to discount the glow and credit these "Knobs as broad as a little silver Peny" (*Journal*, 195) as plausible and articulate vestiges of that world.

It so happens that in Defoe's *Journal* itself, the "mortal Marks" of plague that sooner or later appear on each doomed body express this very paradox. Their status as visible symptoms of contagious disease ("gangreen Spots, or mortified Flesh") can never be completely detached from an occult sense of their relevance to what cannot be "perceiv'd," which is to say to what Defoe comfortably termed the "invisible world" (*Journal*, 195). Defoe's narrator, H. F., attributes both the plague's arrival and its mysterious departure to "the secret invisible Hand" of God; even the physicians, he notes, were "oblig'd to acknowledge that it was all supernatural, that it was extraordinary, and that no Account could be given of it" (*Journal*, 247). And yet at the same time H. F. does intend to account for the plague, and to this end he often supposes it "really propagated by natural Means" (*Journal*, 194). On occasion he even ventures that "there is no apparent extraordinary occasion for supernatural Operation" and that "the ordinary Course of Things appears sufficiently arm'd . . . to execute the Fierceness of Divine vengeance, without putting it upon Supernaturals and Miracle" (*Journal*, 194).

Defoe's famous equivocation between a plague simply "propagated by natural Means" and a plague that supernaturally manifests some "secret invisible Hand" has been interpreted as a kind of Aristotelian compromise with efficient cause. And so it may be.[11] In other circles, that same compromise might well be taken as evidence of the historical encroachment of epistemology on metaphysics, realism on romance.[12] But this equivocation, so central to the *Journal*, is surely more than a sign of the historical mutation of one interpretive (or representational) paradigm into its successor. Rather, we may link Defoe's attempt to adjudicate between the natural and the supernatural dimensions of the plague—between the spectral and material faces of its "Tokens"—to the attempt he made throughout his occult writings to find some "indeterminate" ground between the visible and invisible worlds. For Defoe, this was also to find and chart a representational field halfway "between imagination and solid foundation."[13] And indeed in the *Journal* it appears to be literature, in Defoe's own gerundive sense of "writing," that allows certain pneuma to become visible—and thereby to acquire the efficacious illusion of physical and historical substance. Because the graphemes that compose the literary text behave in this way, misrecognitions such as that by the epidemiologist Hirst are not only understandable but are, in effect, not really misrecognitions at all.

This returns us to the historical difficulty of classifying the *Journal* generically. Walter Scott very typically deposited this work into "that particular class of compositions that hover between romance and history." He also found that in the *Journal* Defoe "drew pictures almost too horrible to look upon," thereby inducing later readers to see Defoe's text as a prototype of gothic fiction (Scott, "Daniel De Foe," 2:256). At the same time, though, Scott's bizarre presumption that the very letters Defoe put down on the page somehow behaved much as pictures do—and with something of their seemingly supernatural urgency—lures us beyond the literary categories that hardened in his own romantic period. The same presumption also entices us to place the "pictures" that seem to fill the *Journal*'s pages in relation to a kind of writing completely devoted to the drawing of such pictures: the apparition narrative.

"The Oeconomie of the Invisible World"

Apparition narratives were registers of what one of their seventeenth-century authors, John Aubrey, called "the Oeconomie of the invisible World."[14] Wending their way through Lucretius and Luther, they reprised incidents from Old Testament times to present English ones in which people claimed, or were said, to have perceived "Apparitions, and other sensible Manifestations of the certain existence of Spirits of themselves Invisible."[15] What we now call séances and poltergeists crowded their pages, along with the "Apparitions of a Man's own self" rumored to appear before death, prognosticating specters, and assorted images of the absent (Aubrey, *Miscellanies*, 75). Flourishing in the last decades of the seventeenth century and in the first decades of the eighteenth, apparition narratives were also cash cows for a burgeoning publishing industry, as their sheer numbers would seem to attest.[16] And although a few of them were written in a debunking spirit, most were at least experimental in their attitudes toward their own subject matter; indeed, many of the most popular spoke with an air of solid conviction.

Though it's tempting to identify them with ghost stories or with gothic fiction, the apparition narratives still so ubiquitous in the 1720s really have no analogue in ours. For one thing, they did not conceive of their subject as necessarily supernatural; as is well known, the category of the natural and its attendant laws were under strenuous negotiation at the time. For another, the strange visitations they narrated were not, in Scott's phrase, considered outright "horrible," owing to the fact that apparitions were seldom taken for ghosts in our sense of the spirits of the departed. That, to many, smacked of purgatory and the coercive chicanery of those "doctors of the Roman church" whom Hobbes so roundly scolded in *Leviathan* (1651). These terrorists of the mind, Hobbes famously charged, had falsely legitimated what are in fact mere "idols of the brain," zealous to make good self-serving "doctrines of hell and purgatory, the power of exorcism, and other doctrines which have no warrant, neither in reason nor in scripture."[17] Writing almost eighty years after Hobbes, at the opposite end of a veritable storm of English writing about ghosts, Defoe was no less eager to indict such "Modern Roguery" (*History and Reality of Apparitions*, 352). Echoing the many self-appointed authorities who had weighed in on the subject since Hobbes, he too maintained that there were few things more "delusive" than the notion that apparitions are really "departed Souls" (*History and Reality of Apparitions*, 26). Without accepting Hobbes's contention that all specters are mere political figments, Defoe nonetheless declared the "absurdit[y] of Souls remaining in a wandring [*sic*], unappointed state after Life" on the sensible or at least peace-appreciating grounds that if they were really able to do so, "the habited Visible World would have been continually haunted with Ghosts, and we should never have been quiet" (*History and Reality of Apparitions*, 95, 98).

Fair enough. But if apparitions were not "departed souls," still bound to the narrative trajectories of human life, what were they? And what might their relationship be to those who still evidently saw, certainly reported, and ultimately read about them? Stuart Clark reminds us that in the wake of the Protestant Reformation, mounting questions about the reality of apparitions bundled themselves up with "more serious problems to do with visual reality"—problems finally resolved within theories of perception that allowed that the eye can "really" be stimulated without any object outside itself necessarily doing the stimulating. If it complicates any knowledge system rooted in ocular proof, this model of perception nonetheless opened the possibility that apparitions were "nothing but phenomena."[18] That possibility in turn has significant implications for the theory and practice of representation, for to the modern eye—of which Scott's may be representative—the apparitions who fill the pages of Defoe and his contemporaries resemble nothing so much as pictures. Anything but revenants, they were frequently taken, literally, to be appearances; in an apparitional event, as Defoe himself put it, "spirit may vest it self so with Flesh and Blood, that is seemingly, so as to form an Appearance" (*History and Reality of Apparitions*, 29). This is "an Appearance" much in the sense we would today intend when speaking of somebody's appearance, except that there is no "somebody" there, and never was. There is *only* the appearance, which is to say only the appearing, seeming without being, a medium that has become its own message. In this sense—one largely obsolete since the early nineteenth century—an "apparition" has most in common with the mechanical details of the Dutch realist canvas.[19]

To rational Protestant observers such as Defoe, the appearances of such "Appearance[s]" thus were not to be confused with visits by angels or devils per se, though both angels and devils had been known to cook them up, and Defoe, for one, was certainly very interested in the recipes they might have used. Nor had apparitions any necessary moral implication, though they could serve justice by, for example, fingering unpunished miscreants or hinting at the whereabouts of lost wills, and, as we've already seen, they could certainly be used as metaphors for insulted conscience. Nor, though they could be faked by frauds, were apparitions often deemed the offspring of natural magic. And though the still-cited Robert Burton had tentatively matched the probability of seeing them to the depth of melancholy, they were only rarely attributed to mental disturbance.[20]

Neither demon nor delusion, nor even necessarily linked to the deceased, the modern apparition was often most exactly painted in the emergent language of modern natural philosophy as a simple configuration of the air. This is not the crude reduction it sounds because, as we've seen, air's material density and attendant mediatory capacities had come under unprecedented scrutiny in the last decades of the seventeenth century, most particularly at the hands of Boyle. As his speculations about the "strange subtility" of effluviums would suggest, the skeptical chemist was in fact extremely curious about what he took to be air's stubbornly occult qualities. He noticed its tendency to assume configurations not necessarily perceptible to all, its residual elusive and incalculable qualities, and even its potential for action as well as for passive "spring." Indeed, like so many of his fellows in the Royal Society, Boyle was keenly interested in apparitions themselves and financed investigations into the second sight.[21] He contributed a preface to the memoirs of the famed Irish "Stroker," Valentine Greatrakes, their acquaintance having been forged in the 1660s at Ragley, home of the esoterically minded (and migraine-suffering) Lady Anne Conway. Conway's circle included the demonologist Joseph Glanvill, the Cambridge Platonist Henry More, and Conway's personal physician, the alchemist Jan Baptista van Helmont.[22] Elements of the Conway circle's inquiries into the possibilities of extrasensory perception and communication found their way into works of natural philosophy such as Boyle's fugitive *Notes . . . about the Atmospheres of Consistent Bodies* (1662). That work in particular sought to quantify what Boyle described as "the cures intimated to be wrought by St. *Peter*'s shadow," so that "to have that pass over [one] was to come within the atmosphere (if I may so call it) which the effluvia of his body made miraculously sanative."[23]

Such speculations aside, the spaces and subtle energies that inhabited even Boyle's physico-mechanical model of springy air left room for speculations, such as Defoe's own, that in an apparitional experience, "moving air describes in (to others invisible) small and exactly shaped Clouds, the very Faces, Shapes, Names and Distinctions of such Persons, Nations, Towns, Cities, and People of whom these Mists, Meteors, and other Clouds represent the Actions. Darkness it self forms the Posture of Things."[24] The northernmost latitudes, from Scotland up to Lapland, were supposedly most conducive to such displays because the air there was held to be exceptionally rigid and hence especially apt to support and preserve apparitional forms.

So conceived, apparitions became the protagonists of a long line of hefty works that fixed matter-of-fact accounts of their manifestation within the frames of Protestant theology and natural philosophy, thereby working a perverse reconciliation between these two discourses on reality. The very title of geologist John Beaumont's *Historical, Physiological, and Theological Treatise of Spirits* (1705) announces exactly this method of exposition and interpretation, one adopted by everyone from contributing fellows of the Royal Society such as Glanvill and Beaumont, to men of the cloth such as Richard Baxter and Robert Kirk, to the more platonically inclined of the Cambridge professoriate such as Henry More. Hand in hand with their weighty works went shorter and more ephemeral popular accounts of apparitional events; indeed, these were often incorporated into longer "Collections of Relations" (as Glanvill called his), which in turn plagiarized freely from one another.[25] Though accounts of local and recent apparitions had obvious roots in oral culture, they thereby also gained a fundamentally literary scaffold.

What we might call the literariness of contemporary apparitions—their manifestation and multiplication primarily via the graphic medium of the printed page—becomes more vivid when we notice that several of the most popular apparitions of the long eighteenth century began their careers as poltergeists, "racketing spirits" who remained invisible to their original percipients.[26] In their company were numbered the so-called Drummer of Tedworth, Old Jeffery of the Wesley parsonage at Epworth, and, later, the celebrated Cock Lane Ghost. Although each of these spirits initially made itself known through fumes, thumps, and other disturbances of the air, none was classified as a bona fide "Apparition" until its story went into print, often accompanied by visual images to supplement a narrative of intrusion into normative experience. The Drummer, for instance, was immortalized as a "Sculpture" in the second volume of Joseph Glanvill's 1665–81 *Saducismus Triumphatus.* Unseen by any of the hundreds of visitors who flocked to the London house she was supposed to have haunted, Scratching Fanny of Cock Lane took the shape of a vaporous maiden at the head of numerous poems and print ballads about her. It was only on such graphic appearances that these and similarly invisible spirits began to be classified as apparitions and became available for ratification or contemptuous dismissal among their several narrators.[27]

As has occasionally been observed, many apparition narratives were partly rhetorical in nature, peddling sensory evidence of a supernatural realm that could double as proof of God's own existence. (No ghosts, no God, as the saying goes.) Hobbes's infamous renunciation in *Leviathan* of all supernatural entity had been linked to his atheism, and most apparition narratives in some way positioned themselves as refutations of Hobbes's seemingly undiluted materialism. Thus Richard Baxter's *Certainty of the Worlds of Spirits* (1691) was, as the title page states, "written for the Conviction of Sadducees and Infidels," and one subtitle of Glanvill's *Saducismus Triumphatus* was "A Blow against Modern Sadducism." In the interest of compelling readerly belief, apparition narratives made conspicuous efforts to verify the good character of living witnesses to the phenomena they described. In so doing, they also committed themselves to formalizing and normalizing—we might almost say materializing—certain practices of linguistic and intellectual exchange. Apparition narratives functioned in these systems partly as tokens whose credit could in turn validate an emerging community of observation and knowledge. The conventional wisdom is that apparition narratives ultimately lost their credit in the economy of knowledge they helped create and ended up in that debtors' prison we call the aesthetic.[28] But if this really was their fate, it had not yet befallen them in the first decades of the eighteenth century. On the contrary, their quasi-aesthetic qualities actually enhanced apparition narratives' epistemological value, along with what we might call their social value as validating tokens within a consolidating (if also ever more widely dispersing) system of linguistic exchange.

One place this spectral economy may be glimpsed is in the standard literary practice of representing apparitions to curious nonpercipients (i.e., readers) as they had first represented themselves to the person who presumably originally perceived them. The person could be anyone of any sex, class, or degree of literacy; individuality was not important, although good character was. What really mattered, though, was possession of the requisite receiving faculty, the second sight. This faculty, still of burning interest to Samuel Johnson when he made his Scottish tour in 1773, was notable in part because it suspended all distinctions between the literal and the metaphorical, and between the historical subject and his demonstrably nonhistorical object.[29] The airy figures available to the second sight often had reference to the future, even as they could impersonate bodies extant in the present and past. Positioned at once outside the history of the senses and entirely within it, apparitions were peculiarly suited to the kind of publicity they received in the printed pages of the day. Indeed, the popular 1707 treatise *Deuteroskopia* (*Second Sight*) by the Scottish clergyman John Frazer defined apparitions as "Representation[s]" linked to "future contingent event[s]"—as representation before presentation, as it were.[30]

If apparitions were anything, then, they were representations of the strangely suspended conditions of their own sightings. Such sightings would resist both the rhetorical and the historical articulation—the narrative—on which they nonetheless depend if they are to enter the system of credible exchange we call collective knowledge and thereby become visible to others. As one Anglican divine put it to Samuel Pepys, "second sight" "is a very proper term of that sight which . . . visions have of things by representation. For as the sight of a thing itself is in order of nature the first or primary sight of it; so the sight of it by any representation . . . is in order of nature the second; or secondary sight of that thing, and therefore the sight of any thing by representation tho first in order of time may properly be call'd the second sight thereof. Thus the sight of a picture in order of nature is indeed the second sight of the thing, whose picture it is."[31]

Defoe was to copy this temporally suspended, seemingly pictorial assessment when he set out to summarize and further popularize decades of English ghostwriting. "An apparition," his *History and Reality of Apparitions* finds, "assumes the shape and appearance of [a] man himself, clothes himself in his likeness . . . as a painter clothes the cloth he paints on with figures, postures, habits, garments, all in colours; while the passive person represented is in no way affected with or concerned in the draught representing him" (*History and Reality of Apparitions*, 132). Defoe's most immediate predecessor, John Beaumont, likewise held that "Genies" can present us "with the Species or Images of those things they would persuade us to, as in a Glass; on which Images, when our Soul privately looks, those things come into our mind."[32] In any event, to the extent that the sighting of an apparition inverts the "order of nature" and supplies representations without originals, clothing without cloth, and reflections without glass, textual records of these sightings replicate their own subject matter and stimulate paradoxical mental states similar to those that characterize the second sight itself. Not to acknowledge these narratives' quasi-pictorial resistance to the very narratives of cause and effect they seem to verify is to miscalculate their value to an emerging practice of exchanging and assigning value to certain forms of knowledge.

For example, McKeon's magisterial *Origins of the English Novel* depicts the late Stuart explosion of apparition narratives as an especially vivid symptom of the transition from metaphysics to epistemology that accompanied the rise of the novel. In their emphasis on empirical verification, he maintains, the apparitionists aimed to impose new criteria of veracity on metaphysical phenomena—a gesture that in itself tolled the death knell of such phenomena as a valid and authentic order of experience (McKeon, *Origins of the English Novel*, 83).[33] Yet it seems that apparition narratives may have been rather more than evidence of metaphysics on the run, if not already overtaken by epistemology. They might also be understood as self-consciously literary structures, too firmly wedded to the metaphysics of apparition itself to liberate themselves. As such, the words in which they were necessarily written could both perform and trigger an intricate mental process. In this process, meaningful and coherent forms take shape within historical and material sign systems only to seek, and gain, autonomy from those same systems. This may be why, in such narratives, the individuated presence of both the (secondary) narrator and the original percipient is so critical. Because the text of an apparition narrative, uniquely, simulates at once both the agent of perception *and* its object, it gains a unique relationship—indeed, a virtual identity—with what it represents. This identity in turn makes it possible for others to "see" what is described by written characters on a page. Unsurprisingly then, apparition narrative seems to have legitimated the specters it represented by exploiting some of the uniquely graphic possibilities of formal textual representation—of "literature."

For someone like Defoe, this was no great leap. He, at least, appears to have seen writing itself as an originally and inherently occult phenomenon, one literally instituted by God's finger as it inscribed the Ten Commandments on stone tablets. In fact, Defoe's 1726 *Essay upon Literature* viewed print as an especially worthy heir to just this occult legacy: not only was printing a matter of "types impressing their Forms on Paper" but the art itself was likely invented by none other than Faustus, who "from thence was taken for a Conjurer."[34] This probability underwrites the faith that many early apparition narratives placed in the unique power of the printed word—a power to validate the specters described both by mimicking the way they had first been impressed on the air and by stimulating the original process by which they had first been seen. Beaumont, for example, identified reading and writing about the "extraordinary Visitations [that had] happen'd to me" with the visitations themselves, confessing to his patron, the Earl of Carbury, that "I could not well with-hold myself from perusing the Best Authors I could meet with, relating to it, in Order to draw an Abstract of what I found most Material in them, and to publish it together with my own Experience and Thoughts in that kind."[35]

In such cases, the page plainly rivaled the air itself. Just so, the most sensational "Relation" in Glanvill's *Saducismus Triumphatus* was the story of the Drummer of Tedworth, a pugnacious spirit allegedly dispatched by one of Cromwell's military drummers to plague a provincial magistrate. "Oral report" had it that Glanvill had retracted his published "assertion of belief" in his own "well-attested narrative" of the spirit's rampages through the magistrate's house, and a new edition of *Saducismus Triumphatus* was rushed to the press after his death in order to "demonstrate [those reports] to be false." Pivotal here were letters of testimony that would "add a very real weight to the value, as of that story, so of the present Edition."[36] Hence it was not the specter itself who both assumed and disturbed the air of reality but rather the previously printed "story" about it. Glanvill even likened the effects of his own "Relations" of modern hauntings to those of the ghosts on their original witnesses: "Not to make much noise to disturb these *infallible Huffers* [who doubt the reality of apparitions]," he wrote, "I softly step by them, leaving only [a] Whisper behind me."[37] Meanwhile, other apparition narratives, such as Cotton Mather's *Wonders of the Invisible World* (1693), drew on Puritan investment in the spirit of the letter; they aimed less to verify supernatural visitations as matters of fact than to realign the experience of letters so it could count as an order of experience in and of itself. Defoe's own commentary on the "Devil's disguises"—his many manipulations of the air to produce apparitions—is exemplary. Complaining that "we dress [the devil] up in more suits of clothes, and more masquerade habits, than ever he wore," the *History and Reality of Apparitions* sets out to help its reader "form such images of the old gentleman in your mind, that you may not be cheated about him" (*History and Reality of Apparitions*, 362).

In the meantime, apparition narratives themselves often breezily announced their inherent literariness by making literary texts germane to the narratives themselves. For instance, in one of the many versions of the still-celebrated *True Relation of the Apparition of One Mrs. Veal*, the image of a dead woman appears to her girlhood friend, and the two spend the better part of an afternoon engaged in literary criticism, discussing "what Books they read, and what Comfort in particular they receiv'd from Drelincourt's Book of *Death*." The latter book, the ghost maintains, "was the best . . . on that Subject was ever wrote."[38] Subsequently, one of the many print accounts of the apparition was attached as an endorsement to a new edition of Charles Drelincourt's *The Christian's Defence against the Fears of Death*. This is more than a story of cross-marketing, as Scott recognized, when, plainly fascinated by *A True Relation*, he turned "to the fact itself" (Scott, "Daniel De Foe," 2:209) and found that fact to consist primarily in the impressive "relation" of it: "In short, the whole is so distinctly circumstantial, that, were it not for the impossibility, or extreme improbability at least, of such an occurrence, the evidence could not but support the story" (Scott, "Daniel De Foe," 2:273). Scott marveled most at the extent to which *A True Relation* boosted Drelincourt's sales volume: "The effect was most wonderful," he writes. "The copies [of *Drelincourt upon Death*] had hung on the bookseller's hands as heavy as a pile of lead bullets. They now traversed the town in every direction, like the same balls discharged from a field-piece. In short, the object of Mrs. Veal's apparition was perfectly attained" (Scott, "Daniel De Foe," 2:274).

That "object" was to *become* an object, in effect if not in "fact." Mrs. Veal is defined entirely in and through the fabric of literary representation—the same fabric in which Drelincourt's assertions about the afterlife were also woven. As Scott so clearly saw—and, justly, believed—writing *about* apparition supplies visual proofs (tokens) that are premised on an immaterial and nonhistorical—an ultimately invisible—body. Its subject is no different from that writing itself. So what we might call the Veal paradox prefigures certain modes of literary representation, particularly those novelistic ones that (like the *True Relation* of Mrs. Veal itself) aspire not just to create the illusion of sense experience but to approximate the very grounds of experience. The same paradox even more wholly validates authors—for example, the unreconstructed Protestant Defoe—who held that apparitions such as Mrs. Veal are not revenants tainted by the purgatorial mists of popish doctrine but pure media: pneumatic impersonations of the absent that both validate and objectify the very faculties that perceive them. Small wonder that *Credulity, Superstition and Fanaticism*, William Hogarth's 1762 attack on midcentury ghost belief, should have shown the most celebrated English specters literally rising from the pages of apparition narratives such as Glanvill's enduring *Saducismus Triumphatus*. Hogarth's Mrs. Veal is even browsing through one such book for herself.[39]

The Face of Things

Defoe himself is sometimes thought to have penned the most popular of the many versions of the Mrs. Veal story in circulation in the early eighteenth century. When his authorship has been disputed, it is typically on the grounds that he "did not believe in ghosts" of the purgatorial sort Mrs. Veal appears to be (Starr, "Why Defoe," 421).[40] But this is to foreclose a view of the text as a kind of intraliterary meditation, one that does not necessarily require belief in returning spirits of the departed, per se. Indeed, in the version of *A True Relation* usually attributed to Defoe, as elsewhere, literature is not just the means whereby apparition is represented. It is also apparition's raison d'être and thus the source of its exchange value—a value that became also its truth value. The same thing is evident in Defoe's *History and Reality of Apparitions*, which explains the proliferation of apparitions in the Old Testament and their relative scarcity after Christ in light of the fact that "we have now a more sure word of prophecy" in the Gospels (*History and Reality of Apparitions*, 21). Modern apparition, in other words, was for Defoe a fundamentally literary event; like weather writing, it seemed to arise from an engagement with letters. *A Journal of the Plague Year*, in its turn, uses literature—"writing"—to image a sensible world now past. The singular nature of this image allows the *Journal* to stand legitimately as the very authentic picture of events it seemingly only "pretends" to be. In this respect, it surely resembles apparition narrative more closely than anything else.

The resemblance holds in other respects as well. One is the form that Defoe's narrative assumes, or doesn't—which is to say that the *Journal* is not a journal at all: it only appears to be. In reality, it is an account shaped out of memoranda presumably scribbled during brief periods of self-induced quarantine by H. F., a London saddler of middle age. Unlike that of Crusoe, H. F.'s "original" journal never appears in the text but is only alluded to from time to time as the hidden source of a narrative that observes no consistent chronological order. In fact, that narrative doubles back on itself many a time, its temporal logic trapped in a maze of puns and obsessive reiteration.[41] In the service of this oneiric idiom, H. F.'s narrative filches all kinds of written signs of the real. It is regularly interrupted by columns of statistics taken from the death rolls, by public ordinances, and by latter-day hieroglyphics—the Jesuit IHS, the printer's flower, the magician's ABRACADABRA pyramidally reduced—which would surely obstruct the narrative's flow from one end of the plague year to the other if that narrative did not so often obstruct itself. Both the *Journal*'s false designation of its own genre and its rampant heteroglossia force us to think of it first and foremost as nothing but writing.

For his part, H. F. never really asks his reader to distinguish him from his writing. Indeed, he actively discourages us from viewing him as a mere eyewitness to the shocking scenes that unfold in his city of death, instead "desir[ing] this Account may pass with [future readers], rather for a Direction to themselves to act by, than a History of my actings, seeing it may not be one Farthing value to them to note what became of me" (*Journal*, 8). If its author is to circulate as a token of any "value," H. F.'s story must itself resist its own presumed historicity, its place in a sequence of perceptible causes and predictable effects. Its task becomes instead to make apparent everything H. F. claims to have "seen." But what *does* he claim to have seen? In fact, he pretends to record little beyond "the Face of Things"—a face from the start so very "much alter'd" by the plague that the appearances of "Things" are as abstract and ideational as they are palpable. Thus at the very moment that he promises to sketch the "Face of Things," H. F. can only tell us that "Sorrow or Sadness sat upon every Face" (*Journal*, 16).

Just so, H. F. is perfectly aware of a strange identification between the words he uses to render "the Face of Things" and the parlous ontological status—the mere apparency, after all—of the world thus rendered: "Were it possible to represent those Times exactly to those that did not see them, and give the Reader due Ideas of the Horror that every where presented it self, it must make a just Impression on their Minds, and fill them with Surprise" (*Journal*, 16). The unfulfilled subjunctive mood of this sentence matters, for "Horror" itself is seldom elicited, only the "Idea" of it. That "Idea" in turn rises less from direct description than from a superficial register of what in fact remains invisible. For instance, we are less shown the city of death than told that "*London* might well be said to be all in Tears; the Mourners did not go about the streets indeed . . . but the Voice of Mourning was truly heard in the Streets. . . . Tears and Lamentations were seen almost in every House" (*Journal*, 16). The pictures the *Journal* paints for us are, by their own confession, no more or less than the shapes of sound. This conflation of showing and telling, if anathema to today's creative writing programs, is so central to Defoe's symbolic procedure that it is only a matter of time before our narrator himself is absorbed into the invisible scene he describes, his place there sealed in one oft-cited editorial insertion near the end of the *Journal*. Here, as H. F.'s eye roves London's now engorged mass graves, an erstwhile silent editor steps in to inform us that "the Author of this Journal, lyes buried in that very Ground" (*Journal*, 233).

This, of course, begs a question: Just who (or what or, indeed, where) *is* H. F.? We're not the first to wonder. No less than his provenance, his immunity to the plague has mystified centuries of readers. The provenance problem is usually solved with reference to Defoe's own uncle, Henry Foe, who obviously shares H. F.'s initials, who was in London at the time of the plague, and of whom H. F. can thus be seen as a kind of literary revenant.[42] The immunity problem, which is more interesting, remains unresolved. H. F. himself wonders why he never gets sick and attributes his stubborn good health to a psalm that seems to promise him protection when his Bible falls open to the relevant page. That the secret of H. F.'s immunity is somehow literary only seems more likely when we stumble across that "Piece of Ground in *Moorfields*" (*Journal*, 233) that is his grave. Under ordinary circumstances, it would be fair to say that the narrator's literal burial in Moorfields coincides with his figural burial in the printed page. But the circumstances of the *Journal* are nothing if not extraordinary; the fact that H. F., as Defoe's construction, only appears to be means that what would normally be considered a literal burial is actually a figurative one. At the same time, H. F.'s literal burial is as literal as it can be: it is his, or at least his voice's, burial in letters. This reversal is the key to understanding the *Journal* as an apparition narrative and as an intervention in that way of writing, narrated as it is not just *about* an apparition but *by* one.

Appreciating this intervention would seem to nudge us toward the emergent English novel, that form that form has none but in whose formation Defoe is always supposed to have played a crucial part. But before we venture further in that direction, we should pay tribute to a third apparitional feature of H. F.'s journal, the plague itself. Like our scribbling witness to it and his text, the plague conflates literal and figural orders of representation. The plague is, of course, invisible in and of itself; it's apparent only through its antagonism to coherent social, moral, and, above all, bodily form.[43] Understood as a "Visitation," it makes many very "dismal Objects" out of all its victims (*Journal*, 66). Yet there is, properly speaking, no "it" for H. F. to see when he speaks of "the Plague," only those dismal objects, the apparent shapes it makes out of the voices and bodies of the living. The pustules that signal the plague's presence are, of course, its quintessential apparitions, "terrifying Particulars" that permit the "Physicians . . . sent to inspect the Bodies" of the dead to ascertain that the cause of death "was neither more nor less than *the Plague*" (*Journal*, 195).

H. F. so obsessively returns to these telltale "Knobs" of "mortified Flesh" that we tend to remember them as having been rendered in scrupulous and graphic detail. But this is not in fact how they are presented: the closest H. F. ever comes is a simile (they are "as broad as a little silver Peny") likening them to the very sign of the presence of an abstraction, a coin. Otherwise, H. F. prefers merely to chart the perception of the "fatal Tokens" (*Journal*, 56), as when a mother "discover[s]" them on her daughter's thighs and realizes that the girl, though still breathing, is a "dead Corpse from that moment." Almost at once, "Fright . . . seiz'd her Spirits," yet this is fright as imagination and projection, aligned with the fact that here as elsewhere the "Tokens" themselves actually shirk both presentation and representation (*Journal*, 56). When the marks of the plague do become grudgingly visible, they tend to knot ordinary chains of temporal sequence, as in the case of the betokened girl who, from the moment of their appearance, is at once already, not yet, and never dead. In turn, though Defoe's reader cannot see these tokens at all, it is by their means alone that the plague becomes in any way visible, historical, or communal—that it becomes, in a word, narratable.

Meanwhile, at the *Journal*'s narrative and apparently historical level, H. F.'s fellow citizens are more than willing to see the plague through the scrim of apparition. Astrologers, "Wizards[,] and cunning People," all perceive dire shapes in clouds and comets and pass the bad news on to a fearful populace (*Journal*, 18). H. F., we note, distances himself from "the Error of the Times" and dismisses such visions as literary manipulations, *effets d'écriture* promulgated by "People who got Money by it, that is to say, by printing Predictions and Prognostications" (*Journal*, 21) He particularly resents the "Books" that, in an eerie evocation of the terrifying tokens of the plague itself, "frighted [their readers] terribly" because they "foretold directly or covertly the Ruin of the City" (*Journal*, 21). If the plague gives an unlooked-for boost to a nascent publishing industry, it also opens a passage into a world of pure media, one in which H. F. need make no functional distinction between the literary texts on which the market confers value and "the Dreams of old Women, . . . Voices warning [people] to be gone" or between the same texts and certain "Apparitions in the Air." The latter include "Voices that never spake," "Sights that never appear'd," and "Clouds" in which people "saw Shapes and Figures, Representations and Appearances," though they "had nothing in them, but Air and Vapour" (*Journal*, 22).

In the *Journal*, then, false apparitions become visible as overmotivated literary constructions. These are figments produced through the material exchange of certain symbolic counters that pretend to be real. But if this is what they are, that leaves literary media themselves open to manipulation into alternative and potentially somehow "truer" apparitional forms. This is exactly how H. F. sees the space of his own page, where apparitions pointing to the plague are replaced by the apparently natural phenomenon of the plague itself. Hence the *Journal* begins with a Dutch outbreak communicated "in ordinary Discourse," and though H. F. concedes that "we had no such thing as printed News Papers in those Days to spread Rumours and Reports of Things" and that everything "was handed about by Word of Mouth only, so that things did not spread instantly over the whole Nation as they do now," the "spread" of plague remains in large part a matter not of "Things" but of lexical appearances (*Journal*, 1). To be sure, public officials initially discover "evident Tokens of the Sickness upon . . . Bodies that were dead," but these mediated sights are immediately "printed in the weekly Bill of Mortality" (*Journal*, 3). "The People's Eyes" fall on these bills, and their "great uneasiness" mounts apace not with actual incidents of illness but with written instances of the plague, each more "frightful" than the one before (*Journal*, 2). H. F. is later at pains to point out that often the bills themselves lied, and he summarily replaces them with his own statistics. But the effect is very much the same.

The other object of horror to "the People" is the very air around them. The air menaces not only because the atmosphere seems to store significant pictures but also because, one way or another, most believe it to be the medium through which the plague itself passes. Some are of the opinion that "the Breath" is full "of strange Monsters and frightful Shapes, such as Dragons, Snakes, Serpents, and Devils, horrible to behold" (*Journal*, 203). Others subscribe to a miasmatic theory of "noxious pestilential Vapours, or any other Thing in the Air" (*Journal*, 219). H. F. himself spurns both of these attributions of agency, figure, and vitality to the air. But in so doing, he also certifies their power in the collective and abstract realm of public "Opinion" to shape perception and action, and on occasion we do find him willing to "join my Experience" to them (*Journal*, 220). "Opinion" may not necessarily have much use value in establishing matters of fact, but like fiction it has considerable exchange value; indeed, it is practical language—such as warnings against conversation with possibly infected outsiders—that H. F. classifies as "talk[ing] to the Air" (*Journal*, 230). Just so, H.F. might well discredit the shape of collective belief along with most forms of certainty about apparition. But once the plague retreats, his words can only confirm the people's collective determination to "depend upon it that the Air was restor'd; and that the Air was like a Man that had had the Small Pox, not capable of being infected again" (*Journal*, 230).

We are reminded here of the only other immune "Man" in the *Journal*, H. F. himself, who at this point again demands our attention less as a narrator or even as a character than as an apparition in his own right—one who plays the same mysteriously mediatory role as the air he maligns.[44] H. F. is certainly the most ethereal figure to appear in his own story: though we never see him, he glides tirelessly through his city's many "common Passages," peeping into burial pits and death carts and admonishing the unrighteous, an omniscient voice trapped in an illusion of the first person (*Journal*, 40). We learn that H. F. often "got Admittance into the Church-Yard by being acquainted with the Sexton" and that he usually "kept [himself] pent up within doors without Air," in effect entombed (*Journal*, 61). When he does venture out into the air, several telling adventures befall him. On one occasion, he joins a crowd "all staring up into the Air, to see what a Woman told them appeared plain to her, which was an Angel clothed in white." The woman "described every Part of the Figure to the Life," and many believe they see it too, but for his part, H. F. insists, she "could not make me confess that I saw it, which, indeed, if I had I must have lied. But the Woman, turning upon me, look'd in my Face, and fancied I laugh'd, in which her Imagination deceiv'd her too, for I really did not laugh, but was very seriously reflecting how the poor People were terrify'd, by the force of their own Imagination" (*Journal*, 23).

While H. F. stresses his own skepticism, what we see is his apparitional kinship with the cloudy angel. Of course, he did not laugh, never having been there in the first place. Thus of course the visionary woman would see him to have done just that. Here, it is through his skeptical rejection of air as a symbolic medium that H. F. himself materializes as just such a medium, onto which an act of perception (verified, we should note, by his own disbelief) is visibly grafted. Likewise, more than once, he appears to his fellow citizens as nothing *but* an appearance. When he tries to chide some "dreadful sort of Fellows" for their drunken revelry, they "ask'd me what I did out of my Grave, at such a Time" and is vindicated when "he who asked . . . was struck from Heaven with the Plague, and died in a most deplorable Manner" (*Journal*, 65, 66). Later, a gang of women attempting to break into his brother's abandoned house fails to register his presence; though he berates them, he confesses, "I could not perceive that my Discourse made much Impression upon them all that while," and they are only disbanded by the sudden arrival of "a Man [who] carry'd . . . all the Dead to their Graves . . . in Form" (*Journal*, 89).

H. F.'s readers are virtually required to wonder who and what he might be. To them he has appeared as everything from a full-blown "character" whom Defoe may or may not want taken ironically to "an intensified and almost abstract version of the ordering self" (Richetti, *Defoe's Narratives*, 236). Speaking of the moment in Moorfields when H. F. unconsciously looks on his own grave, one recent critic says more than he knows when he remarks that "it comes as no shock to learn [H. F.] is dead, for he was never alive" (Ellis, "Defoe's *Journal*," 79). This essentially intermediate quality identifies H. F. not only with the angel in the cloud but also with a ghost another Londoner sees in a "narrow passage." Just as the plague is on the rise, this ghost allegedly "made signs . . . plainly intimating . . . that Abundance of the People should come to be buried in [a nearby] Church Yard; as indeed happen'd" (*Journal*, 24). At the same time, though, H. F. resembles no one so much as the ghost seer himself, who with all the psychotic precision of a latter-day Ezekiel "describ[es] the Shape, the Posture, and the Movement of [the ghost] so exactly, that it was the greatest Matter of Amazement to him in the World, that every Body did not see it as well as he" (*Journal*, 22). The ghost seer, we learn, eventually "persuaded the People into so firm a Belief of it, that one fancied he saw it, and another fancied he saw it" (*Journal*, 24). H. F. can be said to do no less, or more, than this for his own readers.

Yet at the same time H. F. claims to have seen no ghost at all. As with the woman who discerned shapes in the clouds, he confesses that he "could not see the least Appearance of any thing" and that though "that [the seer] saw such Aspects I must acknowledge, I never believ'd; nor could I see any thing of it my self, tho' I look'd most earnestly to see it, if possible" (*Journal*, 24). Far from undermining the seer's authority, H. F.'s skepticism is itself the most valuable spectral token in this literary "passage." As forward-looking assessments of Defoe's narrative procedures are apt to acknowledge, H. F.'s skepticism seems to authenticate him as a narrator of other, less tangible appearances. But, seen at its own moment in what Defoe himself construed as the "History and Reality of Apparitions," this same posture of mind stimulates an economy of perception that seems positively Lockean in its particularity and credibility . . . only, on stricter scrutiny, to resolve into mere abstractions, idealities, the phantasmic shapes of sound.

6

The Dissipation of *Tom Jones*

> There is a cheerful, sun-shiny, breezy spirit, that prevails everywhere.
>
> S. T. COLERIDGE, "NOTES ON *Tom Jones*" (1813)

The earthquake that agitated Crusoe's island was nothing compared to the two that convulsed London and its environs late in the winter of 1749–50.[1] Those "great Shocks" have caught the eye of literary scholars primarily because, at the time, Anglican bishop Thomas Sherlock construed them as proof that God had damned the British book trade. The grievance? Its corrupting truck of "Histories and Romances," not to mention "the Industry used to dispense these Books at home and abroad, and especially to our Plantations in *America*."[2] But even as they appeared to chastise popular print culture for its affront to the Enlightenment's signature principles of consolidation and fixed difference, the London quakes also complemented the world of letters. In their wake, English natural philosophers began to read the geological record *as* a record. Casting the earth as a substantive body with a story that unfolds in historical time, they established a picture of the world that stands still.[3] Not everyone, though, felt compelled to trace London's late "remarkable Shocks" to terra firma. "A few of those Concussions esteemed subterraneous," one onlooker ventured, "may with greater Propriety be referred to the Atmosphere and be characterised by the Name of Airquakes."[4]

This was hardly a novel perception: ancient philosophers from Pliny to Seneca had already supposed that "it is moving air that causes earthquakes."[5] Since then, several fellows of the Royal Society had raised the possibility that air's supposed combustibility could trigger such an event as an airquake. After all (as Benjamin Franklin would still assume), the earth itself might well be a thin, soft shell wrapped around a restive albumen of gases and vapors.[6] And as Kenneth Olwig has recently explored in his speculations on the "imagined scholarly discipline" that felicitously he calls "aerography," seventeenth-century chorographers and landscape artists routinely conceived of the very land as "an illusion of depth stretching into infinite aetherial space."[7]

What, in eighteenth-century terms, might it have meant to "refe[r]" a seemingly grounded phenomenon to the atmosphere? To trade geocentric narrative for a theory of aerial formation? Olwig invokes British astrophysicist Arthur Eddington, who proposed that such a "reversal of the picture" would make "the mere back-ground against which the extension and the motion of matter is perceived . . . as much a performer in the world drama as the matter is."[8] When this "dramatic" perception becomes a textual praxis, as it did in eighteenth-century writing about "airquakes," a peculiar style of attribution surfaces. Instead of cause being imputed, or culpability assigned along a historical trajectory, objects dissolve into their conditions. Just so, the anonymous author of *An Account of Explosions in the Air, or Airquakes* (1750) posited not just abstract equivalence between "the Earth and Air" but a dynamic transfer of properties back and forth between them, one possible because both earth and air "abound with Exhalations of all Kinds." As Olwig implies, this view renders combustible air an incipient theater. Perpetually enlivened by "Lightning and other Appearances," the author of *Airquakes* himself perceived, "the inflammable Substance itself" may be "extended and formed into [many a] Representation," and only different "Disposition[s] of the Air [are] required to produce . . . different Appearances" (*Airquakes*, 4).[9]

For one influential eighteenth-century proponent of a nitrous atmosphere, John Flamsteed, such "Appearances" and "Representation[s]" were less apt to reveal a tangible object of critical knowledge than they were to register the reciprocal influence between such objects and the ambient medium within which they appear. "How simple a body soever our Air seems to be," Flamsteed supposed; it is in reality "a vast Compound." This compound achieves its ever transient meanings not through its ability to hold an inscription, or yet through a legible sequence of internally directed growth and change, but rather by grace of the "Spring and Elasticity" that dramatize its dynamic relationship to whatever it happens to hold.[10] It is, Flamsteed's disciples concluded, the "surprising distinguished Property of Elasticity by which [the air] contracts itself into more or less Space, when an addition of Pressure is laid upon it, and recovers its form . . . when the Pressure is taken off" that may best account for the appearances of seemingly solid "Things." Air's "surprising distinguished Property of Elasticity" gives such "motion to Slabs, . . . Beds and other Things" as will suggest that they are "only relatively and perhaps loosely fixed to the Ground"; it can even "affect the Superficies of the Earth itself" (*Airquakes*, 28). Once air's role in the elastic ordering of things has become apparent, even the most solid bodies are indistinguishable from their passing affectations. Hence arise new manners in which those same bodies' value, significance, and responsibility for their own actions might be judged.

How, then, to judge Sherlock's bookish and delinquent Britain? Surely its tremulous atmosphere compels more clemency than Sherlock could muster. It was true, granted the author of *Airquakes*, that the Mediterranean countries, along with South American ones such as Peru, are inclined to "subterraneous Explosions." But others, if "more thinly furnished with Ingredients productive of these Convulsions, . . . yet are supplied with a sufficient Quantity to replete the Air with Exhalations." And "such," it turns out, "is the Disposition of our fortunate Island" (*Airquakes*, 22). The insular "Disposition" rebukes the earthbound and proportionally *in*clement Sherlock, because very far from corrupting it as Sherlock had charged, "an Explosion" actually "cools the Atmosphere, destroys the noxious Vapours, and perhaps by this means prevents some Epidemical Disorders that might otherwise ensue" (*Airquakes*, 54). If they were indeed airquakes, why might not the "Shocks" that had rocked the capital be construed as "Favours" (*Airquakes*, 54)? In that case, they fall far indeed from the harsh convictions Sherlock said they were. Indeed, Britain itself strikes no one as a definite and delimited body. Quite otherwise, "our fortunate Island" is "replete" with the effects of incessant exchange between form and content, figure and ground. It seems to have become an aerosphere, of sorts—one whose strangely blameless inhabitants are also its readers, and its creators.

Dissipation

Sherlock named no names when he mistook the 1749–50 earthquakes for divine indictments of British literature, but a principal offender is sometimes thought to have been that most English of English novels, Henry Fielding's *Tom Jones* (1749).[11] This fictive history's chief crime has always looked like its untoward appetite for low life, as disapproving readers from Samuel Johnson forward were wont to put it. Fielding brought this appetite to fiction from the notoriously dissipated world of the London theater. Famously driven from the stage by the Licensing Act of 1737, the erstwhile irreverent playwright carried a theatrical sense of space onto—indeed, into—the printed page, thereby lending the latter what William Empson would characterize as "the shimmery mystery of a mirage."[12] Empson wanted this phrase to capture the delicate matter of what he called Fielding's "double irony," but his imagery weds Fielding's writing to the vanished theater of his youth. In more recent times, Jill Campbell has probed this connection with great subtlety in order to make sense of Fielding's slippery sexual politics.[13] Those for the moment aside, in the eighteenth-century theater, as on the social stage, acting was quite literally a matter of assuming an "air."[14] As Fielding's contemporary Thomas Wilkes put it, every actor should "assume the air, look, and language of the person he represents." What is more, the speaking actor's necessary medium of air is held in common with his audience, giving rise to "scenes" that condense, inspissate, dissipate, and reassemble as audiences—and actors—come and go. To Wilkes, such ephemeral effects suggested reflections both generic and local: "For my own part, I would chuse to leave the Theatre impressed by that gloomy pleasure which I feel from the sublimity of Shakespeare, or the tenderness of Otway; and not have it dissipated by Farce," he ventured. "Yet it is perhaps a dissipation necessary in this kingdom, where the temperature of the air inclines to gloom and melancholy."[15]

In Wilkes's words, the walls of the theater themselves dissipate so as to render its interior of a piece with a deliquescent "kingdom" whose very air "incline[s]" its subjects to "gloom" and where bodily experiences double as haunting figures of the mind. One of *Tom Jones*'s own most famous—and self-reflexive—scenes is set in the Shakespearean theater of Fielding's friend David Garrick, an actor praised on his death for performances that, "all perishable, . . . like the electric fire," at last "blend[ed] with air."[16] Equally significant, the scene that Fielding restages in writing is melancholy Hamlet's encounter with his father's ghost. While in Fielding's hands this encounter mounts a comedy of *méconnaissance* possible only in print, Garrick was famous for the fright wig that he adopted in this scene, which prop could easily turn Shakespeare's tragedy to comic effect.[17]

Broadly speaking, Fielding's own comedy owes at least as much to the air as it does to the earth. This debt has subtly shaped the history of English readers' responses to *Tom Jones*. Coleridge duly registered the "cheerful, sunshiny, breezy spirit, that prevails everywhere" in the pages of *Tom Jones*.[18] What Thackeray too would record as a "certain state of the atmosphere" in Fielding's novel may be most apparent in the middle and most "Cervantick" of the twenty-eight books that follow the arc of the lunar calendar.[19] Even as these often moonlit middle volumes are set primarily in the open air, they carry Tom Jones in and out of an array of unstable enclosures—inns, cottages, and barns—that on entry seem less like shelters out of that air than bewitching reconfigurations of it. As such, Fielding's buildings expose distinctions between inside and outside as matters of mere convention, incidental specters of the representational medium. This gives Fielding's omnipresent narrator his own affinity with the air—so much so that, in his invisibility, ubiquity, and ventriloquial agility, "the Author" has struck some readers as the very devil, Defoe's "Prince of the Air" redux.[20]

Such impressions have formed the periphery of critical response to *Tom Jones*, even as the center tends to take the novel as some version of the "History" its full title says it is and thereby emphasizes topical reference or trueness to embodied life. Scott ruled Fielding's "name immortal as a painter of national manners" (Scott, "Henry Fielding," 1:78), and Boswell noted that "he draws very natural pictures of human life."[21] This last was precisely the problem from a moral point of view. But the spontaneity and contemporaneity of these same "pictures," especially insofar as they corresponded to what Johnson disparaged as mere "characters of manners," have also seemed to give Fielding's novel the quality if not the character of a cloud. Thus William Hazlitt's aptly titled 1829 essay "Trifles Light as Air" observes that the "manners and characters which were objected to as low in Fielding have [since] in great measure disappeared or taken another shape."[22] Far from detecting the embodied low lives of mid-Hanoverian English people, Hazlitt's words conjure a recognizably atmospheric condition in which what is "low" inexorably rises, and dissipates. Coleridge likewise mixed his own meteorological reading of *Tom Jones* with reflection on the peculiar intimacy that "prevails" between literary figures and the time and place of their appearance in what we call the real world. "Manners change from generation to generation, and with manners morals appear to change," Coleridge mused. "A young man of the present day who should act as Tom Jones is supposed to act at Upton, with Lady Bellaston, &c., would not be a Tom Jones; and a Tom Jones of the present day, without perhaps being in the ground a better man, would have perished rather than submit to be kept by a harridan of fortune" (Coleridge, "Notes on *Tom Jones*," 2:373).

Coleridge's conception of character has little to do with the "ground," historical or otherwise substantive, that character is supposed to stand on. Whatever it is that names such as "Lady Bellaston" or "Tom Jones" mark, its shape depends first on the environment in which it becomes visible. So is legitimated, in the history of *Tom Jones*'s reception, a nebula of provisional thought, one that includes "suppos[ition]," speculation as to what "would have" been, and a willingness to attend to "manners" and the apparent change of "morals"—to what perhaps merely "appears" to be. Perversely enough, it is through this charitable cloud that the bold lines of Fielding's original composition are most clearly perceived. Once so perceived, however, they can no longer be definitely judged, only reflected on indefinitely. It was, predictably, that master of indefinite reflection, Henry James, who suggested that *Tom*

Jones's "author . . . has such an amplitude of reflection for him and round him that we see him through the mellow air of Fielding's fine old moralism."[23]

If James discovered that Fielding's "moralism" was more an effect of "mellow" medium than a formal reaction to definite objects, Fielding's own method for fixing his "History" involved something besides strict reference to relatively current events such as the Jacobite rebellion of 1745 (whose influence is anyway felt only from afar) or the incorporation of contemporaries from Hogarth to Garrick (both of whom Fielding registers only through their own transient representations of bodies that never existed). We can also position Fielding's own narrative within a "History" of belief that does not, by the Author's own admission, ever actually progress. So while the Author insists that *Tom Jones* was not "writ in the Days of Superstition," in the next breath he vows to "burlesque the superstitious Faith of his own Age and Country."[24] And even if we were to apply ourselves strictly to Fielding's plot—the very one that Coleridge so famously regarded as one of "the three most perfect . . . ever planned"—we may detect something rising from it that is, properly speaking, no thing at all and so can be ruled neither perfect nor imperfect.[25]

Fielding's Victorian biographer Austin Dobson captured this dimension of Fielding's fiction even as he marked its "carefully-arranged progression towards a foreseen conclusion." Especially in *Tom Jones*, Dobson noted with appreciation, "little details rise lightly and naturally to the surface of the narrative, not more noticeable at first than the most everyday occurrences"; eventually, though, "we look back and discover a hundred little premonitions which escaped us at first."[26] Dobson's airy "rise" of "little details," if ironically, makes it possible for this narrative's surfaces to present as depth. In this manner, *Tom Jones* fosters a style of reading that seems to look through the causal chains of forward movement that are prerequisites of judgment. Here "premonitions" become objects of curiosity and delight in their own right, even as they lodge neither inside nor outside the material signs that precipitate them.

When Thackeray wrote of "a certain state of the atmosphere" in *Tom Jones*, then, the word "certain" conveyed something besides certitude, namely, the passing impression of temporary location "in" an undulant field of unranked phenomena. The Victorian novelist's atmospheric reading of *Tom Jones* was evidently inspired when Charles Lamb wondered how often the laughter of its protagonist "clears the air." It is the kind of wonder possible only when a reader, rather more wonderfully still, senses that she may be comprehended in the very air in question. Thackeray, clearly laboring under much the same impression, added only that the "air" Tom's laughter clears must be "in a certain state of the atmosphere." Tom's laughter, Thackeray granted, "might clear the air when such personages as Blifil or Lady Bellaston poison it. But I fear very much that (except until the very last scene of the story), when Mr. Jones enters Sophia's drawing room, the pure air there is rather tainted with the young gentleman's tobacco-pipe and punch" (Thackeray, *English Humourists*, 171).[27] Readers of his day, as of our own, have often claimed familiarity with the world of *Tom Jones*. In resorting to the nimble idiom of atmosphere, Thackeray managed to turn this sense from a ground of judgment into a paradoxical source of intimacy with the glancing impressions left by letters . . . and, in turn, with other readers.

Fielding's contribution to that sphere of British insular fiction whose matter is manners lies in precisely the atmospheric sense that conviction itself is conditioned, and rendered conditional, by the medium it shares with its own objects. After Fielding published *Tom Jones*, his work as a magistrate attracted him to the one legal construct that entertains the possibility of this kind of knowledge: spectral evidence, which renders the testimony of apparitions admissible in courts of law. Fielding's seldom-read *Examples of the Interposition of Providence* (1752) would not rule out that kind of evidence. On his way to *Tom Jones*, however, Fielding had something else before him: his day's popular astrology manuals. These, like the spectral *Examples* of providential interposition that he would later collect, treated appearance as a legitimate form of evidence and presented judgment as itself a transient figure of the mind. So perhaps it was by drawing on the suspended forms of estimation encoded in the overlapping graphic and almanac cultures of his own day that Fielding developed that aspect of English fiction that, in James's words, "we see . . . through," yet also see.

The Disposition of *Tom Jones*

What is it like to be in *Tom Jones*? A reader might begin to say by considering the shapes that Fielding's fiction assumes—or does not.[28] When Coleridge held that "there is a cheerful, sun-shiny, breezy spirit, that prevails everywhere" in *Tom Jones*, that is, where was "everywhere" (Coleridge, "Notes on *Tom Jones*," 2:374)? For that matter, where was—or is—the "there"? It's possible that "there" might lie round about the same place as the emphatic "there" with which Fielding apparently locates Paradise Hall's "Air of Grandeur" in the fourth chapter of the first book of *Tom Jones*. "There was an Air of Grandeur in it," he allows, "that struck you with Awe" (*Tom Jones*, 1.4.42).[29] But where *is* "there"? In pursuit of some answer to this question, the enterprising reader naturally seeks refuge in Fielding's plot, in his habits of characterization, in his stated themes and topics, or even in his theoretical pronouncements. All of these express doubt as to the possibility of gaining access to interiors of any kind, and it is this very skepticism that Johnson put beneath contempt when he famously likened Fielding's fiction to a "dial-plate" that the author merely "look[s] on" so that he can "tel[l] the hour" (Boswell, *Life of Johnson*, 389). All too were surely in the back of Coleridge's mind when he switched the metaphor and gauged the "prevail[ing]" conditions of *Tom Jones.* But Coleridge's "breezy" estimation of Fielding's outward-bound art draws less on any of that author's overt propositions, or on any of his famously superficial habits of proposition and exposition, than it does on a cluster of prepositional assumptions that we still make about what it is like to be "in" the world of a literary fiction. Without ever directly acknowledging them, we still betray the presence of these assumptions whenever we speak of a novel's atmosphere.

We might expect to discern, and even quantify, such an atmosphere by means of the sort of inductive data sifting that Fielding recommended to those who seek "Knowledge of the Characters of Men," as the title of his oft-cited 1743 essay put it.[30] "The Author" of *Tom Jones* himself appears to invite this breed of curiosity when he declares his intention to "take us behind the Scenes of this great Theatre of Nature" (*Tom Jones*, 7.1.327). But it soon becomes apparent that he conducts the tour from within a preconceived figural design and therefore in a manner irreducibly ironic, self-reflexive, and indifferent to the open-ended unfolding of evidence in time. Empson's observation that *Tom Jones* "builds up like Euclid" captures just what it is that defies inductive, penetrative reading practices: the writer and the geometer are alike "didactic in the propositions they have been taught already" (Empson, "*Tom Jones*," 19, 26).[31]

One way to account for Fielding's Euclidean bent would be with reference to what some have suspected might be an underlying providential argument.[32] But for Empson the visibly prefigured nature of Fielding's writing was implicated in its elusive "double irony." Its clear-eyed analysis of this brightly mysterious quality—in which interpretive possibilities hover in suspense—made Empson's 1982 reflections on *Tom Jones* a belated coda to *Seven Types of Ambiguity* (1930). Indeed, Empson remarks at the outset of his essay on Fielding that he had "been meaning to write about [him] before" (Empson, "*Tom Jones*," 17). *Seven Types* would have been the perfect place to do so, given that its penchant for isolating "types" and identifying literary "coefficients" seems to align it with the "'scientific' mode of literary criticism."[33] Empson's peculiar science, however, was not one that derived symbolic notations from empirical phenomena; on the contrary, its first premise was that symbolic notation grounds and generates the forms within which phenomena appear not just to be meaningful but, simply, to be. On this premise's infirm foundation Empson based his sense of literary "atmosphere" as "an undifferentiated mode of being" that is "conveyed in some unknown and fundamental way as a by-product of meaning."

Empson went on to identify this "mode of being" with an "infantile state before [sensations] had been distinguished from one another" (Empson, *Seven Types*, 13). No doubt relevant to any eighteenth-century novel that revolves around a found infant and his oft-censured commitment to promiscuous sensation, this association also brought the modern reader into sympathy with a notable eighteenth-century one. When the talk turned to Fielding, James Boswell, in contrast to his friend, could but "wonder at Johnson's excessive and unaccountable depreciation of one of the best writers that England has produced." He himself preferred to shift attention from objects and surfaces (that infamous "dial-plate") to the less calculable impressions of volume, emphasis, and even "bright[ness]" that Fielding's writing leaves. Boswell further found that if Fielding's "characters" "do not expand themselves so widely in dissertation," their compression itself makes for exceptionally "just pictures" of human nature, and indeed for positively "striking" ones, virtually seismic registers of "the touches of the pencil" that set them down (Boswell, *Life of Johnson*, 389).

Granted, it takes imagination to detect compression in *Tom Jones*; few novels, even of the long-winded eighteenth century, take up more pages than it does. And yet there *is* something compressed, even evacuated, about the way Fielding sets his scenes and strikes out his characters. He appears to do so via uncomplicated acts of reference: "There was an Air of Grandeur in it"; "The Shadows began now to descend larger from the high Mountains"; "*Allworthy* was of a cooler Disposition than [Mrs. Miller]" (*Tom Jones*, 1.4.42, 8.9.435, 13.5.931). Such statements drift from their own pretenses to reference or proposition even as they seem to take shape on the page. The effect is not simply to call attention to *Tom Jones*'s fictionality. Nor is it to make our narrator's rhetorical designs any more apparent than they already are. On the contrary, Fielding's marking gestures register the conditions under which persons and their surroundings press on one another, and under which the language that would account for such effects is moved to abandon all distinction between its literal pretensions and its figural reality. Superficially declarative and referential sentences thus naturally turn on temperature, season, or climate. These are atmospheric tropes that, we've seen, were conceived to be as literal as they are figural, as much impressions as expressions. "Cooler," "warm, "an Air of" this or that: such lexical turns automatically implicate definite objects in their indefinite environments.

Boswell was hardly the first of Fielding's readers to attempt to capture the ambiguous type of interiority that atmospheric effects such as these can foster. In his 1762 essay on Fielding's "Life and Genius," the playwright Arthur Murphy notes that as his "imagination [is led on] through scenes of prodigious variety," the reader of *Tom Jones* is liable to find himself "in the situation of a traveler to a distant town, who, when he perceives but a faint shadowy glimmering of its walls, its spires, and its edifices, pursues his journey with more alacrity than when he cannot see any appearances to notice the place to which he is tending."[34] The impression of nearness is also one of an illusory and "shadowy glimmering." "Notice" belongs both to the attentive mind and to the destination that gives it. Meanwhile, *Tom Jones*'s own primary conceit of itself as a "public Ordinary" both sets a reader apart from it and insists that she look on herself as having somehow entered it (*Tom Jones*, 1.131). The novel's partition among twenty-eight books whose opening chapters invariably place us outside the fiction proper thereby produces the impression that we are otherwise inside it. But once Fielding's reader is within doors, so to speak, some critical proposition or other is bound to put her without them once again. So the very distinction between inside and outside on which this work apparently relies would seem always to be dissipating. Very much as Murphy apprehended, it is precisely this dissipation that produces the sense not just of observing Fielding's writing but of moving, really moving, in and through it.

To the reader who enters such a state, Tom Jones himself will seem to be very naturally conveyed in the idiom of "Disposition." "As Tom had the Vices of a warm Disposition," we are told, "he was entirely free from those of a cold one" (*Tom Jones*, 12.13.677). What does it mean to know that Tom's "Disposition" is "warm"? Presumably, it means that he merits pardon for any associated lapses. But do we in fact know that his disposition is warm in the first place? If so, how do we know? After all, the Author doesn't say that Jones actually *had* a warm disposition, merely that he had the "Vices" of one. Consequently, we register Tom's disposition only obliquely—though surely this is also the way we register the dispositions of those who really are around us all the time. Then too Tom's disposition is conveyed simultaneously in the forms of syllogism, causal statement, and preposition: that pivotal "as" could mean "if," "because," "like," or "while." All four connotations must be upheld if we are to register Tom's "Disposition" in its entirety, if we are to seem to sense it. That sense itself proves a condition of Fielding's irony and so of his freedom of expression—a freedom bound up with Tom's own freedom from the vices of the colder disposition that he so palpably lacks.

The interesting mist of "Disposition" shows Tom Jones to be of a piece with his native home. The Author devotes an entire chapter to Paradise Hall: "The *Gothick* Stile of Building," we're told, "could produce nothing nobler" than that magnificent edifice. Yet this "Description," as the Author calls it, shows us next to nothing. Is this a "*Gothick* Stile of Building," or is it not? We are never told. The house simply disappears into a comparison with that "Stile." At the same time, though, what rises from an evacuated simile is an unmistakably firm impression. Whatever might be said, or not said, of Paradise Hall, "there was an Air of Grandeur in it, that struck you with Awe, and rival'd the Beauties of the best *Grecian* Architecture; and it was as commodious within as venerable without" (*Tom Jones*, 1.4.42). Persisting in disregard for the difference between "within" and "without," the Author goes on to place Paradise Hall where "the Country gradually rose into a Ridge of wild Mountains, the Tops of which were above the Clouds" (*Tom Jones*, 1.4.43). Even Coleridge puzzled over the discrepancy between those aspiring mountaintops and apparent references to the tamer terrain surrounding Fielding's own native Glastonbury, itself mystically charged with the remnants of the Grail legend.[35] The most authoritative of Fielding's twentieth-century biographers worried about the same thing.[36] And both echoed contemporaries such as the one who carped that "to reconcile this Description with Probability will be the Difficulty; for, unlucky for our Author, the Counties of *Devon* and *Dorset* stretch out between *Somersetshire* and the Sea; and if we place this Seat in the *North-West* part of that Shire, the Hills of *Devon* will intercept the View to the Sea; and if we should imagine the Scite to be in the *South East* part of that County, then *Dorsetshire* interferes."[37]

Thanks to so much interference, the Author's "Description" of Paradise Hall precipitated a rumor reported with satisfaction by Richardson: that the novel that surrounded *it* was but "a rambling Collection of Waking Dreams."[38] "A most extensive Ken indeed!" one reader scoffed on finally arriving at the end of Fielding's chorographic tour de force, "and shows the accurate Author endued with more than a *Second-sighted* Mind" ("Aretine," *Old England*, 167). Yet, in the event, it's hard to judge this scene a flagrant violation of the chorographic contract. For one thing, chorographies of the period, such as Robert Plot's enduringly popular *Natural History of Staffordshire* (1686), often slipped surreal images of whirlwinds and tornadoes between pages that seem to us designed to indicate stable features of the human and natural terrain (fig. 7). Such chorographic techniques replicate in textual space the

FIGURE 7. Detail from Robert Plot, *A Natural History of Staffordshire* (1686). Photograph: The William Andrews Clark Memorial Library, University of California, Los Angeles.

experiments with light, space, and aerial perspective that Olwig shows to have been conducted in the stage designs for seventeenth-century masques such as those of Inigo (no relation) Jones. In both settings, landscapes mixed appearance with physical reality, ethereal speculation with sense experience.[39]

Fielding's own natural history of the terrain about Paradise Hall may not come with pictures of the air. Nonetheless, it conspicuously grounds itself in the half-sublimated forms of graphic language—forms that present space itself as a composed scene created by the performances that arise and dissipate within it. In the "Description" we're looking at, these forms are intimated in the prepositions that so often displace them ("above," "into"), and they are actively figured in the clouds that at the referential level mark the last things we see even as, at the lexical level, we are asked to look through and beyond them to the last thing of all: "the wild Mountains."[40] Through a reversal of the eye's natural progress—a reversal that draws attention to the mediating presence ("the Clouds") of Fielding's verbal art—the "History" of Tom Jones apparently rises from the marks that pretend to point toward the place where the story takes shape. Precisely by accumulating so many terms of reference that it is impossible to see through them without also seeing them, Fielding summons an order of experience that his words cannot simply dictate. So even as their pretensions to historical grounding evaporate, the figures of the fiction rise, seemingly unencumbered by their mediating forms. It is a freedom possible only, and ironically enough, because those forms are so immediately visible as perception's (and its objects') requisite background.

Wolfgang Iser's classic study of the reader's role in Fielding's fiction was perhaps the first to specify Fielding's use of background as a constituent of social knowledge. Iser tracked Fielding's manipulation of shifting backgrounds and so of "negative form" and its corollary, contrast, as these evoke in the reader an "explicit conception of phenomena" that for Iser was exclusively human.[41] But when the Author lays his claim to "open a new Vein of Knowledge" by "demonstrat[ing] every quality" through its "Reverse," he turns from the language of sociocultural norms to the example of unpeopled season: "Thus the Beauty of Day, and that of Summer, is set off by the Horrors of Night and Winter" (*Tom Jones*, 5.1.212). To "reverse" the literal and metaphorical aspects of this pivotal statement of intent is to reveal a narrator as preoccupied with physical nature as with its "human" counterpart, but most preoccupied of all with the mediating forms that just perceptibly shape both.[42]

"Under . . . precarious circumstances," Walter Scott was thus to write, "the first English novel was given to the public, which had not yet seen any work of fiction founded upon the plan of painting from nature" (Scott,

"Henry Fielding," 1:103). It would seem that the "circumstances" of *Tom Jones* are "precarious" not because of the strict plan of "painting from nature" but rather because of the "not yet seen" nature of Fielding's "fiction." So may the world of that novel appear to be at once natural and supernatural, seen and second-seen, and at last experienced most completely through the impression it leaves, the airs that it theatrically assumes. Then, of course, *Tom Jones* required of its reader "a more than *second-sighted* Mind."[43] The second sight was popularly supposed to give the "seer" access to shapes in or of the air, to apparitions out of place and time. Likewise, even if the shapes made visible in *Tom Jones* do not override ordinary markers ("it was now the Middle of *May*"; "in the latter end of *June*"), they nonetheless make such markers visible as temporary forms. However temporary, these formations remain the immediate conditions of personal presentation, experience, reflection, and (ever provisional) interpretation.

This may be what prompted contemporary readers, however sardonically, to commend "Mr *Fielding's* Proficiency in the Royal Science of Demonology."[44] Ironically cultivated though it may be, this "Proficiency" is responsible for much of the atmosphere of *Tom Jones.* In much other fiction of the period (including what Coleridge called Richardson's "close, hot, day-dreamy" *Clarissa*), dislocated sensation is conspicuously mediated by letters (Coleridge, "Notes on *Tom Jones,*" 2:374). It's mediated, that is, by figures that aim not just for reference to objects that anyway never were but for the kind of labile cross-reference that can make their own conditions apparent. Unlike Richardson's tragic ones, Fielding's mannered comic characters thus express metonymically the behaviors of the graphic characters that compose his pages. Early identified with the biblical conjurer of apparitions, the Witch of Endor, for example, Tom's mother, Bridget Allworthy, ultimately dwindles "from black to grey, from grey to white" (*Tom Jones,* 3.1.117). In the same spirit, Fielding identifies Bridget with a figure drawn by "Mr *Hogarth* himself, to whom she sat many Years ago, and hath lately been exhibited by that Gentleman in his Print of a Winter's Morning, of which she was no improper Emblem" (*Tom Jones,* 1.11.66).[45] Evasive representations such as these make it easy to see wintry Bridget as the hypermediated condition of Tom Jones's—and thus *Tom Jones*'s—own being. His invisible, but ubiquitous mother, she naturally sucks air currents at keyholes and hovers over most of Fielding's "History" in the invisible form of an unread letter.

Bridget finds her reverse in Benjamin Partridge, the melancholy and superstitious schoolmaster who appears to be Tom's father until, at the eleventh hour, one Mr. Summer takes those honors. Critical history tends, from a historical perspective, to judge Partridge as either a superstitious dupe or a

Jacobite straw man. But this consistently peripheral figure has a more vital function: to make apparent the novel atmosphere that Bridget, seemingly from within it, partly creates. We have Partridge's own literary genealogy to thank for this. His name and *modus cognoscenti* drawn from the popular astrologer John Partridge, Fielding's Partridge places *Tom Jones* within the vital context of the English astrological almanac.

Before examining this relationship, however, we need to consider Partridge's own peculiar "Disposition" in Fielding's novel. His place is always just to the side of most of its significant scenes. Halfway through *Tom Jones*, for example, Partridge and Tom are dithering at the foot of the most baffling of Fielding's several hills: while Tom can talk only of its hypothetical summit, Partridge complains that his companion has "made my Blood run cold with the very mentioning the Top of that Mountain." At the same time, however, "frozen to Death," he sensibly pleads for "a place under Ground, to screen ourselves from the Frost" (*Tom Jones*, 8.10.443). Suspended between two pictures of reality, Partridge's thermostatic engagement with "the Hill" realizes that terrestrial formation not as an object that Fielding's novel posits but rather as a shape or contour that is expressed in a figure in it. Appropriately enough, Partridge's hill-induced chill is itself expressed in and through a double entendre. When the schoolmaster's blood runs cold, literal meanings dissipate into figural ones, a unique individual's bodily state into a conventional figure of speech. So in turn may a mental plight (fear), a communicative act (mere mention of the hilltop), and a self-negating physical sensation ("frozen to Death") become apparent as the elements of an otherwise imperceptible condition—an "Atmosphere."

Partridge's complaints about the temperature produce something very like what Elaine Scarry calls "vivacity," in that they allow the graphic field in which Fielding's writing consists to present as a perceptible environment that can be negotiated as if it were really there.[46] For Fielding, the possibility of vivacity is linked to the process of judgment. In fact, Partridge first emerges from the undifferentiated background of his writing in a scene that looks back to the witchcraft trials in the days of Matthew Hale—trials that infamously revolved around evidence that was as circumstantial as it was spectral.[47] The melancholy schoolmaster stands falsely accused of adultery and battery. While his ruthless wife claims "Marks enow about my Body to shew of his Cruelty to me" (*Tom Jones*, 2.6.99), the presiding magistrate—Allworthy—believes that he has seen not just the incriminating marks but the lines that link them to the accused. Hence he charges Partridge as he would a witty witch, with "lying backward and forward" (*Tom Jones*, 2.6.100). It is suddenly obvious why Partridge should be the only character in *Tom Jones* to "trembl[e] in the firm Belief" that there may be such a thing as a witch; to doubt it would be to doubt himself, at least as he first appears (*Tom Jones*, 8.10.446). And of course that is what, if anything, the scene of his judgment by Allworthy is really about: how—in what manner—things appear.

Partridge sinks into depression and gives up his school; later, he lets himself be led by dreams of worm-gnawed barrels and white mares. Having forgotten he's in a history where people cause things to happen, not a romance where things befall them, Partridge blames sorcery for a tumble from his horse. Tom charms him. When the two are caught in "Darkness, Rain, and Wind," Partridge is naturally convinced that time has stopped and that "tho' the Horses seemed to go on, they had not moved a Step forwards during at least the last half Hour" (*Tom Jones*, 12.11.661, 12.12.664). To the extent that Partridge seems to see himself as plotted on a fixed grid, he stands out from the boisterous, autonomous, and forward-thrusting line of Fielding's "History."

Yet again and again the "History" itself compromises that line. Is Partridge really so far outside it? Chapter headings ("In which the History is obliged to look back"; "Wherein the History goes back") are forever backtracking or hovering in one place too long. Homer Brown points out that throughout *Tom Jones* "the nonrecurrent, arbitrary intersection or co-incidence of separate and unrelated, irrelevant causal chains" can occasion whole chapters of "unintended effect." "Side effects," Brown calls them (*Institutions*, 113). But why might not the "side" effects actually lie in "back" of the main event, invisible only because but one character, and he an outcast, possesses the expressive and analytic styles adequate to it? If his superstitions alienate him from the common sense by which Tom, for all his faults, seems to be guided, Partridge remains the figure in Fielding's novel who is most "in touch" with his own literary surroundings. From this perspective, he is Fielding's the most complete realist, the key to each (real) page's legibility from the inside out.[48]

In its Latin form, "superstition" simply means standing outside, over and above. It means assuming more or less the position the Author pretends to. Partridge's own superstition, with its burden of omens and demons and weather-turning witches, gives him a similar status with respect to the outline of Fielding's fiction proper. Unlike the Author, however, Partridge is free to drift back and forth across the edges of its silhouette. He might be the designated magical thinker in their two-man traveling party, but when Tom waxes eloquent on "the Beauty of the Moon" and imagines Sophia contemplating it at the same time he does, it's his presumably loony sidekick who sensibly avers that "if my Eyes were fixed on a good Surloin of roast Beef, the Devil might take the Moon and her Horns into the Bargain" (*Tom Jones*,

8.9.437–38). As a matter of fact, this unlikely skeptic adds, "If I believed [his wife] was in the Moon, according to a Book I once read, which teaches that to be the Receptacle of departed Spirits, I would never look at it for fear of seeing her" (*Tom Jones*, 8.9.438).[49]

Is the "it" "the Moon" or "a Book I once read"? Once we've put ourselves in his literary frame of reference, the incongruities among Partridge's different beliefs—his positions on things—melt into air. If he's still discernible as a character in Fielding's fiction, then character itself must be something other than a stable object of perception or a reliable sign. It looks much more like an indeterminate set of aspects, masks less donned than improvised by the (here literary) environment in which they are seen. Each of these masks is potentially revealed through the influences and pressures at work in any given moment. And each, as it is revealed, likewise indicates that those same influences and pressures are the work of the literary medium. Partridge's otherwise puzzling adoption of so many positions in the world, from schoolmaster, barber, and tailor to amateur theater critic, builds up this conception of character.

One professional guise that Partridge doesn't assume, however, is the very one with which the contemporary reader would have been most likely to associate him: that of astrologer. Partridge, of course, shares his name with the maligned astrologer John Partridge. As both Partridges would have agreed, this can be no accident. Or rather it is an accident *only* in the astrological sense, where an "accident" is the conjunction of two symbolic forms. Fielding is known to have used an almanac to plot the scrupulous chronology of *Tom Jones*. But the so-called celestial science of astrology that was perpetuated through eighteenth-century almanacs also conditioned the mental dispositions that turn chronology into experience and persons into impressions no less vulnerable than vivid.[50]

Celestial Science

According to one practitioner at its seventeenth-century zenith, astrology is the "Art which teacheth by ye motions, Configurations, & Influences of the Signes, Starrs, & Celestial Planets to prognosticate of ye naturall Effects and Mutations."[51] This definition suggests that astrology's communicative and informative intentions are not easy to distinguish from its truth claims. As a result, those claims were not destined to be honored (by those wishing to be seen as learned anyway) for very much longer. The prolific John Partridge, whose star at the time of Fielding's own birth had just risen as high as it was going to, was one of the last English astrologers to argue explicitly that astrology is a science.[52] Since his time, it has looked more and more like the kind of magical thinking at which Benjamin Partridge excels. Such thinking compels juxtaposed appearances (the sun in Capricorn, the king in chains) into meaningful conformity with one another.

Fielding's Partridge "was, in Truth, a *Jacobite*" (*Tom Jones*, 8.9.436). By contrast, the historical Partridge was a mildly rabid Whig whose title pages often advertised his ironic determination to unify the heavens with his own aversion to "Popery, and Arbitrary Government."[53] Here, surely, he was guilty of little more than upholding a national custom. In the middle of the seventeenth century, English hands on both sides of the civil war had turned astrological reckoning into a political weapon, further undermining astrology's authority as a legitimate language of nature, in contrast to the apparently ideology-free discourses of astronomy and physics. Besides its fall into rhetoric, many other precipitants of astrology's eighteenth-century "decline" have been adduced. They range from the proliferation of cheap, formulaic almanacs that lacked the individuality, charm, and snob appeal of their predecessors to the growing acceptance of new scientific descriptions of the cosmos. (The two most influential of these descriptions would seem to cancel each other out, because they promote a heliocentric model of the cosmos on the one hand but, on the other, establish heavenly bodies as no more fixed or permanent than sublunary ones and so no stable platform from which to "prognosticate of ye naturall Effects and Mutations" here below.[54]) The autopsy report also mentions enlightened anxiety to deny that "geniture" is an implacable determinant of individual human fortunes—an anxiety nowhere more fully registered than in Fielding's own *History of a Foundling*.[55] Then too there was high literate culture's weakness for ridicule, especially when it could be leveled at any form of knowledge that resembled superstition.[56]

For our purposes, however, English astrology's most significant transformation in modernity was into the quasi-aesthetic form of the popular almanac. Earlier astrologers from William Lilly to John Gadbury had published their readings of the stars in complex, layered, and geometrically sophisticated pamphlets that were often as devoted to explaining their own hieroglyphics as they were to displaying them in the first place. Beginning in the 1690s, however, astrology began to dissipate into cheap, mass-produced almanacs such as John Partridge's and so into a symbolic form whose ephemerality and superficiality paradoxically both figured and grounded its authority. While historians have often documented astrology's "decline" over the eighteenth century, they have also had to acknowledge the almanac's ascent into one of the most popular and ubiquitous literary genres of the period.

In the new almanacs that saturated the anglophone world from the beginning of Fielding's own century to far beyond its end, an evidently dumbed-down astrology fell into two distinct subcategories. "Judicial" astrology, which survives in today's online and newspaper horoscopes, interpreted (and infamously projected) collective and individual history in relation to heavenly configurations. Little different from provincial weather journals or the data charts of the day's virtuosi, "natural" astrology, since largely vanished, sought to coordinate terrestrial phenomena—"great Earthquakes," say, or an early "Thawing"—with the positions of celestial bodies.[57] While natural astrology would seem to be more obviously related to the question of how English readers experienced their own physical atmosphere, of greater interest than what either of these astrologies did is how they seemed to do it. Most obvious, instead of assigning meanings and correlations, both natural and judicial astrology "cast" them. The interesting vagueness of this verb captures astrology's true flexibility and latitude as a system of meaning. Because it dealt in spheres and angles, with relatively little interest in linear connections or one-to-one correspondences, astrology enjoyed a secret kinship with irony, the most oblique of all symbolic modes. Possibly this kinship was too close. Certainly the eighteenth-century satirists for whom irony was a stock in trade were among the quickest to exaggerate astrology's rigidity, often exemplifying it, as did Swift, in the almanac's signature pretension to "fi[x] the weather."[58]

Fix the weather: How does one do that? To fix could mean to rig or determine, as well as to repair. But in an astrological context it's more likely to imply pinning something down in a stable and self-evident symbolic form. From an astrological perspective, of course, the heavens themselves take a similar shape, one that thus physically encompasses our immediate environment. Yet horoscopes were never "cast" directly from the stars. From astrology's earliest hours in ancient Babylon, they relied instead on a graphic representation of the heavens in the form of a diagram, or ephemeris, not just to communicate information (or the next thing to it) but to gather and organize it in the first place. Unlike the astrological texts of the seventeenth century, however, an eighteenth-century almanac seldom if ever actually printed the ephemeris in back of it. If anything, its margins were given over to the remotest of side effects: jaunty couplets, math problems, aphorisms, and even pocket histories.

This left the ephemeris itself an invisible intermediary between the vertical rows of dates and conditions that the almanacs did faithfully reproduce and the natural, political, and moral phenomena to which such notations referred. The relatively elite readers of a prominent Stuart astrologer such as Lilly could have inspected the coordinated imagery of houses, signs, and planetary bodies on which his knowledge was based and the meaningfully wide or narrow angles that related them to one another. But for the common readers of an eighteenth-century almanac—for example, Partridge's *Merlinus Liberatus*, or the physician Francis Moore's *Vox Stellarum* (which sold 107,000 copies in 1768 alone) —the ephemeris had become the invisible pretext of a print impression and thus an unspoken determinant of the information that presumably flowed from it.

What at first looks like deception, however, might rather be a sign that certain reading strategies had been internalized. Some degree of mediation was now taken for granted as an aspect, even condition, of the kind of experience—the experience of existing within a meaningful whole—that astrology makes possible. Not to be confused with the Puritan conviction of interiority (where one harbors an autonomous interior), this experience of interiority *to* some encompassing configuration is temporal as well as spatial. With regard to Swift's charge of fixity, almanacs did appeal to a craving for stasis; as Maureen Perkins remarks, even the aphorisms that often studded their edges—not unlike Pope's true wit—pandered to the assumption that one wants most to hear what has been said before. As for astrology itself, at least when considered as a kind of writing rather than as a branch of knowledge, it would appear to influence (fix) even as it arranges (again fixes), subordinating time to space in its own symbolic form, in its internal logic, *and* in the world it describes. Consequently, all these domains are in effect if not in fact the same.

As for specifically fixing the weather, the astrologer William Ingpen confirmed that astrology "foretelleth the variable state of the air, and other elements at every moment."[59] Perhaps more significant than the contents of this claim are the ways it was made good formally. The almanac pages that displayed relationships between "the variable state of the air" and the heavens above did so in vertical columns that had to be read horizontally—more from within the page and in relation to points of reference there established than from outside it (fig. 8). In Partridge's *Merlinus Liberatus*, popular astrology also fixed reading time within literary space, subjecting supposedly autonomous movement across a graphic surface to the visibly artificial parameters of that surface. In this way, almanacs conditioned English readers to be in mediated worlds. It scarcely mattered whether those worlds consisted in letters or in island air, as witness perhaps the latter's heavy representation throughout almanacs such as Partridge's. Here, however, this same air is a construct that is at once manifestly relative to other such constructs and fundamentally imaginary. That is, if on a given day an almanac page could

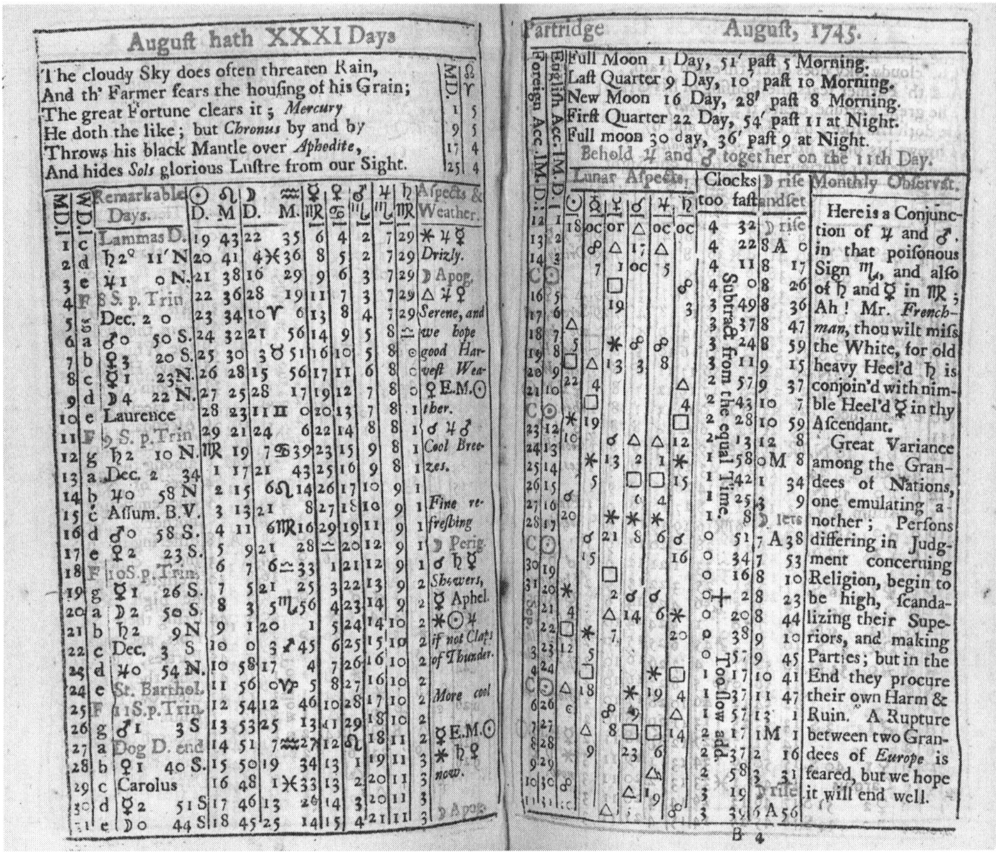
August hath XXXI Days

The cloudy Sky does often threaten Rain,
And th' Farmer fears the housing of his Grain;
The great Fortune clears it; *Mercury*
He doth the like; but *Chronus* by and by
Throws his black Mantle over *Aphedite*,
And hides *Sols* glorious Lustre from our Sight.

Partridge August, 1745.

Full Moon 1 Day, 51′ past 5 Morning.
Last Quarter 9 Day, 10′ past 10 Morning.
New Moon 16 Day, 28′ past 8 Morning.
First Quarter 22 Day, 54′ past 11 at Night.
Full moon 30 day, 36′ past 9 at Night.
Behold ♃ and ♂ together on the 11th Day.

Monthly Observat.

Here is a Conjunction of ♃ and ♂, in that poisonous Sign ♏, and also of ♄ and ☿ in ♍; Ah! Mr. *Frenchman*, thou wilt miss the White, for old heavy Heel'd ♄ is conjoin'd with nimble Heel'd ☿ in thy Ascendant.

Great Variance among the Grandees of Nations, one emulating another; Persons differing in Judgment concerning Religion, begin to be high, scandalizing their Superiors, and making Parties; but in the End they procure their own Harm & Ruin. A Rupture between two Grandees of *Europe* is feared, but we hope it will end well.

FIGURE 8. From John Partridge, *Merlinus Liberatus* (1745). Photograph: The William Andrews Clark Memorial Library, University of California, Los Angeles.

only posit "Aspects of the Weather" ("overcast," "Cold Winds"), the day itself was given, much as the first step in a Euclidean proof is given, as a symbolic formation rather than as an absolute description of reality.[60]

On this very principle is any day in "Partridge" marked, nonhierarchically, in more than half a dozen different ways: as a day of the month; a day of the week; a lunar phase; a "remarkable Da[y]" (which could as easily be a saint's day as one devoted to a Roman god); or a position in relation to the sun and the ascendant planets and constellations. These are all figural designations. Time, whether measured in minutes or shaped by seasons, cannot be conceived or experienced outside of them. Astrology's distinctive idiom reinforces this impression of containment. In Partridge's own almanacs, for example, a point of reference was not an object that could presumably be found somewhere; it was instead an active sign of orientation within a meniscular space of relative, and ever shifting, positions. Every year the title page of Partridge's *Merlinus Liberatus* thus promised an almanac "calculated and referr'd to the Meridian of *London*, whose Longitude is 24 degr. 20 Minutes" and "whose Latitude is 51 degr. 32 Minutes."

This frame of "refer[ence]" exists in symbolic coordinates of meridian, longitude, minute, and degree. A word such as "when" can function as a temporal marker only after it designates a coincidence between two points. Similarly, the word "in"—as when a planet is "in" a "house"—doesn't describe a body's change from having previously been out. Rather, the word "in" marks a superficial position in space that is unapproachable outside symbolic mediation. Not only is there no way *not* to be "in," but when an almanac indicates that the sun is "in" a sign or house, it means this no more figuratively than literally. "In-ness" is cast in graphic form at exactly the same time and place that it's defined conceptually.

The vagaries of its relational terms (especially the prepositions) tell us why astrological writing could be maligned as cryptic, occult, or manipulative. However, on repeated exposure, this same deviance begins to look remarkably straightforward. Ann Geneva supposes that the picture of the world that astrology builds became increasingly "illegible" in the transition from William Lilly's era to Fielding's (Geneva, *Astrology*, 263–82).[61] This happens to have been a transition over which John Partridge presided, and it occurred in inverse proportion to the rise of the quasi-referential languages that the English novel is also supposed to speak. Yet the opposite is also true: astrological writing—even or even especially in almanac form—was *too* literal. Whereas astrology's apparent rival, natural philosophy, systematically refers to what cannot be seen or experienced by a reader—and that thus craves validation by as many eyewitnesses as possible—astrological writing replaced narrative reference with a set of manifest relationships. Consequently, not to say ironically, it required much less anxious verification. Popular astrology's internal world was a remarkably self-sufficient one, even as the individual constituents of that world could look, well, overdetermined.

Now, from within this frame of reference, what does writing look like? It does not, we might suppose, look all that much like writing at all. Swift's Bickerstaff groused that astrologers "do not so much as understand Grammar and Syntax; that they are not able to spell any Word out of the usual Road; nor even in their Prefaces to write common Sense, or intelligible English" (Swift, "Predictions," 427). Worse yet, astrological writing substitutes "Pot-hooks for Planets," and even "foretelling the Weather" happens only at the hand of the printer who "takes it out of any old Almanack as he thinks fit" (Swift, "Predictions," 433, 435). Bickerstaff could have made an even more rigorous case, because besides astrology's cryptic "pot-hooks," almanacs might incorporate roman and Arabic numerals, print, cursive handwriting, roman characters, and gothic typeface. What was the difference among any of these typefaces or, for that matter, among the claims to differentiation made by the "Word[s]" they presumably formed? In their virtually automated shuffling and reshuffling, moreover, astrological "characters" replicated the alleged actions of celestial bodies, moving them from their old place "out there" to a new and disconcertingly close one "in" between a reader's hands.

Alison Chapman proposes that seventeenth-century astrology preserved the concrete specificity of medieval and non-Protestant modes of orientation in space and time through the cultural shift brought about by the Puritan ascendancy in the middle of the seventeenth century. To the (considerable) extent that the Puritan movement coincides with the rise of print culture, we can expand Chapman's premise and suppose that this was to carry non-print-based styles of orientation into a new medium.[62] And indeed, over the next century, many almanacs disdained the shape of bound books to take the form of sheets that could be pinned to walls or fashioned to conform to the crown of a hat. In contrast to leaves that could be turned, *their* other side was thus a real boundary to which contiguity nonetheless lent the status and charisma of a symbolic form (Perkins, *Visions of the Future*, 16). Other almanacs, once exhausted of their pertinence to the present year, apparently got twisted into kite tails and sent into the skies. All these fates build the sense that the world to which almanacs referred was materially and symbolically of a piece with them. So astrological writing cannot be described as palimpsestic, any more than it appears to have denoted objects that ultimately differed from it in kind. Its layers are all transparent. Rather than locate marks on a blank page, almanac astrology places them in an influential field that like that page is potentially visible as a form in its own right. Indeed, the blank spaces in almanacs could be experienced as a field of this "nature." This may be why readers felt free—or possibly compelled—to write on them.

Finally, what does such writing make of reading? Certainly reading would not seem to involve penetrating layers of meaning any more than it would seem to promise access to knowledge of objects in the external world. Rather, astrological deciphering is the other side of ciphering. And of the modernizing English astrologers, John Partridge had done the most to establish astrology as a kind of writing inseparable from common English reading practices. The high literate mockery that Swift spearheaded reminds us that "Partridge" was a writer at exactly the same time he was a reader of the heavens. And he was an inexhaustible one at that, his fertility as tied to time, tide, and climate as any arable field. While this might have naturalized his compositions, Partridge ultimately seemed to exist nowhere else—so much so that "Partridge" continued to be printed in his name for many years after his death. It was a fate ironically predicted in Swift's notorious April Fools' Day attacks on him in 1708. Having forecast Partridge's death, Swift's Isaac Bickerstaff went on to "prove him to be dead, out of his own Almanack," solemnly vowing that "if an uninformed Carcass walks still about, and is pleased to call it self Partridge; Mr. Bickerstaff does not think himself any way answerable for that" (Swift, "Predictions," 439).

Integral to Swift's joke was the impossibility of anything—anything at all—occurring anywhere "out of [the] Almanack." In turn, in a temporally suspended world that consists entirely of graphic impressions and the formal background that is the condition of their legibility, the "uninformed" form popularly known as "Partridge" can never be either living or dead. Furthermore, Partridge's almanac—known simply as "Partridge"—continued to be published for decades after the astrologer's bodily death. As late as 1745—the "Year of Our Redemption" at the center of *Tom Jones*—"Partridge" bore the epigraph "*etiam mortuus loquitur*" (even now the dead may speak). At any rate, the content of Partridge's writing, ideological or otherwise, clearly mattered far less than its formal and phenomenological implications. What most rankled Swift must have been something closer to home than a radical Whig who bent his readings of the stars until they could not be told from coercive representations of them. It must have been the enthralling second nature of the English print medium made manifest.

Almanacs such as "Partridge" made that nature apparent without making it any less captivating. They thereby mocked the enlightened illusion that information automatically nurtures freedom. At the same time, conspicuous figuration emerged as a legitimate condition of experience. This possibility haunts Swift's perception that almanacs were themselves figural, albeit in a manner far harder to detect, let alone resist, than the kind of rhetorical flower repudiated in the day's mania for plain speech. Almanacs were implicated in a diffuse state of mind indistinguishable from communicative form. The rhetorical flower's own most zealous cropper, Thomas Sprat, clucked over their "wild amuzing of mens minds, with Prodigies, and conceits of Providences." It was "a vanity to which the English seem to have bin always subject above others," and Sprat lamented the reputed "disposition of our Countrymen, that they us'd to order their affairs of the greatest importance according to some obscure Omens, or praedictions that pass'd about amongst them, on little or no foundations."[63]

When a native "disposition" is said to incline toward "obscure Omens," it's not just that *things* threaten to become interchangeable with *signs* ("praedictions"). It's that they thereby "pas[s] about amongst" people, "on little or no foundations." From an astrological standpoint, omens are contagious, though not in the sense that they are things that passed from individual to individual. Rather, omens are found, if anywhere, at once in and between individuals, both "about" and "amongst" them. It is in turn this unfounded and yet ubiquitous location that puts omens necessarily outside the limits of a proper knowledge even as it lodges them immovably inside English "mens minds." Astrological writing, especially in the form of the almanacs that formally shaped daily life at once from inside and from without, paradoxically gave this ominous phenomenon a place to "be." That writing consequently informed a collective mind that could manifest itself only in the figural modalities that had informed it in the first place.[64]

The contents of this mind certainly cannot be described as the ideational traces of sense impressions that one finds in the Lockean "Understanding." The boundaries between it and what's about it are too porous for that. So in its eighteenth-century incarnation, the astrological mind may be best captured in our present-day idea of the "airhead"—a comic stereotype that all too easily applies to Fielding's Partridge. English print astrology treated the layered air to which he is so attuned as both a transparent entity and a figure of both communication and influence. For instance, one prominent interregnum almanac, John Dove's *Speculum Anni* (1653), held that worm-eaten oak casks accompany dearth, wars, and plague. But while "some perhaps may think this to be but a superstitious observation," the conjunction in fact "proceedeth from the constitution of the air; which (whether wholesome or unwholesome) appeareth more subtly in such things as these, than to the sense of man it either doth or can." The shared medium of the air not only produces a sensible impression (porous wood) but links it with a second display of its own "subtl[e]" influence, human catastrophe. In this way, air also mediates a correspondence between mental and bodily states: "If the air be corrupted, putrefaction abounds . . . and diseases equivalent produced, as also the mind of man inclined to such and such actions as agree thereunto."[65] Air here mimics the communicative medium of the print astrology that communicates *it*. Thence arises the print almanac, now a strangely exalted figure for what Swift satirically termed air's mechanical operations, if also its corruptibility.

Judging Partridge

As it turns prepositions such as "in" and "before" into self-evident propositions, almanac astrology's peculiar literalism lent it an affinity with at least one figure of speech that was making the arbiters of English literary culture more nervous by the day: the pun. It is by way of a pun that the Author of *Tom Jones* purloins astrology's lexicon the better both to dispose his "History" and to dispose his "Reader" to move through it. In the English almanac, such terms as "Aspect," "Coincidence," "Accident," and "Geniture" at once designate substantives and coordinate proximal states with remote ones, spatial configurations with temporal order. Despite—indeed, perhaps because of—Fielding's ironic distancing of it, the vocabulary of aspect and accident directs his reader through the novel notional space to which such concepts visibly give rise. The sense of direction that it transposes from "Partridge" and its kin in turn makes *Tom Jones* look less and less as if, from a secure position within a different symbolic system, it simply mimics an astrological description of reality. Fielding's work now appears as a point of transference between the notional environments of the English almanac and English fiction.

By dint of his name alone, Partridge is obviously the visible agent of transference. Pressure points within Fielding's plot emphasize his role as such. Its appearance of linearity aside, that plot returns Tom and Partridge to their original position at Paradise Hall, thereby retracing the astrological model of the Great Year in which planets assume their original relative positions.[66] The travelers' path is easily charted across a standard map of England, whereupon it may be seen as a triangle. In Ptolemaic astrology, the background against which that triangle appears itself appears as a triangle, with Britain "in the north-western triangle" of Sagittarius, Aries and Leo. It was its "general characteristic, . . . by reason of the predominance of the triangle and the stars which join in its government, to be independent, liberty-loving, fond of arms, . . . cleanly and magnanimous."[67]

Ever susceptible to Tom's liberty-loving magnanimity, Fielding's Partridge knows how to read the embedded angles that dispose both characters in figural and physical space. Consider the manner in which, having disappeared, he resurfaces. Tom, exiled from Paradise Hall, has fallen in with the British army and been wounded by Ensign Northerton (whose name works along the lines that Western's does). Now a barber who deals in "Suds" (*Tom Jones*, 8.4.414) and symptoms, Partridge turns up in the "House" where Tom is convalescing. The woman of the "House" has just been "moved by" a "Motive, I cannot tell" to put him "in the Sun"—which spot happens to be "the worst room in the House." The woman then retreats to "entertai[n] a Circle she had gathered about her, with the History of poor *Jones*" (*Tom Jones*, 8.4.416). Partridge eavesdrops on these entertainments and makes a plot out of them, easily measuring the distance between "accidents" such as the one that Tom's "Name be *Jones*" and the one that he is "a poor Parish boy." He thus greets

Tom himself "with a seeming Surprize, as if he had recollected a Face he had seen before" (*Tom Jones*, 8.5.417). Partridge isn't exactly dissembling here; he is responding appropriately to the set of notional figures that contains a "Circle" of relation.

Because Partridge immediately experiences the counterpoise of appearances as an expression of what binds them into a meaningful whole, he is well equipped to navigate the formal conditions of the fiction that contains this scene. Within that fiction, one given is that human characters such as Jones himself are not autonomous and self-determining. Nor, though, are they automata; rather, they are figures moved, shaped, and conveyed by forces ("[the] Motive I cannot tell") inherent to "the Representation" (*Tom Jones*, 8.4.416). It's exactly this literal dimension of Fielding's novel that Partridge is able to negotiate and interpret with such singular, if ironic, accuracy. Tom Jones himself might omit "many material Details" of his own plot to date. But Partridge still manages to ascertain the truth that Tom has been a victim of appearances. His "Curiosity, cold as it was," even drives him to discover the "principal Cause of all [the] Mischief" that has so far befallen Jones (*Tom Jones*, 8.5.420). His own "Air and Aspect" perpetually subject to change (*Tom Jones*, 8.6.423), Partridge expertly correlates the transient "Aspect[s]" of others with their characters. Nor is it any accident (better yet it is only one) that Partridge does his best work when he reads—without judging—Tom Jones himself.

What, then, are we to make of Partridge's "Aspect" as "the most superstitious of Men" (*Tom Jones*, 8.7.426)? Most obvious, that aspect supports Fielding's parody of the English Jacobites: its jones for cryptology and superstition had long marked this political subculture, both for the insiders who could identify one another by it and for the outsiders who could thereby satirize them.[68] But from precisely this point of view, Partridge's superstition is better described as a disposition than a belief. As such, it allows him to gauge his own surroundings and to realize them for Fielding's creative—yet formally contained—Reader. For example, in a famous passage, Tom and Partridge, having "missed their way" in a storm, approach a barn full of reveling gypsies, where "Lights and different Noises appea[r]." These in turn call to mind "the whole Furniture of the infernal Regions [as] appropriated by the Managers of Playhouses." Partridge's lascivious manners with respect to the "Set of Beings" who appear on this nebulous stage land him before their king, charged with adultery for the second time. Although, like Allworthy before him, "his *Egyptian* Majesty" judges by appearances, he is for that very reason wholly uninterested in punishment. His stance has prompted critics to read the scene in which he assumes it as political allegory, albeit one that takes the shape of an unresolved debate between what Fielding's Author himself describes as the relative virtues (or evils) of "absolute Power" and "the limited Form of Government" (*Tom Jones*, 12.12.672).

In Martin Battestin's influential sounding of this debate, Fielding's atmospherics are made to parse its grammar: "We may find in the atmosphere of utter darkness and superstition which introduces Tom Jones's Egyptian adventure an emblem of the condition of England in the stormy winter of 1745, an emblem of what Fielding considered to be the perilous fascination of a free but benighted people with the false lights of absolutism."[69] Battestin was right not to overlook the weather that precipitates this "adventure." Noting the "hesitant and indistinct" frame around penetrating analyses like his, however, Manuel Schonhorn places the gypsy encounter inside the frame of art that it, to his eyes, seems to generate. Shifting from narrative and allegory as this scene prompts makes it "possible to read *Tom Jones* itself backwards as well as forwards" and to relate this particular event "to events antecedent to it, not subsequent to it."[70]

Schonhorn happens to be the one judge of this scene to have measured its numerological and astrological underpinnings, though even he does so only in a footnote by definition peripheral to his formal aesthetic judgment of the book. As for the "Set of Beings" whose "*Egyptia*n" aspect links them to the birth of astrology, most readers have made the Author's claim that "these People are subject to a formal Government and Laws of their own" (*Tom Jones*, 12.12.667) the foundation of their varying decisions as to whether Fielding's own political sympathies lay more with the "limited Form of Government" or "absolute Power." This kind of political and historical—and allegorical—analysis cannot be ruled irrelevant to any understanding of *Tom Jones*. Nonetheless, its relevance arises from an underlying indeterminacy. Without saying what it is, Fielding's Author thus gestures toward "the only absolute Power which can by Scripture be derived from Heaven" (*Tom Jones*, 12.12.672). The "scripture" of astrology, understood as such, derived that power contingently as it involved the outer sphere of the stars and planets with the inner one that encompasses the human world. How arbitrary or fixed was the authority of the celestial bodies? What role did human judgment—reading—play in upholding that authority?

The eighteenth-century almanac's solution to such questions was less to answer them than to make reading part of the world of appearances that they disposed. Consider John Partridge's overt animosity toward "Popery and Arbitrary Government." Many of the astrologer's judicial readings were bent on justifying the Hanoverian succession and doubled as diatribes against "Romish" authority. But they also responded to "Arbitrary Government" at the level of symbolic practice. Between the positions of celestial bodies and earthly events, "Partridge" found merely correlations and at most influences, never evidence of implacable design. Yet design played a crucial role in the communication of these correlations and influences; it was, if anything, their condition. Far from presuming to secure certain knowledge, "Partridge" the almanac thus represented, and rewarded, states of mind that could remain suspended among alternative possibilities. Such states of mind sponsor clement appraisals, not assertions of causality from within a haze of appearances. It is this state of mind that disposes "his *Egyptian* Majesty" to pardon the evidently adulterous Partridge, even when appearances are against him. They must be, because *everything* is appearance. And it is so not just in the playhouse of the gypsy barn but in the world outside that barn—a world that Fielding shows to be nothing more, or less, than the gypsy world's obverse.

He does so through the reversible figure of clemency. The weather that prompts Tom and Partridge to take refuge in the barn is, as Battestin politely remarked, frightful: "The Night was dark, and it just then began to rain very hard" (*Tom Jones*, 12.10.660). Partridge suspects a witch. If risible, this is also—and rightly—to suspect some formal condition of the experience that he is about to have. In this sense, the "utmost Inclemency" of the weather forms the natural obverse of Partridge's experience of "Clemency" inside the gypsy barn (*Tom Jones*, 12.13.672). This shaping background is also a temporal power: Tom and Partridge are in the space that it limns but "for a few Minutes, whilst the Rain continues" (*Tom Jones*, 12.12.664). The rain's "whilst" presumably marks the real world, with its laws of separation and duration, as the other side of the undifferentiated and atemporal space of the Egyptian "Playhouse." In training to be Fielding's own reader, Partridge is asked to assimilate both worlds as aspects of one experience that belongs to no single person, fictional or real, judge or judged. His Egyptian majesty's clemency is in turn the other side of the inclement conditions that surround the display of it. The transition between these two sides, tellingly, is thus almost imperceptible: all of a sudden, "when the Storm was over," Tom and Partridge simply "took Leave of his *Egyptian* Majesty." They recover from their "Deviation" in a twinkling, having corrected it entirely but "a little after Noon" (*Tom Jones*, 12.13.673).

Some two hundred pages later, Partridge is experiencing "so violent a Trembling, that his Knees knocked against each other" (*Tom Jones*, 16.5.853). He is now in London, attending a performance of *Hamlet* along with Tom Jones and Mrs. Miller, and though he declares that "I know it is but a Play," who can tell where a naïve and visceral response to the play (knocking knees) stops and a sophisticated grasp of theatrical illusion (the knowledge "it is but a Play") begins? As one of this chapter's liveliest readers, John Stevenson, has suggested, it is also hard to imagine that it fits logically into Fielding's novel as a whole; for his part, Stevenson salvages the night at the theater from irrelevance by treating it as part of an intricate web of allusion to the Jacobite threat with which Partridge has been identified all along.[71] Yet the Jacobite reading remains a referential and allegorical one, so committed to unearthing a subtext that it misses what lies on the surface of this chapter—which opens not at the theater at all but rather with Tom reading "a Letter from Sophia"; "a Child who hath just learnt his Letters, would have spelt this Letter out in less Time than *Jones* took reading it" (*Tom Jones*, 16.5.851). Tom's naïve parsing reduces writing to its elements even as it magically materializes marks on a page into objects of sensuous knowledge. Indeed, Tom "spend[s] three Hours in reading and kissing the aforesaid Letter" until he is, "at last, in a State of good Spirits" (*Tom Jones*, 16.5.852).

With Tom transported out of historical time and into the mediated illusion of immediacy, this chapter's frame forecasts in literary terms what is about to happen in theatrical ones. For once we're in the theater, it is Partridge who is first brought to mind of "the Man at the end of the common-Prayer Book, before the Gunpowder Treason Service." This image is stimulated by the sight of a man lighting the upper candles in the theater, which is to say by someone who, neither outside nor inside the frame of the play, is part of its illuminating, even animating condition. When Shakespeare's most famous ghost appears on stage "in a strange Dress," Partridge again notes that this is "like what I have seen in a Picture" (*Tom Jones*, 16.5.852). So does Partridge appear to be blessed with the heightened attunement to peripheries that we find in those readers who have, by definition, seen it all before.

And yet what most of Fielding's readers remember from this byzantine chapter is that Partridge mistakes the stage ghost of Hamlet's father for a reality: he's scared to death of it by the sixth paragraph. The more confidently we advance through narrative time, the likelier we are to miss the fact that Partridge is more skeptical than anyone at first ("Ghosts don't appear in such Dresses as that, neither" [*Tom Jones*, 16.5.853]). He changes his tune only when he sees how frightened Garrick's Hamlet is: "If that little Man there upon the Stage is not frightned, I never saw any Man frightned in my Life" (*Tom Jones*, 16.5.854). "That little Man" less expresses than proleptically mediates Partridge's own fear. Fielding's epithet is usually taken for a historical reference to the diminutive Garrick. But now it opens onto the sphere of homunculi and faerie lore, even as it shrinks "Garrick" to the dimensions of the "little" characters—the alphabetical ones—that convey *him* to Partridge's own reader. In short, Partridge looks far more like a reader than he does a playgoer. And as a reader his habits of attention are completely appropriate. How else would he spend the rest of the performance but "with his Eyes fixed partly on the Ghost, and partly on *Hamlet*"?

Within the strict symbolic parameters of Fielding's own text, Partridge is surely right to speculate as he does: "Perhaps, it is the Devil." The visual image on which Partridge's eye is (partly) "fixed" is demonically compounded with the words Garrick's Hamlet speaks ("Hush, Sir, don't you hear him!"). This occult transaction between sounds and visual signs is one that best pertains to the written characters that literally convey it. Immersed in this second-order transaction, Partridge very naturally loses himself in Hamlet's response ("Is he frightened now or no? As much frightened as you think me"). Nor do his words mark any distinction between himself and the reader outside the book in which he appears: "As you think me," is what he says, not "as I am" (*Tom Jones*, 16.5.854).

So does Fielding's rendering of a performance look less *at* than *with* the question of what it is to have a literary experience. In the context of such an experience, unsettling questions may arise: What, finally, *is* the difference between Hamlet and the ghost? Between Garrick and Hamlet? Between Partridge, the literary character, and either one of them? Between any of these figures and Fielding's own reader? Partridge's literal-mindedness allows all that is solid on both stage and page to melt into a common air. The resulting world is atmospheric: dreamily paced, softly lit, and poorly differentiated into discrete sensory registers. But Partridge's critical interiority to this evocative sphere allows us to see how literary worlds come not just to seem but to *be* as if they are all about us. He should be judged—or not—accordingly.

7

Glanvill's Ghost, Cold Sociability, and "the Cure of *Arabella*'s Mind"

> It is one of the most pregnant facts of mental life that, if we weld certain elements taken from the whole of being into a realm of their own, which is governed by its own laws and not by those of the whole, this realm, if completely cut off from the life of the whole, can display in its inner realization an empty nature suspended in the air; but then, often altered only by imponderables, precisely in this state of removal from all immediate reality, its deeper nature can appear more completely.
>
> GEORG SIMMEL, "Sociability" (1910)

> Tho' I banished Mr. *Glanville*, I did not desire his Death; and, questionless, if he knew how I *resent* it, his Ghost would be satisfied with the Sacrifice I make him.
>
> CHARLOTTE LENNOX, *The Female Quixote* (1752)

What, then, to make of Arabella, Charlotte Lennox's famed female Quixote? She appeared on the London literary scene just three years after *Tom Jones*, and Fielding himself made haste to praise the "Coldness of the Narration" that delivered her there.[1] If "Coldness" implies remoteness, Fielding isolated something besides a quality of Lennox's narrative voice: Arabella lives, to echo Georg Simmel's evocative phrase, in a realm of her own, one governed by its own laws and not those of the whole society.[2] The latter she notoriously rejects in favor of her private library and the unshared phantoms of her mind, and when at last she comes to her senses and forsakes those, the English novel is supposed to have reached a climacteric of its own. In Catherine Gallagher's justly influential reading, Arabella's conversion to the credo of common sense at the hands of a mysterious "Divine" marks her submission not, as would appear, to reality but rather to a distinctively modern conception of fiction—a regime, governed by "probability and illusionistic realism," where "'credit' is solicited only conditionally, on the revived Aristotelian terms not of 'belief' but of 'suspended disbelief.'" For Gallagher, Arabella's conversion comes at the price not just of other forms of fiction but of material embodiment too. Gallagher's Lennox herself paid that price as, for all the popularity of this novel, she was ultimately left to vanish, impoverished, from the increasingly abstracted marketplace of mid-eighteenth-century English letters.[3]

Gallagher's persuasive reading redresses decades of criticism in which "the Cure of *Arabella*'s Mind" was understood as an allegory of the English novel's break with its own nebulous past, and especially with the archaic figures of romance that could ever after only haunt it. Just as it once supported confident accounts of British literature's arrival at the age of reason, this allegory also underwrote counterarguments for romance's subliminal persistence in the kinds of literary "History" that pretend to certify its death.[4] Yet I want to suggest that "the Cure of *Arabella*'s Mind" might have little to do with transition and that the solids that seem to melt into air over the course of Lennox's novel (or vice versa) are nothing more or less than social facts—facts that Lennox tests via a supernaturalized account of what happens in the air. Because what happens in the air turns out to differ little from what happens in the mind, this account is indifferent to the stipulations of either "Romance" or "History," "fiction" or "reality." It is invested instead in mediated environments that mix material and immaterial, interior and consensual orders of experience.

Here, the first thing to be observed is this: what precipitates "the Cure of *Arabella's* Mind" is also a symptom of her malady. She believes in apparitions. Very early in Lennox's novel, Arabella imagines that she has killed her long-suffering lover, Mr. Glanville, and her anxiety about his vengeful "Ghost" conditions her first impulse toward rational sympathy—a habit of mind that potentially rescues her from ridicule and thereby integrates her into a community whose laws she has held in contempt. Conveniently, Lennox's apparent endorsement of that integration allied *her* with such respectable, if motley, champions of her writing as Samuel Richardson, Samuel Johnson, and Fielding. This alliance is formalized at the end of *The Female Quixote* when Arabella's "Imagination, bewilder'd . . . in the Follies of Romances," stimulates a second vision: believing that she is about to be "seiz'd" by "three or four Horsemen riding along the road," Arabella throws herself into the river Thames with the intention of eluding these apocalyptic forms. A fever ensues, then a civil debate with the Divine about the nature of fiction, and finally a lecture on the importance of embracing convention—in this case, a realist convention that guarantees community by regulating how fiction's nature may be understood. At last, Arabella's "Heart yields to the Force of Truth," if not to "Truth" itself, and her "History"—a "History of Nothing" anyway, as her maid ruthlessly points out—abruptly ends. Apparently the female Quixote's insular imaginings have been exorcized by "the Divine who had the Cure of *Arabella*'s Mind."[5]

Or have they? It is significant that Arabella's first experience after her apparent rehabilitation is one of mortification. "Wholly absorb'd in the most disagreeable Reflections on the Absurdity of her past Behaviour," she is now horrified by the figure that she must have cut in the social eye as long as she appeared to be seeing things (*Female Quixote*, 383). Just possibly, Arabella's newfound ability to see herself by seeming to stand outside herself is not simply a cause of her plunge to earth and into the company of others. It is also of a piece with the kind of experience she has been having all along. And to the extent that it turns out to ground social being, that experience shows real sociability to be what Simmel would call a "play form"—a "castle in the air," as he would also put it, that is every bit as threatened by "entanglement with real life" as it is by what, in Lennox's novel, threatens *it*—the self-contained sphere, "suspended in the air," of a young lady's romantic imagination.

Gallagher rightly insists that that imagination resists modern (realist) fiction, not—as is commonly supposed—reality. But why is this supposition so common? One answer is that Arabella's apparent conversion provides a medium through which her living readers can see *her* as a modern literary character, which is to say as one that they can collectively pretend to find in any way believable. Nonetheless, Arabella herself is (literally) in a position to be brought into a community of shared perceptions because she has been fished from the Thames "senseless, and to all Appearances dead" (*Female Quixote*, 363). If she is to stop believing in apparitions—and if her readers are to begin doing so—it appears that she must herself become one. Lennox is interested in this necessity from two perspectives: that of sociable forms and that of thought forms. In the end, these forms converge within a novel sphere of sensory possibility. As it tests the edges of that sphere, Lennox's satiric novel of enlightenment engenders the "romantic" dimension of modern literary realism that now goes by the name of atmosphere.

Silent Emphasis: From Glanvill to Glanville

Lennox's only modern biographer, Miriam Rossiter Small, found that "atmosphere" is actually the one thing that *The Female Quixote* lacks. "We do not get many touches of eighteenth century life," Small regretted. "The fact that Arabella occupies the center of the stage almost continuously accounts for this lack of eighteenth century atmosphere."[6] Lennox's contemporaries may or may not have agreed. What *did* Henry Fielding mean when he referenced "the Coldness of [Lennox's] Narration"? That question for the moment aside, two things about Arabella that we get instead of "atmosphere" will surely impress any reader: Arabella's aversion to touch and her intractable "Coolness." She is, questionless, the coldest of the mid-eighteenth century's popular literary heroines, surpassing even that "charming frost-piece," Samuel Richardson's Clarissa, in the attraction she holds for the most frigid of adjectives and nouns.[7] As it whispers of formally composed, inconceivably beautiful air,

Arabella's very name evokes the same sphere of associations. Yet the cold, untouching, and untouchable properties of Lennox's "Narration" are also what give her reader an immediate experience of her heroine—immediate precisely insofar as it arises from nothing but the pages in that reader's hands.

Gallagher has stressed that Lennox's words try very hard never to refer to things in the "real world," further linking this resistance to the wholesale "rise of 'pure fictionality'" as a cultural idiom that excludes female embodiment from the literary marketplace (Gallagher, *Nobody's Story*, 90). Lennox's contempt for referential language appears to have troubled her own contemporaries from time to time; as Small points out, late scenes set at Bath, London, and Richmond teased them with the possibility that a few names at least might be linked to living persons. By significant contrast, though, Arabella's singular remoteness, her assumption that experience can be grounded in the lexical signs that only seem to mediate it, and the lofty mental habits that let her feel at home in the world that those signs project all *do* appear to have absorbed eighteenth-century readers. "I had some very odd adventures since I saw you," Susanna Highmore wrote to Hester Mulso, "not unworthy of a Lady Arabella, which happily I may relate, when we have the pleasure of meeting again, when in return, I shall, questionless, claim your history."[8] Mary Delany likewise professed not just to "like the design" of "the Female Quixote" but to be "glad to get into good company again" when she sat down to read it.[9] In this last instance, the company of *The Female Quixote* cannot be told from that of the female Quixote herself, and both are less joined than, however facetiously, gotten "into."

The plot of the novel that brought her "into [such] good company" is, of course, driven by Arabella's settled aversion to "company," especially that of her suitor, Mr. Glanville. She systematically evades his persistent "attempt[s] [upon] her Lips" and "pressing[s] [of] her Hand," and only on the last page of *The Female Quixote* is he permitted to "kis[s] the Hand she gave him with an emphatic Silence" (*Female Quixote*, 28, 32, 383). It is a question whether this "emphatic Silence" is Mr. Glanville's or the Lady Arabella's. A reader's answer to the question will hinge on her view of the sexual politics of the novel, not to mention the kind of pleasure, if any, she takes in the story it tells. All the same, this hard-won point of contact matters to far more than the mechanics of narrative desire. It also reveals a dimension of Lennox's fiction very different from what we find in her plot, in her characters, or, indeed, in any of this novel's empty referential gestures. This dimension—which I will, in spite of Small, be calling atmospheric—can be sensed where the words "emphatic Silence" converge with the phenomenon of soundless impression that they evoke. It may be mapped through the literary rapprochement that the female Quixote's emphatically silent contact with Mr. Glanville both represents and accomplishes: that of *The Female Quixote* with Joseph Glanvill's soundless yet emphatic "Blow against Modern Sadducism."[10]

Even if Arabella's social legitimacy were not confirmed by her eventual marriage with someone who shares his last name, her fate should bring Glanvill to mind. That notable demonologist might have been dead for well over half a century, but even in 1752 his oft-revised trove of "full and plain Evidence concerning Witches and Apparitions," *Saducismus Triumphatus* (1665–81), remained a fixture in Britain's supernatural library. Arabella's swimming at the end of *The Female Quixote* follows standard procedure for evidencing a witch, whom Glanvill took to be anybody whose quixotic "powers of *imagination* may form . . . shapes" that are "onely *illusions*."[11] By 1752, the witch tests that *Saducismus Triumphatus* seemed to legitimate had been outlawed for more than a decade. But Lennox herself, in her capacity as editor of the midcentury periodical *The Lady's Museum*, was soon to print a lament for the recent "execution of poor Thomas Colley, only for stifling Ruth Osborne, the witch of Tring, in a pond of water."[12] Several layers of irony shade the fictitious letter, by one "Grace Viper," which lodges this particular complaint. Still, irony is not necessarily incompatible with conviction. Nor is Arabella's own "Wit," even at the height of her visionary "Absurdities," something that her own Mr. Glanville ever doubts, even after she is nearly stifled in the water herself.

When Mr. Glanville's seventeenth-century namesake laid out his "full and plain Evidence concerning Witches and Apparitions," he identified the verbal forms that such "Evidence" took with the literary genre of romance that defines Arabella's reality. Revised just after Glanvill's death by the Cambridge Platonist Henry More, the 1681 edition of *Saducismus Triumphatus* begins with the complaint that, any more, "stories of *Witches*, *Apparitions*, and indeed everything that brings tidings of another world, are but *melancholick* Dreams and *pious Romances*" (Glanvill, *Saducismus Triumphatus*, 9:3). The identification of ghost stories with melancholy dreams and "*pious Romances*" bears on *The Female Quixote*, whose anachronistic obsession with literary romances "long since exploded" has perplexed readers from Clara Reeve on down. It has never been very clear why Lennox should have wished to beat such a dead horse: "This book came out 30 or 40 years too late," Reeve estimated, recognizing *The Female Quixote* as a revenant of the very genre that it appears to write off.[13] The romances that Lennox satirizes first appeared twenty or thirty years earlier still. In short, the era of romance's last gasp also gave birth to distinctively modern apparition narratives such as Glanvill's. In evoking the one, Lennox breathed new life into the other.

Their natural affinity not just with apparition narrative but with apparitions themselves is evident in Lennox's treatment of romance as an environment. The books that line her heroine's walls are "not in the original *French*, but very bad Translations" (*Female Quixote*, 7), ghostly simulacra that cross symbolic boundaries and zones of time. Arabella, for her part, does not merely—or at all—look at these spectral forms. She immerses herself in them, at one point "so extremely interested in the Fate of the Princess *Melisintha*, whose Story she was reading, that she could not stir till she had finish'd it" (*Female Quixote*, 281). Arabella initially experiences her romances as a "Store" that fills her father's library, and on moving to London her first act is to distribute a "moderate Quantity of them" about the room (*Female Quixote*, 332). Her romances thus appear less as bad copies of vanished originals—or as shrapnel from the "exploded" past—than as elements of her immediate environment. Finally, we saw with respect to Defoe's *Journal of the Plague Year* that eighteenth-century spirits could be vital markers of the sphere of commonly sensible effects that arise from representation. Glanvill's ghost "stories" were meant less to prove that spirits existed than to show that truth claims that rest on nondiscursive proof might be mere propositions that monger demons and delusions of their own. As she engaged the "*pious Romances*" that Glanvill classified with those stories, Lennox in turn seems to have been less interested in *re*viving romantic apparition than she was in *de*riving it from the collective *folie* that bound not just the living literary culture about her, but the sphere of social relations about it.[14]

The group referenced in Glanvill's "Blow against Sadducism"—the Sadducees—were Jewish doubters of Christ's resurrection. It took Glanvill some time to settle on a formal title for the ever-expanding book of "Evidences" he first published in 1665, but when he did, *Saducismus Triumphatus* foretold "triumph" over the Sadducees' latter-day avatars, the materialists and mechanists who demanded empirical evidence of the physical reality of spirits. Glanvill himself was far from hostile to the yoking of belief and sense impression. He even began his literary career as a prolific apologist for the methods of the Royal Society, of which he was a fellow, and to whose annals he contributed several experiments. (One concerned the minerals at Bath, of whose Abbey Church Glanvill happened to be rector.) Professionally anxious to reconcile Christian revelation with the empirical protocols and models of community that the Royal Society exemplified, Glanvill regularly warned his readers to beware the lure of self-deception, the insular mind's powerful inclination to objectify its own illusions. "The *apparitions* of our frighted *Phancies* are real *sensibles*," he wrote in *Scepsis Scientifica* (1665). "But if we translate them without the compass of our Brains, and apprehend them as external Objects, 'tis the unwary rashness of our *Understanding* deludes us." Indeed, we are so "impatient for *Truth*" that the mind will continually "take up short, of what is really *so*, and please it self in the possession of imaginary appearances, which offering themselves to its embraces in the borrowed attire of that, which the *enamour'd Intellect* is in pursuit of, our impatient minds entertain these counterfeits, without the least suspicion of their cousenage."[15]

If the impatient mind too readily puts itself "in possession of imaginary appearances," however, it doesn't necessarily follow that there are, external to it, no real appearances whatsoever. Glanvill's skepticism was not an intellectual habit of simple negation. This made it possible for him to suppose that formally ordered disruptions of the air might be something other than counterfeits minted by "the *enamour'd Intellect.*" Glanvill's dogged pursuit of this possibility has moved modern parapsychology to hail him as the "Father of Modern Psychical Research." To phenomena we now classify as supernatural he seemed to apply rhetoric and rules of evidence—inherently social standards of collective witness and empirical verification—virtually indistinguishable from those found in, say, the transactions of the Royal Society. Such applications provided an investigative and discursive model for everyone from Andrew Lang at the end of the nineteenth century to the academic constituency of the Oxford Psychical Society throughout the twentieth.[16]

But Glanvill's massive *Saducismus Triumphatus* was not just a record of his efforts to make supernatural "Evidence" a common object of knowledge that could in turn accredit the thought communities that surrounded it. His eulogist Anthony Horneck considered that Glanvill's own "Soul seemed to be spun of finer Thread than those of other Mortals, and Things look'd with another Face, when they passed through the quicker Fire of his Laboratory."[17] No less than *New Experiments* was Robert Boyle's, *Saducismus Triumphatus* proved Glanvill's "Laboratory." In its pages the demonologist experimented with typographical design and narrative voice to make his invisible subject matter "loo[k] with another Face." That matter had often been as invisible to Glanvill and his informants as it was to his own reader. But Glanvill treated this coincidence as a point of enlivening convergence that turned his writings about the invisible world into "proofs [that] . . . come nearest the sence [*sic*], [and] indeed strike our minds fullest, and leave the most lasting impressions."[18]

The pages of *Saducismus Triumphatus* made especially effective "proofs" because, with respect to the phenomena they represented, "these things were not done long ago, or at a far distance, in an ignorant age, or among a barbarous people." For that matter, "they were not the passages of a Day or Night, nor the vanishing glances of an *Apparition*; but these Transactions were *near* and *late, publick, frequent,* and of *divers years continuance*" (Glanvill, *Sadducismus Triumphatus,* 9:116–17). Their claim to immediacy—albeit immediacy already mediated through "publick" affirmation—lent Glanvill's "Relations" special resonance with the lexical marks that conveyed them. As present "Transactions" in their own right, those signs made splendid conscripts in the demonologist's war on the "dull Lethargy" of the modern Sadducees, who may well have been his primary audience.[19] We have seen how much it pleased Glanvill to report that "this sort of sensible Experiments cuts them and stings them very sore, and startles them" (Glanvill, *Saducismus Triumphatus,* 9:12).

No literary "Experimen[t]" was more likely to startle or sting than Glanvill's rendering of the so-called Demon of Tedworth, who soon emerged as the undisputed star of his "Collection of Relations." The Demon—or Drummer, as he was more familiarly known—was an invisible spirit believed to signify its presence by disturbing the air: in April 1661, the justice of the peace he plagued, one John Mompesson, thus found his Wiltshire home troubled with "a Thumping and Drumming on the top of his House, which continued a good space, and then by degrees went off into the Air" (Glanvill, *Saducismus Triumphatus,* 9:91). With the Demon's arrival came "an hurling in the Air over the House, and, at its going off, the beating of a Drum like that of the breaking up of a guard" (Glanvill, *Saducismus Triumphatus,* 9:92). Before long, the Demon turned up inside the house itself, whose air thickened with a "bloomy noisome smell" and grew "very hot, though without fire, in a very sharp and severe Winter" (Glanvill, *Saducismus Triumphatus,* 9:99). If air was the Demon's preferred medium, writing came in a close second. It liked to flip the family Bible into the ashes—the good book fell open to Jesus casting out demons in the Gospel of Mark—and a favorite trick was tracing, again in ash, "some Letters, . . . which [their readers] could make nothing of, besides many Circles and Scratches" (Glanvill, *Saducismus Triumphatus,* 9:100).

As writing, Glanvill's "Relation" of the Demon's misadventures could register this spirit's effects only indirectly, on open pages through the circles and scratches of typographical marks. But this necessity merely identified the experience of reading with that of perceiving the signifying demon himself, the experience of writing about it with that of investigating Tedworth directly, as Glanvill relates he did. Samuel Pepys pronounced Glanvill's book a "strange story of spirits, and worth reading indeed."[20] Indeed, in 1663 he recorded a dinner with the Earl of Sandwich "both at and after [which] we had great discourses of the nature and power of Spirits, and whether they can animate dead bodies: in all which, as of the general appearing of spirits, my Lord Sandwich is very scepticall. He says the greatest warrants that he ever had to believe any, is the present appearing of the Devil in Wiltshire, much of late talked of, who beats a drum up and down; there is books of it, and they say very true."[21]

Aiming to stimulate the speculative social exchanges that could make it possibly "true," the 1681 edition of *Saducismus Triumphatus* added still "other passages at my being in *Tedworth*, which I published not" (Glanvill, *Saducismus Triumphatus*, 9:104). In one such passage, Glanvill roams the "passages" of Tedworth in hopes of seeing the Demon for himself, but can perceive only the mysterious wriggling of a linen bag. When he "caught it by the upper end with one Hand," the demonologist confides, he "found nothing at all in it," though "there was no body near to shake the Bag, or if there had, no one could have made such a motion, which seemed to be from within, as if a living *Creature* had moved in it" (Glanvill, *Saducismus Triumphatus*, 9:103). As if to echo Boyle's experiments with efficacious but invisible air, this passage aligns primary "Evidence" of the spirit's presence with the secondary "Evidence," likewise the effect of an insensible cause, that the self-defined social medium Glanvill's writing offers. But the "passages" in which Glanvill's demon appears are also striking because of the hybrid (thus inclusive) belief state into which they coax their own readers. The telltale bag, for example, moves "as if a living *Creature* had moved in it" (Glanvill, *Saducismus Triumphatus*, 9:103). Whereas another kind of book would want us to see that a living creature moved in the linen bag, Glanvill's pages ask the slanted analogy to define the reality: this was something that moved "as if" it were living. Glanvill lets his readers see the "as if" they see through, and so the physical marks of his "Relation" mark also the provisional nature of his readers' emergent conception of reality. Indeed, Glanvill aimed to provoke "a true, genuine and consistent Notion of the Nature of a Spirit, such as will not beget misbelief of their existence" and intended his apparitional "Experiments" to "rub up and awaken . . . benummed and lethargick Mindes into a suspicion at least, if not an assurance that there are other intelligent Beings besides those that are clad in heavy Earth or Clay" (Glanvill, *Saducismus Triumphatus*, 9:16).

"Suspicion," not necessarily (or at all) "assurance." No misbelief, not positive belief. These were the postures of mind that Glanvill's written words stimulated, confirmed, and strove to render communal. The result was a new category of sensation—the grounds of knowledge in any empiricist framework but now a source of community because it mixed material effects with mental causes and conditions. This expanded sense of "sense" made it possible to embrace the possibility of "a Spirit" within an agnostic framework.

What is more, Glanville's own satire depended on it. When it came to the "brisk and confident Men of our days" who were his implied readers, he expected his writing to "affront them, greatly provoke their rage and contempt, and to raise the Devil of their Wit and Buffoonery" (Glanvill, *Saducismus Triumphatus*, 9:A3r). This image of the written voice as a kind of abrasive that could stir up "the Devil" seems to have motivated even Glanvill's notoriously bizarre and distracting typography, which could move through four different typefaces in a single sentence. But all of them actively secured a mixed mental state as an order of sense experience in its own right. Because that state was heterogeneous and elastic, the objects that evoked it melded disparate groups into unified subjects of common experience. "Yet why," Glanvill legitimately could wonder, "should [we] not believe that the *Air*, and all the *Regions* above us, may have their invisible intellectual Agents of nature like unto our *Souls*, be that what it will. . . . This *Hypothesis* will be enough to secure the possibility of *Witches* and *Apparitions*" (Glanvill, *Saducismus Triumphatus*, 9:8).

By identifying "*Hypothesis*" with the shared maintenance of a possibility, Glanvill wrote his own readers, as a notion-holding community, into his pages. But he himself went missing from that community. The Demon of Tedworth was irreverently incarnated as "the Drummer" of Joseph Addison's 1716 farce of that name, and the final revision of *Saducismus Triumphatus* was edited after Glanville's death by his friend More. Part of *The Female Quixote* too has always been suspected of having been ghostwritten; besides promoting it in the popular press, Samuel Johnson not only composed its strangely evasive dedication to the Earl of Middlesex but may well have drafted Arabella's long debate with the Divine who presumably cures her mind.[22] In possibly playing Henry More to Lennox's Joseph Glanvill, Johnson left but one of many traces of affinity between *Saducismus Triumphatus* and *The Female Quixote*; another is the barrage of italics that dramatically impresses each page with silent, often arbitrary emphasis.

Above all, however, Glanvill and Lennox made each such page less show *that* than, through silent emphasis, project *how* mental states, provoked by notional objects, might condition the experience of reality. Lennox's Arabella absorbs the same sensory impressions that those around her do. But the concepts within which she makes sense of those impressions have been previously adapted from a second-order reality. Her Glanvillean interpretations of events are for this reason both extraordinarily provisional ("it may still be as I say") and, as Gallagher also notes, even more extraordinarily coherent; even Richardson stood ready to praise Arabella's "Absurdities" for this virtue.[23] Glanvill too challenged those to whom "*the* Notion *of a* Spirit *is im-*

possible and contradictious" on the grounds that the materialist "Proposition, *that the Notion of a Spirit is absurd*" goes hand in hand with the notion that "a God and a Soul distinct from Matter, and immortal, are likewise Absurdities" (Glanvill, *Saducismus Triumphatus*, 9:5). Indeed, though *Saducismus Triumphatus* vividly rendered scenes of physical contact with a world not itself physical, it is ultimately the mental effects of any "Proposition" as they register in graphic space that matter, *are* the matter, and finally make up what counts as the matter of *Saducismus Triumphatus.* The only real "Impossibilities and Absurdities," as far as Glanvill was concerned, are ideas *about* our "Conceptions."[24] In and of itself, any "Conception" is fair, though the more elastic—thus potentially shared—a conception is, the better.

Lennox obviously shares this interest in conception as the condition, rather than the trace, of our experience of sensory objects and thus as a paradoxical foundation of sociable community. So "the Adventures of Arabella" very naturally end with her incorporation into the Glanville family. Up to that point—and much as Joseph Glanvill tracked spirits through sensible disturbances of invisible air, replicating this pursuit in lexical space—Lennox's upright and long-suffering Mr. Glanville struggles to figure out the elusive object of his desire. As the percipient narrator of Glanvill's evidentiary "Relations" had done with his aeriform "spirits," Lennox's own Mr. Glanville tries to do so by feeling Arabella out. His first, near fatal act is thus to "salut[e] her with the Freedom of a Relation," a freedom whose accompanying "Freedom of Expressio[n]" he exercises (and literalizes) through regular efforts to "press" some part of Arabella "to his Lips" (*Female Quixote*, 28, 124). But once Lennox's Mr. Glanville has realized that Arabella is not "in jest" (*Female Quixote*, 34)—that her meanings and motives cannot be made out through ironic indirection—he has little choice but to try to discern them by other means.

In turn, Mr. Glanville's precious moment of reciprocal contact with Arabella in the last paragraphs of Lennox's novel rewards him for managing to adopt the mixed epistemological attitude—"suspicion" without guarantee of reassurance—that Joseph Glanvill charted in *Saducismus Triumphatus.* In this way, Mr. Glanville has turned his intelligence from Arabella's interrogator into her medium, for at the end of the day, it is surely through Mr. Glanville that Lennox's living readers move into the sphere of Arabella's influence. On the page, this sphere presents naturalistically as an atmosphere. It is registered in the intermittent "Gloom" and "Brightness" of Arabella's face, in her damp eyes, often "swimming in Tears" (*Female Quixote*, 35), but above all in the visible effects of all these elusive yet seemingly ambient phenomena upon her earthly lover.

Arabella's Air

Mr. Glanville, we are very often told, has "Vivacity." Arabella has none. For all her many references to Arabella's "Figure," Lennox never really fills that figure in.[25] Instead, she swathes her heroine in the trappings of women of yore, hiding her behind not even a veil but the "Form of a Veil." Lennox takes care that the one physical feature to which she does refer directly—Arabella's "fine black Hair"—has only "the Appearance of being artless" so "that all but her Maid . . . . imagined [that her tresses] were so" (*Female Quixote*, 9). The attentive reader can discover that Arabella's eyes are "Black." But that reader can do so only indirectly, through Mr. Glanville's debate with her about an "express[ion]" ("those fair Eyes") that a rival has worked into a love letter. As David Marshall has shown, Arabella herself would evidently prefer to exist only in a reverberant chain of simulacra, bound by the obscure chemistry of emulation to the idealized heroines she holds in her "Mind." Meanwhile, at the level of Lennox's plot, Arabella can only reflect others' impressions of her.[26]

Then how do we know that Arabella is there at all? The most obvious answer is, of course, that we don't, because she isn't. But she *is* notionally present to other characters in the novel. How do we know that? There is certainly a set of print impressions (apparently they copy Arabella's personal black-and-white color scheme). Almost as certainly, there are what those impressions evoke: ours of Lennox's heroine. The act of forming impressions is modeled by Lennox's characters, and that act itself takes various forms. *The Female Quixote*'s running joke is that the female Quixote's own impressions are almost always false, but they are also internally coherent. As such, they double as throwbacks to the standards of "Evidence" that Glanvill had set—standards finally justified, which is to say realized, as "possibilities" in the experience of conspicuously literary representation. Within the premises of Lennox's own narration, the qualities that should damn them thus lend Arabella's false impressions significant reflexive and satiric force.

If we can doubt how much it matters to Lennox's narration that Arabella's quixotic impressions are false, then how are we to value other characters' impressions of her? Is it possible to say that anyone in this novel is impressed by her at all? Another of Lennox's favorite jokes is that the female Quixote affects—and effects—next to nothing. Arabella has a story only because she listens to those of others, and because she is a champion magical thinker who believes her wishes can make things happen, as naturally they cannot. But what Arabella *can* do is put people "out of Temper." This must not be confused with affecting things, because moods—even "bad" ones, states of being "out of Temper"—are not themselves affections. Their essence, if essence they can be said to have, lies in their cloudlike suspense between affection and cognition, feeling and thought, sensibility and sense. Still, it is this suspense that comes closest to producing sensible experience of Arabella. Even the good "Doctor listen'd to her with mix'd Emotion, between Pity, Reverence, and Amazement" (*Female Quixote*, 368). The "mix'd" mental and bodily conditions that Arabella seems to precipitate in others also give rise to a sense, ultimately communal and to that extent verifying, of her presence as an object of perception.

Here Lennox appears to grant Arabella the kind of influence that was supposed to belong to Arabella's own medieval heroines of romance: the power to create what we call ambience. In his suggestive history of the idea of ambience, Leo Spitzer attends to the "concept" of female divinity "emanating from the love mysticism of the Middle Ages"; this concept, Spitzer perceives, induced the poets to "surround [her] with a halo, an 'atmosphere,' which served to veil her in mystery, to suggest that she dwelt in a divinely harmonious world of her own—which her lover, nevertheless, sought to penetrate."[27] Spitzer's vital point is that the "*aria*" of the so-called *donna angelica* (whose "sigh . . . is thus both air and spirit") cannot be told apart from its linguistic expression. *Aria* simply could not be without being mediated through formal language. At the same time, *aria* confirms language's own capacity to operate as an expressive medium, its potential to leave residues, shadows, and shades of meaning as well as to mean through its more typical gestures of proposition and reference. Every time the tempers around the "out of Temper" Arabella ruffle, flare, or are lost outright, Glanville's "Angelic Cousin" (*Female Quixote*, 125) summons this oft-hidden capacity of the literary forms in which she also consists.

Here too lies Arabella's difference from the atmospheric women of medieval lyric poetry. Unlike them, she is at last not just a linguistic but a fundamentally and conspicuously *lexical* composition, something that she herself seems to recognize. (Indeed, from this point of view, Arabella is the only character in Lennox's novel who grasps that her own genetic code is nothing but code and who may thus be said truly to comprehend who and what she is.) Between the Lady Arabella and the medieval lady haloed in *aria* lie the French prose romances that grounded evocative and charismatic objects in the material signs that mediate them. Unless this ironic grounding could be made apparent, the effects that Spitzer charted could not have been maintained in a modernity committed to empirical and consensual models of reality verified in writing. But Lennox does make them apparent, through the witty screen of female quixotism.

In this way Lennox discovers the possibility of such things as immaterial effects and taps into the other side of consensual reality. Virtually everyone in Arabella's world puts on "Airs" of some sort or another. But *only* Arabella is able to see these airs. If the impression she draws from, say, the "Person and Air" (*Female Quixote*, 22) of a conniving young gardener is by conventional standards false, it nonetheless reminds us that the communicative medium that binds the world of Lennox's novel together is, like all printed matter, literally but figural.

Consider one "surprising Adventure." Arabella, her cousins, and uncle are all on their way to Bath when they are accosted by highwaymen whose imminence (or is it immanence?) is first "whispered to Mr. *Glanville* in the Ear" (*Female Quixote*, 257). Arabella, by contrast, looks out "to see what was the Matter" and thereupon appears to form the impression that the brigands are actually "valiant Men" who "falsly suppose" that she and her cousins have been "violently carried away." Her subsequent "fine Speeches," as usual, strike all who hear them as nonsense. "What is the Meaning of all this?" demands Sir Charles. Even the highwaymen "could not distinguish her words." At the sight of Arabella, however, the ruffians inexplicably think "fit to abandon their Enterprise." Hence Arabella concludes (correctly) that they "had some bad Design" and (incorrectly) that they must be "Persons of Quality." Arabella's guardian, Sir Charles, declares that he "can't possibly understand" what she has in mind. It is Mr. Glanville who recognizes that Arabella "has not yet, possibly, believed them to be Highwaymen" and attempts to set her straight. For her part, Arabella will concede only that, though "I have been strangely mistaken it seems" about appearances, she may still be correct about motive: "I apprehend there is no Certainty, that your Suspicions are true; and it may still be as I say, that they either came to rescue or carry us away." And in this indeterminacy she finally rests, so that Mr. Glanville can only "chang[e] the Discourse; having observed with Confusion, that Sir *Charles* and his Sister, seemed to look upon his beloved Cousin as one that was out of her Senses" (*Female Quixote*, 257–59).

It is difficult for anyone, Lennox's reader perhaps least of all, to "see what [is] the Matter" here, and Lennox's relation of this "surprising Adventure" is in danger of not making sense for that very reason. That relation appears to settle things on the side of the senior and female Glanvilles only when we forget that, in this scene, everything "seem[s] to look" but nothing "is." Ordinary "senses" are useless: it hardly matters whether one is in or "out of" them. In turn, the appropriate interpretive orientation must resemble the singularly antisocial suspense that Arabella assumes ("I apprehend there is no Certainty, that your Suspicions are true; and it may still be as I say, that they either came to rescue or carry us away"). Besides Arabella, only Mr. Glanville can sustain this state of suspended mental animation, and only he expresses it in the social form of a "Confusion" that drives him to "chang[e] the Discourse." Mr. Glanville's "Confusion" is literally a synthesis of various positions that make sense only in themselves. Because it is a synthesis, Mr. Glanville's confusion does not enforce isolation. To the contrary, it motivates politeness. Indeed, we could say that Mr. Glanville's "Confusion" institutes a novel form of sociability even as it rebukes the irrelevance and possible cruelty of old ones. And for the time being, Mr. Glanville's confusion permits Arabella to ride on unperturbed, her composure justified by what we might call the visible air of Lennox's writing, as it speaks the language of pure appearance.

Such subtle affinities between Arabella's supposedly ridiculous "Figure" and the potentially sociable ground against which it appears to be so become most apparent when Arabella, as here, is out of doors. For such a bookish person, Arabella spends a surprising amount of time in the open air. Very often she is on horseback. Spitzer mentions in passing that "[the] word air, conveying also 'moral health'; a well-tempered climate," and so central to the evocative figure of the romantic heroine, was "also applied to the gait of thoroughbred horses" (Spitzer, "Milieu and Ambiance," 260). Closer to home, Johnson's *Dictionary of the English Language* (1755) notes in the last of its sixteen definitions of "Air" that "*Airs* denote the artificial or practiced motions of a managed horse." Arabella's equestrian bent may or may not express this embedded association; Lennox anyway supplies a more immediate reason for it insofar as "the only Diversion [Arabella] was allowed, or ever experienced," is that of "riding out" (*Female Quixote*, 23). Arabella's "Airings" are the only occasions on which what others would count as "experienc[e]" coincides with what Arabella herself would consider to be so. It is during her airings, moreover, that Lennox's own reader is most explicitly charged with forming some impression of Arabella's "Figure." As in the "Adventure" of the highwaymen, this is possible only through the screen of ironic narration. As is true for the reader of *Saducismus Triumphatus*, by definition eager to see an apparition, persistent awareness of this screen is what makes it possible to see Arabella for the only thing she can finally, with any accuracy, be said to be: the evocative stimulant of her readers' own reflections. Once the air about Arabella becomes visible, it no longer matters *what* we think of her, only *that* we do.

And what of the romantic literary signs in which Arabella herself believes? These don't seem to correspond to any contemporary empirical reality, as verified through the performed perceptions—the common sense—of others. Resisting that sphere of perceptions as they are realized in behavior,

Arabella displays a "most happy Facility in accommodating every Incident to her Wishes and Conceptions." So she "examine[s] [a] Matter so many different ways, drew so many Conclusions, and fansied so many Mysteries in the most indifferent Actions" that she is inevitably "confirmed in [her] Opinion[s]" (*Female Quixote*, 25). But Arabella's way of being in the world is only superficially at odds with the relatively open-ended experimentalism that held together the community of natural philosophers in which Joseph Glanvill, for instance, participated. "Questionless" Arabella may be a "strict Observer of romantic Forms" (*Female Quixote*, 13), but she *is* an observer. Some of her favorite subjects even include the elements of fire, air, and water, whose surprising commonalities obsessed the explosive pneumatic chemistry of Lennox's day. Unlike your standard-issue empiricist, however, Arabella does not stand apart from her sense experience of these elements to produce a consensually valid, replicable physico-mechanical account of any of them. On the contrary, she immerses herself in them, and in this way withdraws into Simmel's suspended "realm of [her] own" (Simmel, "Sociability," 139).

Lennox makes this marriage of abstraction and experience possible, and even erotic, by killing Arabella's mother before the daughter's story properly begins. The premise of idealizing romance is lack of sensation. But for Arabella this absence evokes an experiential reality—that of her own mother's absence—and so an objective breach in the history of transmission that has made Arabella herself possible. It is this objective breach that makes Arabella the kind of reader who can "conceive, without being told" (*Female Quixote*, 239–40). Her power of so conceiving means that Arabella herself is possessed by her books at least as much as she possesses them. Just so, the "strange" if not downright "absurd and ridiculous Notions" (*Female Quixote*, 254) that this heroine harbors seem equally to harbor her.

Hence arises Arabella's aversion to "Presumption" on one side and to the "Pressure" of physical touch on the other. As if to prove that they are, in effect, the same, Lennox periodically makes her own reader feel the air about Arabella. The female Quixote has trouble breathing on more than one occasion, and she leaves London because "the gross Air of that smoaky Town affected her Health so much" (*Female Quixote*, 340). Arabella's difficulty with the air makes her a comic distillation of the romance heroines of yore. At the same time, though, Lennox subtly modernizes an archaic sense of atmosphere: the concept of atmosphere that both surrounded and mediated the medieval *donna angelica* is what Spitzer called "not a 'cold' abstraction, but a warm one" (Spitzer, "Milieu and Ambiance," 187). Lennox cools this abstraction into the shapely, brittle "Airs" that rise from her purely literary characters and to which only Arabella is properly attentive.

Spitzer would have given Lennox little credit for strict originality. He maintained that Newton had already done her work for her when he calculated the ambient medium, so entirely dislocating it from physical space that the word "medium" itself for the first time assumed its purely mathematical meaning of "average." The same shift can be seen, if less starkly, in natural philosophy's embrace of what Lorraine Daston, with respect to the then-fashionable study of phosphorescence, calls the "cold fact."[28] Daston maintains that over the course of the eighteenth century, "scientific" facts chilled as they came to describe classes of events as opposed to singularities. We might add that this perversely divorced them from their referents; the pun that Daston uses to account for the solidification of the "cold fact" draws attention to exactly that linguistic, hence social, fact. Arabella's own coldness, however, perversely reheats the airs that float through Lennox's fiction (and, presumably, her culture) precisely by calling attention to their purely figural nature. The "cold Salute" with which Arabella checks Mr. Glanville's "Vivacity," the "very cool Compliments" she deigns to give him, and the extra "Coolness" she reserves for his sister (*Female Quixote*, 29, 42, 108)—even the "fact" that she almost dies of what seems to be a cold—all are double negatives that almost magically reverse the curse of abstraction by building the possibility of experience back into it.

Arabella's obstinate coldness, even when she is gripped by fever, also defies the contemporary social conceit of the female thermometer, so wittily explicated by Terry Castle, to remind us of what lay behind midcentury fascination with new instruments for detecting "alterations in the atmosphere." Such devices, Castle shows, carried special resonance for the popular woman of sensibility, whose ever-visible, notoriously unstable body underwrote a whole new generation of thermometric and barometric metaphors for "mysterious processes of political, social, or economic change."[29] But Arabella—who simply cannot get warm, reasonably fears love, and could not care less about "today"—will not be turned into an icon of ambivalence and inconsistency. She resists her own conversion into either an instrument or a joke and instead actively composes the man-made atmosphere that verbal or visual instruments pretend to make visible.

No one appreciated this better than Henry Fielding, who in comparing *The Female Quixote* to its most obvious model, *Don Quixote*, found one vast difference between them. "Don Quixote is ridiculous in performing Feats of Absurdity himself," Fielding observed. By contrast, Lennox's airy heroine "can only become so in provoking and admiring the Absurdities of others" (Fielding, review of *Female Quixote*, 1:280). This difference reminds us that Arabella's absurdities produce an unflattering picture of the "others" around her. But they themselves possess an even more subtle and pervasive effect: that *of* a subtlety and pervasiveness that, like Arabella's own "Imaginations," are "too quick for Language" (*Female Quixote*, 370). Thus, whereas in the "Case" of *Don Quixote*, "the Ridicule hath all the Force of a Representation; it is in a Manner subjected to the Eyes; in [*The Female Quixote*] it is conveyed, as it were, through our Ears, and partakes of the Coldness of History or Narration" (Fielding, review of *Female Quixote*, 1:280). This synesthetic and ambient quality of Lennox's "Narration" rises from Arabella's own function as the precipitant—not at all the measure—of the grotesquely mixed conditions known as "Absurdities." In her own coldness, she prefers "convey[ance], as it were," over conveyance pure and simple, "pleasing Images" to "horrid Sound," and *images* of sound—otherwise known as letters—to everything (*Female Quixote*, 25–26). Why else write so many notes to so many people? Lennox seems to share her preference. The distance that writing both bridges and makes visible justifies Lennox's pervasive irony, naturally the coldest (but also the most conspicuously social) of all communicative styles.

The coldness of Lennox's narration accompanies the happy reversal of spirit and letter, figure and ground that realizes narration for her own reader. At such points of realization, things naturally tend to go up in flames. Early on, for instance, Arabella's incensed father has "resolved to cure *Arabella* of her Whims, by burning the Books that had put them into her Head." Mr. Glanville diverts the marquis's attention to the purely metaphorical register occupied by "the Incendiary *Statira*" and persuades him to spare the books in order to support the figure of "Merit" he himself hopes to make with Arabella. He is materially rewarded with a play on words: Arabella's "Countenance brighten[s]" (*Female Quixote*, 57), and Lennox's "Narration" makes an impact without sacrificing its kinship with silence and cold abstraction.

More telling still, the tenth chapter of book 7 finds our heroine "so extremely interested in the Fate of the Princess *Melisintha*, whose Story she was reading, that she could not stir till she had finish'd it" (*Female Quixote*, 281). Arabella's engagement with this story is figured in it, because the princess she is reading about is herself "at present in a most terrible Situation. She has just set Fire to the Palace, in order to avoid the Embraces of a King who forc'd her to marry him." Arabella is in such "Pain to know how she escapes the Flames" that she begs her cousin Miss Glanville to "sit down and listen to me while I read what remains" (*Female Quixote*, 282). Though her desire is for narrative, Arabella's "present" tense interest in Melisintha's "Situation" works against that very drive in the interest of immersion in narration. The fact that the narration consists in nothing but abstracted lexical signs would obstruct the tracking of plot or the inference of character. But it only enhances the sense of immersion. "Fire" here both represents and effects an otherwise absurd confusion: to burn is at once to transcend sensation and to be engulfed by it. Hence, Melisintha's "set" flames satisfy that princess's burning desire "to avoid" the sense of touch itself.

Arabella realizes this desire by reading aloud, thereby turning the material signs that compose the romance in front of her back into air. This act of sublimation reminds us of what is distinctive about all literary fiction, both the written characters that compose it and the characters (for example, the here indistinguishable Melisintha and Arabella) who arise from them. As Arabella's "Divine" will later emphasize, such characters matter not because they refer to matter, or because they designate a particular species of fictionality, but because they make the sense of immersion in multiple modes of matter possible. The irony is that they can do so only as long as the dangerous element—like Arabella's "History," and at last like Arabella herself—consists in "Nothing."

The Devil in Miss Arabella

Fielding's remarks on the "Coldness of [Lennox's] Narration" are penetrating and sympathetic. But in the critical essay where they appear, some strange mistakes are made. When the pretended narrator of Fielding's review, his eidolon Sir Alexander Drawcansir, "perused a Book called, THE FEMALE QUIXOTE," he naturally perceived it to be a "Copy" not of the romances of Scudéry or La Calprenède but of "the famous Romance of Cervantes," *Don Quixote.* Fair enough. However, to Drawcansir *Don Quixote* appears to have been copied from Milton's *Paradise Lost,* whose author had, like Cervantes, kept some "great Cause . . . before [his] Eyes when he [planned]" his work. In Milton's theodicy, Fielding's "Censor" explains, this sacred aim was the "Cause of Christianity"; in Cervantes's work of "Ridicule," it was the expurgation of a "vicious Folly" (Fielding, review of *Female Quixote,* 1:279). In both cases, though, literary composition involved catching—and holding—sight of some invisible object (be it the Christian "Cause" or the "Folly" of romantic delusion) that could be seen only to the extent that it was mediated through a visible one.

While this apparently shared method sounds a little, well, quixotic, the real trouble is that in a historical sense Cervantes could not possibly have copied Milton. The latter was not yet born at the time of *Don Quixote*'s first appearance in the world. Fielding suggests that he *knows* that "Imitation" is a temporal phenomenon but apparently *believes* it to be otherwise, for only through such mixed and unstable epistemological styles may copies be perceived to be contemporary with their originals, and vice versa. Fielding implicitly saw *The Female Quixote* in the same way, because "the Humour of Romance, which is principally ridiculed in this Work, be not at present greatly in fashion in this Kingdom" (Fielding, review of *Female Quixote*, 1:279).

Clara Reeve was also to comment on the anachronism of Lennox's novel, whose preoccupation with literary figures "not at present greatly in fashion," we saw, finally yielded insight into the immanent quality of its supernaturalism. Arabella validates that supernaturalism, without naturalizing it, through her uncanny resemblance to Joseph Glanvill's possible apparitions. Indeed, Arabella's first appearance to a society of eyes is at Glanvill's own cure of Bath. Long before this, however, Arabella has struck us as one of the sullen spirits who fill the "Relations" of that seventeenth-century Divine. At first, "she took all Opportunities of avoiding [Mr. Glanville's] Conversation; and seemed always out of Temper when he addressed any thing to her" (*Female Quixote*, 30). So before her lover learns to absorb her temper, he fails to grasp her at all. All he can manage is to "ris[e] at her Appearance" when she joins him at the bedside of her languishing father (*Female Quixote*, 35).

Meanwhile, Arabella's own desire for death, expressed more than once, often seems to be beside the point, not least because the minute her father dies she seems to join him on the other side. Just as "his Eyes never remov[ed] themselves from her Face, till they were closed in Death," so Arabella's "Spirits now failed her at once; and she fell upon the Bed without Sense or Motion, as soon as she saw him expire." Indeed, Arabella "continu[es] so long in her Swoon, that [others] apprehended she was dead." She eventually revives, but only to renew her sense of kinship with the "dear and sacred Shade of my loved Father" (*Female Quixote*, 60). On her Bath debut, Arabella's "Appearance" in a veil thus naturally approximates two ghostly stereotypes: some "supposed her a *Scots* Lady," while others "concluded she was a *Spanish* nun" (*Female Quixote*, 263). "The Phenomena of the Veil" alone not only seems like "a Cloud" but guarantees that Arabella's "singular Appearance" will, meteorically, create "a great Disturbance" (*Female Quixote*, 205, 263). Few literary heroines are ultimately more unnerving than Arabella, and Lennox does nothing to mitigate or explain her spine-tingling effects on those around her. "Bless me, Madam," cries her cousin Miss Glanville after Arabella calmly announces that her tormentor, Sir George, is only not quite yet a ghost. "What horrid things come into your Head! I vow, you terrify me out of my Wits, to hear you" (*Female Quixote*, 194).

Miss Glanville's terror is not, perhaps, unwarranted. Arabella is soon to fail a witch test, and the direction of her uncanny "Wit" will change only after she has been dumped into a "warm Bed" (*Female Quixote*, 353) and faced interrogation by the "Divine," who has heavily forearmed himself against it. Apparitions of the sort that Arabella seems to be need not be quite this unhinging. Even Joseph Glanvill's Demon of Tedworth aimed more to entertain than to frighten; the "Relation" of Mrs. Veal was a little sad, never scary. But Arabella gives those who see her the chills. Indeed, her Divine's success with her "Cure" hinges on his ability to make her see in her romances the "Evil which can strike a good Man with Horror." It is an "Evil" that, Arabella confesses, she "dread[s] the more, as I do not feel it" (*Female Quixote*, 390).

Much the same shadows fall on contemporary depictions of Charlotte Lennox herself. John Hawkins's record of the launch party for Lennox's first novel, *The Life of Harriot Stuart* (1750), mentions that "the place appointed was the Devil tavern," and his depiction of the all-night "phenomenon" ends with Hawkins remembering "at the instant of my going out of the tavern-door, the sensation of shame that affected me"—a "sensation," he insisted, "occasioned not by reflection on anything evil that had passed in the course of the night's entertainment, but on the resemblance it bore to a debauch."[30] Gallagher accounts for Hawkins's uneasiness by invoking the scandalous "apparition of the author-whore [who] continued to haunt [a] site" that excludes her (Gallagher, *Nobody's Story*, 147). And it is true that Hawkins barely mentions the living "Mrs. Lenox" herself. Still, we need not look to what his words might be repressing to catch sight of the specters there: they cling to the surface of this passage, every bit as obvious as the Devil in the tavern sign. The frisson that "evil" once induced is now stimulated by nothing but "resemblance." And Hawkins's guilty response to mere "resemblance" is downright quixotic. In a manner worthy of Arabella herself, he responds to the mere appearance of a "debauch" not just *as if* it were real but so as to make it experientially so. In fact, insofar as Hawkins wrote down the story of Lennox's first appearance to literary society a good fifteen years after that of *The Female Quixote*, he stands as another example of the uncanny fluidity with which Arabella's style of ideation passed from Lennox's fictional world to the one inhabited by her living readers.

The word "evil" comes up more than once in contemporary representations of Lennox, and always in relation to the idea of groundless yet prolific appearances. Johnson went out of his way to vouch that he had "heard no evil" of her, thereby raising the specter whose existence he denied.[31] Later, Clara Reeve would finally trace her own discomfort with *The Female Quixote* less to its unaccountable revival of long-dead literary forms than to the fact that it had given rise to a "swarm of imitators" that in Reeve's view amounted to a "public evil" (Reeve, *Progress of Romance*, 7). By "public evil," Reeve seems not to mean some endangered social good so much as the demonic "swarm of imitators" themselves—or itself. As Defoe had influentially proposed, one problem with the devil who lingers here between the lines is that of whether he is many or one. How, then, is he to be represented? After all, "*Satan*'s Name may well be call'd a Noun of Multitude," and any close consideration of this fact only causes us to encounter him in the language of paradox that we inevitably use not just to speak but to conceive of him.[32]

That the devil appears less in the details than in the ways we collectively think those details into conceptual forms was of course one of Joseph Glanvill's pet ideas. And it would be hard to underestimate this idea's relevance to the apparitional fabric of Lennox's novel—a novel no less preoccupied than Glanvill with how states of "Mind" (its last word, after all) become apparent to themselves through representational objects. In his own "Blow" against strict materialism, we saw, Glanvill insisted that the materialist "Proposition, *that the Notion of a Spirit is absurd*" goes hand in hand with the notion that "a God and a Soul distinct from Matter, and immortal, are likewise Absurdities." Glanvill's "no ghosts, no God" syllogism included the possibility that the ghosts themselves may be the handiwork of the devil. But the real proof of the latter's existence emerges not from the illusory forms for which he may or may not be responsible but rather from the absurd certainty that he cannot possibly exist. To those who insist that all ideas are "but the Thrusting of one Part of Matter against another," Glanvill submits that "that *evil Spirit* whose *influences* they will not allow . . . hath a greater hand and interest in their *Proposition* than they are aware of" (Glanvill, *Saducismus Triumphatus*, 9:2). Like his attendant "evil" in the commentary of Hawkins, Johnson, and Reeve, the devil is just an expression. Here, he expresses the influence of materialist thought forms. In Glanvill's book, that is exactly what makes him demonstrably present in letters.

It would seem that the devil is very much out and about in *The Female Quixote*. The "Devil" is a figure of speech elicited four times in the space of little more than seventy pages. In a fiction that tests all propositions about reality by showing what kinds of experience might arise from them, this may not be so remarkable. More so, though, is the extent to which what should be nothing but an expression nonetheless opens new dimensions of the scenes in which it appears. Take, for instance, the adventures of the insufferable beau Mr. Tinsel. "Devil take me," curses he, "if I can tell what all this means." "All this" is as follows: Arabella's maid, Lucy, has just become "possess'd with the Thought" that Tinsel is about to ravish her mistress; Arabella has consequently fallen "back in her Chair in a Swoon, or something she took for a Swoon"; and Tinsel himself has found that "he had got into her Chamber he knew not how" (*Female Quixote*, 300).

The devil *has* taken Tinsel. Possession, telekinesis, curses, and trances are, of course, nothing but figures of speech through which certain misprisions are expressed. Here, however, they also support the scene, lending it what coherence it possesses. As Mr. Glanville walks a still-dazed Tinsel to the door, the latter pronounces Arabella "a most accomplish'd, incomprehensible Lady" and yet once more invites "the Devil [to] take me, if I think there is her Fellow in the World" (*Female Quixote*, 303). He says more than he knows. So, perhaps, does Mr. Glanville himself when he later responds to Arabella's stipulation that he "was[h] in Blood . . . and not appear before the Nation" by demanding to know "what the Devil does the Girl mean?" (*Female Quixote*, 350).

Conventionally, the prince of the powers of the air appears wherever acts of thinking and meaning are revealed to be figural, conventional, and precipitate. So it's no accident that Lennox's Devil makes an appearance that is never quite—and yet is only—one in that part of Lennox's fiction that is set in and around Bath. "The Situation of the City" causes Arabella, as she approaches it, to "mak[e] a Comparison between the Valley in which it was placed (with the Amphitheatrical View of the Hills around it) and the Valley of Tempe," which she believes to be in Turkey. Her uncle points out that in that country "the People are all Pagans, they say, and worship the Devil," and anyway Tempe is in Greece. The dispute as to its location is never resolved, leaving a reader to wonder whether geography is not just a state of mind but the representation of one. Be the Valley of Tempe in the devil's own territory or no, Arabella makes it clear that she would have no objection to being "forcibly carried thither"—a possibility that she regards as "not unlikely," given the "Similitude of Destiny" she believes herself to share with the "many illustrious Ladies" who have been similarly spirited away (*Female Quixote*, 259).

No less than her belief in and through "Similitudes," Arabella's Glanvillean thinking in terms of what is "not unlikely" helps to explain why she might be very much at home in a valley of devil worshippers. Indeed, Bath may justly be regarded as one. If reality is, as Joseph Glanvill insisted, the impression that rises from a temporarily coherent set of propositions—and if one grants that Bath is projected, not referenced, by Lennox's literary language—then Lennox's Bath could easily be the work of the Devil too. Since Defoe at least, the old gentleman had marked the intersection of two media—air and writing—and the possibility that the one may be transferred into the other. In Defoe's *Political History of the Devil*, he is "the Prince of the Air, having, at least, the whole Atmosphere to range in." Defoe's sociable history makes that "Atmosphere" inseparable from its own graphic body: it begins by teasing "reading friends" whom it aims to "amuse" with the likelihood that they will "make a pause, perhaps as they do at a witch's prayer, and be some time a-resolving, whether they had best look into it or no, lest they should really raise the *Devil* by reading his story" (Defoe, *Political History*, 35, 67). And the "story" itself is a tongue-in-cheek explication of a number of literary stories that feature this "*Tyrant of the Air*" (Defoe, *Political History*, 35). The devil is thus a witty instrument for Defoe's Glanvillean narrator to test the realizing potential of his own voice: "Seeing we are not to expect he will write his own History," the supposed historian thus concedes, "I shall see if I cannot write it for him" (Defoe, *Political History*, 45).

Doesn't Arabella herself subscribe to this order of demonic authorship? "I shall not write it," she declares to Miss Glanville when the subject of her personal "History" comes up, "tho,' questionless, it will be written after my Death." Arabella does compromise with her destiny sufficiently to appoint the luckless maid, Lucy, her biographer. Some "curious Instructions for relating" Arabella's "History" ensue, though perhaps the least curious of them is Arabella's stipulation that this history will include all of her "Thoughts, however instantaneous." The registry of these immediate "Thoughts" (whose quality is later confirmed in the Divine's sense that Arabella's "Imaginations are too quick for Language") would seem to be at odds with Arabella's anxiety that Lucy also "number all my Smiles, Half-smiles, Blushes, Turnings-pale, Glances, Pauses, Full-stops, Interruptions; the Rise and Falling of my Voice; every Motion of my Eyes; and every Gesture which I have used for these Ten Years past" (*Female Quixote*, 121–22). Arabella's haptic catalogue accumulates abstraction with each successive item until it is apparent that the finished product will describe nothing and "number" everything. It will not exist except as the multitude of virtual impressions whose unifying form Arabella outlines here. It will be damned pure writing.

Arabella's demonic authorship metonymically figures the fictional source of Lennox's own cold "Narration." One reason the narration is cold is that Lennox's narrator makes virtually no appearance in it; indeed, there is no more purely third person to be found in the fiction of the mid-eighteenth century. When a token "We" appears in Lennox's final paragraph, it distracts attention from the double wedding that this paragraph supposedly references to the manner in which it is (or is not) possible to "expres[s] this Circumstance." This is an incredibly cold conclusion. The author of Arabella's "History," whoever that might be, proves at least as elusive as she is, detectable only in expressions that themselves remain invisible, and condensing only provisionally in the mixed states of mind to which "the Adventures of Arabella" give rise.

The Cure of *Arabella*'s Mind

Just what, then, *are* "the Adventures of Arabella"? In his study *On Human Conduct*, the political and social scientist Michael Oakeshott distinguishes between natural and social knowledge: the latter, he suggests, is unique in its awareness of the conditions, circumstances, and precipitants of action. Action consciously so conditioned, circumstanced, and precipitated is behavior. Lennox's novel is social science to the extent that it analyzes the conditions of conduct, first making them visible as fictions. In ordinary society, Oakeshott proposes, such conduct takes two forms. In one, actions are "adventures to procure wished or substantive satisfactions in the responses they receive"; in the other, they are "exploits in self-enactment."[33]

It is significant that this distinction does not apply to Arabella, whose "Adventures" are also exploits in self-enactment. But did they even apply to the mid-eighteenth-century social world, urban and novel, in which Lennox and her novel took shape? Here social stratifications were increasingly realized in formal behaviors. The "airs" that individuals assumed formally marked them as members of groups; and as Richard Sennett proposes, such marking took place in and at the moment an impression was made, thereby detaching individuals from histories and subjecting them to the common air of contemporary London. The challenge embedded in this shared yet insular adventure, Sennett writes, is one of "how to arouse belief in one's appearance among a milieu of strangers"; in such a "milieu, the people who witness one's actions, declarations, and professions usually have no knowledge of one's history, and no experience of similar actions . . . in one's past." Sennett here offers an account of how an assumed air becomes a milieu and of how managed impressions become a context.[34] "The Adventures of Arabella" reflect this context back to those who perhaps less created than were it.

Those adventures end with the "Cure of *Arabella*'s Mind." How we see Arabella's "recover[y] to the free use of all her noble Powers of Reason" (*Female Quixote*, 382) is supposed to reflect—perhaps even decide—our interpretation of Lennox's novel as a whole. This interpretation will in turn decide (or at least reflect) the novel's place in a broader critical geography: as pro-romance or anti-romance, patriarchal or proto-feminist, pro-fiction or anti-fiction, for one kind of fiction and against another. All these critical frames assume that Arabella's cure *is* a cure; the only question is the degree of irony with which Lennox treats it. But "cure" in the sense of healing, even of temporal shift from one state to another, may not be what is meant at all: the word "cure" appears in relation to "the Divine, who had the Cure of *Arabella*'s Mind," and in the environment of this phrase the word "Cure" means merely sphere of influence. A Divine's cure is nothing but his territory, and this particular Divine's cure is not *of* but *is* whatever is meant by the expression "*Arabella*'s Mind." Here that mind's celebrated "Cure" has nothing to do with recovery from illness, mental or otherwise; as our seventeenth-century divine, Joseph Glanvill, had grasped, it is all about the position and disposition of a mind shaped by, in, and through the airy and ultimately social forms of writing. Whatever else is true, the other side of "Cure" guarantees that Arabella's "History" will end not with a bang but a pun.

This is no doubt what cold Arabella would have wanted. Yet what does it mean for Lennox to have left us in a suspense so unsettling that few of Lennox's readers have let themselves realize that they are even in it? One who does is Gallagher, whose acute reading of the warm and specific materiality of the body eliminated in the "marketplace" of modern fiction sets her against a rising theory of fiction as "nobody's story" even as she explains its power. It is worth noting, however, that Arabella's arrival at "Reason" and even common sense—and so her apotheosis into a social and novelistic nobody—does not happen here for the first time: Mr. Glanville has "fancied [it] to himself" before, and with "the most ravishing Delight" (*Female Quixote*, 382). Likewise, when Arabella, supposedly cured, "expatiate[s] upon the Follies her vitiated Judgment had led her into," those follies continue to present as an environment in which she immerses herself (*Female Quixote*, 383). Lennox thus frames Arabella's "Cure," if cure it even be, less as a transition from one state to another—insular to sociable, imaginary to sensible, flighty to grounded—than as a making visible of the no less delusional conditions of representation. *The Female Quixote* offers social critique, even social science, insofar as these conditions apply to the social presentation of persons as much as they do to the representations of literary romance.

Here Lennox seems to anticipate the theory of the modern German sociologist Georg Simmel. In contrast to the fact-based or historical and institutionally oriented sociology of his contemporaries Max Weber and Emile Durkheim, Simmel's sociology is singular in its commitment to the aesthetic and formal dimensions of social life—dimensions that he captured in the same idiom of air once cultivated by such enlightened theorists of manners as Shaftesbury and Hume. Simmel's modern urban persons famously travel inside closed spheres that are meant to keep others at a distance, and in "Sociability," his most pertinent essay, he posited that "within this constellation called society, or out of it, there develops a special sociological structure corresponding to those of art and play"—the "frictionless" and "suspended" world of pure form commonly known as "sociability" (Simmel, "Sociability," 128).

Upheld in Simmel's view by the coquette—thus by the Belindas, not the Arabellas of the world—this was the sphere of sociability that Pope had limned in *The Rape of the Lock*. That poem, of course, ends with Arabella Fermor's curl—having passed through the solvent of literary fiction—idealized into a "constellation" of Pope's own making. We have also seen that Pope at once embodies and disembodies the "play form of association" in his sylphs, at the same time linking it to both natural knowledge and the experience of art. And Pope's sociability, like Simmel's, is threatened by any shadow of individual desire, the earthly lover lurking at the heart. Sharing Pope's idiom of air, Lennox takes this form of sociability a step further by containing it in her own Arabella's mind. It is thus only in the "Cure" of Arabella's *un*cured mind that the world of empty social forms, of fictions, can actually be inspected. This involves less the refutation of Glanvillean skepticism than its rediscovery in the forms of fiction that Arabella negotiates with her own Divine. Once Arabella, "to all Appearances dead," has been "wafted . . . over in a few Moments to the other Side" (*Female Quixote*, 363) and put in a "warm Bed" for colloquy with "the Divine," who is to say that she is even there? As David Marshall implies, Arabella seems rather to have been folded back into the ghostly hall of mirrors already inhabited by her romance heroines (and by women of the "real" world). Their desire to be imitated she herself had copied when, "possess'd with [a] Belief," she jumped into the Thames (Marshall, *Frame of Art*, 168). When Arabella is dispossessed of that belief, her debate with the Divine takes on the air of an exorcism. This air is only enhanced by something that Gallagher emphasizes: their conversation pivots not on superficial questions such as that of whether romances actually correspond to reality, or even on that of their inherent absurdity, but rather on questions of moral value (Gallagher, *Nobody's Story*, 193). In a word, the Divine says they are "evil." While Gallagher rightly links these moral categories to *ideas* about what fiction is and does, "Wickedness" and "evil" also attach to *experiences*, and especially to social experiences. In this case, the experience is one of "Fright." Arabella's imperviousness to the "Evil" in her books seems to demand that the doctor "harras[s] [her] with innumerable Terrors and Perplexities" until her own "Fears" are directed toward a proper object (*Female Quixote*, 371).

For his part, the Divine brooks only the kind of "Fright" that is based on the kind of linear prognostics we find in assessments of the weather, as when a sailor sees "Clouds rise" and "experience bids him expect a Storm" (*Female Quixote*, 372). But what is special about that kind of fright? How different are these clouds from the cloudy "Phenomena" of Arabella's veil? Both illuminate the horizon where experience meets conception—a horizon that turns out to be marked and maintained by social convention. Just so, much of Arabella's debate with the Divine is "really" a dispute about geography. They quarrel, specifically, over the accuracy of the authors of romance. Those "French Wits of the last century," the Divine charges, "disfigure the whole Appearance of the World and represent every Thing in a Form different from that which Experience has shown" (*Female Quixote*, 375, 379). For her part, Arabella maintains that, though the castles that romances figure may have disappeared, "Lakes and Caverns, and Desarts [*sic*], must always remain" (*Female Quixote*, 374).

It may be possible to reconcile the "Form" that Arabella values with the "Experience" her Divine does. At the point of their intersection Lennox forges a fictional geography that is valuable not as a proposition about a world external to itself but as a reflexive environment in its own right. Indeed, this sense of romantic habitat echoes Joseph Glanvill's famous surmise that, "as things are at present, the LAND OF SPIRITS is a kinde of AMERICA, and not well discover'd *Region*; yea, it stands in the Map of humane Science like *unknown Tracts*, fill'd up with *Mountains*, *Seas*, and Monsters: for we meet with little in the *Immaterial Hemisphere, but Doubts, Uncertainties, and Fables*" (Glanvill, *Saducismus Triumphatus*, 9:115). Glanvill's words ground "the LAND OF SPIRITS" in a mode of expression that is also a frame of conception, one he calls a "Map of humane Science." Glanvill's "Map," however, is anything but an abstraction. Even if purely figural, it is nonetheless what we think through and with, and in turn it makes experience of Glanvill's "LAND OF SPIRITS" at least possible. For her part, Arabella contrasts the "*Immaterial Hemisphere*" of romance with "*Fables*," albeit only those of Aesop. Those are distinguished—on the basis of the rule that in Aesop's fables "the Absurdity discovers itself"—from romantic "Tales which are told with the solemn Air of historical Truth" (Female Quixote, 377). In contrasting these two aspects of fiction, one obscuring and the other elucidating, Arabella also reconciles them. Neither is real. Both are true.

As it yields social knowledge and dictates conduct, Arabella's debate with the Divine is itself a fiction in this mixed sense, at once granulated and diffuse. A malleable chorography thus arises from, and through, it. The Divine rebukes romance writers for building ethereal landscapes of desire, charging that "they have distributed Mountains and Desarts [*sic*], Gulphs and Rocks, wherever they wanted them, and whenever the Course of their Story required an Expedient, raised a gloomy Forest, or overflowed the Regions with a rapid Stream" (*Female Quixote*, 378). But the Divine cannot do without streams altogether; he disparages "the Brave and the Coward, the Sprightly and the Dull" who "suffer themselves to be carried alike down the Stream of Custom", and it is Arabella who takes the part of the literalist, pointing out that in her case a woman did in fact "fl[y] from a Ravisher into a rapid Stream" (*Female Quixote*, 379). The streams that feed Arabella's conversation with her "Divine" are shared because they are no more or less literal ("[the] rapid Stream") than they are figural ("the Stream of Custom"). Their suspense between these two conditions briefly allows Lennox's "Streams" to present her literary fiction itself as a layered, and thus shareable, experience—as common nonsense, a potential basis of community.

In Glanvill's view, nature itself moves "by *subtile streams* and *aporrheas* of minute *particles*" even as a witch's "*vigorous imagination*" can transmit "*pestilential spirits*" so that "*infection* should pass in certain *tenuious* [*sic*] *streams* through the Air from one house to another" (Glanvill, *Saducismus Triumphatus*, 9:23, 24). Inseparable from communication, ambient "streams" evaporate into possible worlds as well as sensible ones. The common foundation of both turns out to lie not in putatively unmediated sense experience but rather in the sort of agreement about how things *might* mean, given the mediated nature—and so the inherent lightness—of all of them, that Arabella reaches with her Divine. And because Arabella and her Divine are finally constituted out of elements of the same lexical stream, it is natural that in the end their positions should reverse. The female Quixote generously professes to "wonder how the Blaze of Enthusiastic Bravery" could have blinded her to her own crimes" (*Female Quixote*, 381). The Divine politely withdraws so that she can "compose her Spirits" (*Female Quixote*, 382) and is not heard from again. The agnostic figures of speech that buoy their common separation limns an ideal of insular sociability that Simmel would capture in another tongue and time—"that simultaneous sublimation and dilution, in which the heavily freighted forces of reality are felt only as from a distance, their weight fleeting in a charm" (Simmel, "Sociability," 39).

8

In Factitious Airs: Radcliffe's Priestley

The circumstances in which any kind of air is decomposed tend to discover its composition.
JOSEPH PRIESTLEY, *Experiments and Observations on Different Kinds of Air* (1774)

Keats called their author "Mother," thereby securing Ann Radcliffe's wildly popular romances of the 1790s a place at the start of many things, most famously the widespread literary practice of gothic fiction.[1] And a key gothic signature is atmosphere. This, appropriately, can mean more than one thing: in Deidre Lynch's illuminating view, mist crept onto the gothic page so as to mark a break with the Enlightenment's presumptions of transparency.[2] From another point of view, however, gothic fiction is atmospheric because of the weather it obsessively specifies: lightning strikes, gales shriek, and rain picks up as in no other novelistic subgenre. Anticipating Mark Twain's quip that everyone talks about the weather but no one does anything about it, Defoe and Fielding had invoked the elements to limn a determining horizon implicitly at odds with "the" rising novel's since-vaunted ideology of autonomy and inwardness. Later writers from Radcliffe and Matthew Lewis to Charlotte Smith and Clara Reeve brought this horizon forward, and the unfailingly exceptional weather in their novels made its presence apparent.

Yet there is a more immediate way of describing the rise of atmosphere in the gothic fiction of the 1790s. Since Radcliffe wrote, that is, we always look for the words that compose such fiction to chill the blood, tingle the spine, and make the "hair stan[d] on end the whole time"—the electrical effect of Radcliffe's most acclaimed fiction, *The Mysteries of Udolpho* (1794), on Jane Austen's susceptible Catherine Morland.[3] Such formulations posit the kind of totalizing aesthetic environment that seems possible only in the wake of the romantic revolution.[4] But scrutinized *as* formulations, they also bind gothic atmosphere to the rise of the literal into the figural that underwrites such turns of phrase and renders them communal.[5] At the same time, and via these same expressions, a plainly artificial composition—the marked page—seems to give rise to an undifferentiated "sphere" of effects. This sphere strikes us as unified and immediate even as we can plainly see that its lexical precipitants are particulated and factitious. It is only for the moment anyway—which is to say for "the whole time" we are reading—that we willingly believe it to be otherwise.

To understand Radcliffe's gothic atmosphere as a shared sphere of habitable beliefs—potentially enlightening precisely because it is so visibly the effect of mediating forms and intimate in its very drift toward abstraction—is to remember that "Mother" Radcliffe was a daughter too. And her forebears are not just Fielding, Lennox, and Defoe, but pneumatic analysts, such as Boyle and Arbuthnot, who were committed to making "the Effects of Air" visible through the notional screen of the written word.[6] Radcliffe came of age just as popular exhibition science began to supplant the page, mediating natural philosophy through show and spectacle for bourgeois and often female consumption. Staples of the circuit were recapitulations of Boyle's air pump experiments. As these found themselves, in turn, remediated in oil on canvas, paintings such as Wright's 1768 *Experiment on a Bird in the Air Pump* managed both to probe and to profit from the interfusion of hedonic and aesthetic experiential modes with scientific knowledge. To like effect, in 1781 the painter and stage manager Philippe de Loutherbourg's celebrated eidophusikon—a cycloramic re-creation of stormy weather—attracted English painters from Gainsborough to Reynolds to the installation's venue inside de Loutherbourg's house.[7]

Three years earlier, an artist and scientist devoted to the study of electricity, Benjamin Wilson, had simulated lightning storms inside a different building—the London Pantheon—the better to display the virtues of round as opposed to pointed lightning rods. Wilson's published account of the proceedings ended with a warning to anyone inclined to restage the lightning rod experiment at home. This was not because Wilson intended his ninety pages to bear any special burden of proof: he fully expected his air study to be restaged and worried only that on such occasions interference from "much smaller circumstances than have been here specified, as the intervention of a single hair, a fibre of down or even a little vapour arising from perspiration, or otherwise, will, where great exactness is required, sometimes prevent the success of an experiment."[8] Wilson's performance space in the Pantheon had been made to exclude other "circumstances" as well. Besides needing to "construct a substitute for a cloud" out of which "real" lightning could be simulated by way of an electrostatic cylinder, for example, Wilson required some sort of structure for said lighting to strike. To this end, he thought it best to "move [a] building indoors," leaving out "windows and doors . . .

because they were immaterial on this occasion" (Wilson, *Account of Experiments*, 5).

Wilson's *Account of Experiments* anticipates the architectural forms and models of inwardness, at once claustrophobia-inducing and "immaterial," that gothic romances such as Radcliffe's would contain (and that, within those romances, would themselves contain gothic plots and characters). Less proleptically, in its preoccupation with technological display over states of air, *Account of Experiments* seems to abstract the English aerographies of the earlier eighteenth century. It splits writing about the air from frequently tendentious experiment with air's effects and returns that kind of experiment to a theater demarcated as such. Still, writing such as Wilson's was hardly "immaterial": many more people read about the experiments made at the Pantheon than saw them. Of these, Radcliffe may well have been one, and in the 1780s, she began reproducing the effects of artificial air in her own writing; by 1794 the word "air" and its variants would appear 197 times in *The Mysteries of Udolpho* alone. Yet the discipline and fascination with which Radcliffe investigated air's properties in the airless space of the page finally identifies her less with stage managers and visual artists such as Wilson, de Loutherbourg, and Wright than with the leading "aerial philosopher"—and perhaps the most compulsive writer—of her generation, Joseph Priestley (1733–1804). More than the influence of exhibition science on her work, Radcliffe's affinity with Priestley is what urges a reconsideration of the kind of language that early gothic writing speaks. This in turn entails a reconsideration of the kind of natural philosophy that Priestley was doing at a transitional moment in the inevitably cultural history of air.

The Blue Tinge of Air

Air is often the first thing a reader sees when looking into any book that Radcliffe wrote. Her most popular romance, *The Mysteries of Udolpho*, opens onto a view of "partial vapours" suffused with "the blue tinge of air"; the last that she published in her lifetime, *The Italian* (1797), ends in the "airy hall" of a villa that appears to have been "called up by enchantment."[9] In retrospect, Radcliffe's ubiquitous air seems to negotiate a transition between the past and then-future of British literature. Chapter epigraphs and inset verses compulsively reproduce the "aerial forms" (*Udolpho*, 84) of dead English poets from Milton and Thomson to Collins and Gray even as they forecast the Aeolian breezes of a Coleridge, a Wordsworth, or a Keats. Likewise, even as the damp and moldering air of Radcliffe's castles and abbeys proves, in Coleridge's phrase, at least as "close" and "daydreamy"—if never as "hot"—as a Richardsonian bedroom, it is still the same air, exhilarating in its freshness, that blows from her mountaintops in the clear direction of Waverley—or of Pemberley for that matter: once airborne, the place names sound very much the same.[10]

Readers of Radcliffe's time naturally drew on the language of air to decipher the spell that her fiction seemed to cast over them. Summarizing her celebrated vistas, Walter Scott remarked that "there is, as it were, a haze over her landscape, softening indeed the whole, . . . but without conveying any absolutely precise or individual image to the reader."[11] As for Radcliffe's other trademark, the so-called explained supernatural, an irked Mary Wollstonecraft likened the novelist's quasi-scientific "mode of accounting in a natural manner for supernatural appearances" to the manufacture of "glistening bubbles, blown up in air, only to evaporate more conspicuously; leaving the aching sight searching after the splendid nothing."[12]

Thanks *for* nothing, Wollstonecraft appeared to say, linking the conspicuousness of Radcliffe's air to its absolute fictionality. With Scott, Wollstonecraft also used air to mark the spot where all that's solid melts into . . . well, into what? We might as well say that it vanishes into the literary medium that conveys it. In this way, Radcliffe's "glistening bubbles" grant her reader intimate knowledge of her own experience *as* a reader. Indeed, this is probably the most common effect of Radcliffe's air: it makes apparent the literary forms that communicate it. For instance, the air that Radcliffe's presumably French and Italian heroines breathe is never English. The five novels that she published in her lifetime all build up in "the air of Languedoc and Provence," of Venice, Florence, Naples, the Apennines, and the Alps. Scotland looms briefly, early on, but avoided altogether is the "thin" air of her native England.[13] And as the "blue tinge of air" suffuses her landscapes, it also encodes the "softer graces" of the non-English-speaking landscape painters, from Nicolas Poussin to Claude Lorrain, whom Radcliffe is supposed to have copied with her pen. At the same time, Radcliffe's air looks toward the abstracted meteorology of her contemporary J. M. W. Turner, the first English painter who appears to have taken air as his sole subject.[14] In both cases, however, Radcliffe's air visibly marks the presence of English lexical mediation. It lets us know that her improbably English-speaking heroines really *aren't* breathing (or speaking with) continental air. Nor, by the way, are we looking at an old French painting: we are reading a modern English novel. Radcliffe's air makes the facts and elements of mediation apparent as a condition of experience.

Another distinction of Radcliffe's air is that it comes freighted with emotion. For this reason, it also, and again distinctively, *attracts* emotion: ambivalence, anxiety, trepidation, melancholy—in essence, the very feelings that it is also charged with drawing in a different sense, which is to say with expressing. These feelings, which Radcliffe's air itself can seem to have, tend to be mixed, albeit not so much with each other as with thought. We should therefore call them moods. And anyone compelled to breathe Radcliffean air is bound to be moody. "Gloomy ideas" lower on her heroines like clouds (*Udolpho*, 49). As this same gloom begins to saturate the living reader, Radcliffe's "atmosphere" at once tenders and depicts that diffuse element of literary experience that we know to be partly an ideation—that we know, in fact, may not be there at all—but that holds us all the same.

In all these ways, Radcliffe's air graphically exaggerates—almost parodies—the kinds of air we have weighed so far. But as those kinds of air did not, hers snaps shut like a trap. She herself succumbed to spasmodic asthma in 1823, more than a quarter century after *The Italian* was published. Radcliffe's silence in the intervening years has always posed a mystery worthy of Udolpho. Perhaps even she had begun to feel suffocated by the symbolic conventions within which her own romances were compelled to take shape. Nor does breathing come easy to anyone *in* those romances: "Give me air," gasps Emily St Aubert in the bowels of Udolpho (*Udolpho*, 258). She and her Radcliffean sisters—most notably Adeline de Montalt of *The Romance of the Forest* and Ellena di Rosalba of *The Italian*s—are always "breathless" from fear or the want of fresh air, if not from their loquacious servants having sucked all of it out of the room. Even the "spirits" of Radcliffe's minor characters "suffe[r] extremely from want of air" (*Udolpho*, 459).

With air so abundant, why do they struggle? The breathing problems that seem to arise from Radcliffe's aesthetic program might simply register the formal truth that by the time we get to her—or to the end of her anyway—English air was too much with the literature that had lent it something of its modern form. Radcliffe was quite literally sick of it. At the very least, "air" had perhaps done all it could to animate the complex transaction between form and content that had come to distinguish modern fiction in English. What Radcliffe's aeriform romances marked, in this view, is not the sublimation of English air into what we postromantics still mean by "atmosphere" but instead its death. Then again, perhaps these are the same thing. It is gothic writing, after all.

No matter its moral, Radcliffe's "recollection" of more than a century of airs may be seen as an analytic process necessarily conducted from within the very medium that her fiction places under observation.[15] And the possibility of just such experiment and observation uncovers a vital and regenerative power in that medium—a kind of oxygen detectable not at the level of what

Radcliffe's writing says, or even strictly at the level of how it says it at any specific juncture, but rather at the very point where that writing at first seems least original, even dead. This is one reason that Radcliffe's fiction asks to be read alongside Priestley's. As a pneumatic chemist, Priestley was notable not just for what he discovered about the air but for what he didn't, quite, for all that he exhaustively tracked its character and effects. That would be oxygen, the as-yet-unrevealed secret of the candle's flame, the blood's efficacious crimson, the last-minute reanimation of the half-suffocated form. Knowledge of oxygen promises to demystify the air for good: in a chemical anticipation of Radcliffe's own supernatural *expliqué*, it is to bring air as literate eighteenth-century Britons had known (and un-known) it to its end. Priestley's finally ambivalent relationship to this project seals his affinity with Radcliffe, even as it helps to account for the rapidly institutionalizing differences between their kinds of writing.

Unsettling

Priestley was a rabble-rousing Unitarian minister in an age of revolution. His books, in Samuel Johnson's view, "unsettled every thing and yet settled nothing."[16] Despite the conservatism that literary critics have often deduced from Radcliffe's fiction, her husband, the journalist William Radcliffe, was of Priestley's liberal political stripe, and Radcliffe herself may well have met him during girlhood visits to the home of another dissenting intellectual, her uncle and Priestley's friend, Thomas Bentley. In his recent biography of Radcliffe, Rictor Norton speculates that thence she may even have absorbed Priestley's heterogeneous and democratic conception of the sublime long before she encountered the version of it that we find in Edmund Burke's *Philosophical Enquiry* (1757), and that is usually assumed to have shaped her celebrated landscape aesthetic.[17]

What matters here, however, is that both Priestley and Radcliffe were Britons who wrote about the air. Of course, Radcliffe used the word "air" without once referring to anything that ever was there. Priestley's air, by contrast, certainly appears to have been the real thing. His *Experiments and Observations on Different Kinds of Air* (1774) is a landmark text in the history of British research into the air. In effect, *Experiments and Observations* closed the chapter of that history that begins with Boyle. With Boyle (and Radcliffe), Priestley was also a prolific person of letters. *Experiments and Observations* eventually took up four volumes, Priestley's *History and Present State of Electricity* (1767) three. There were also books on history, perspective drawing, politics, and theology. And Priestley was a professor of literature and language at dissenting Warrington Academy; he taught there for more than a decade, producing hefty studies of grammar, rhetoric, and language while he was at it. Priestley's "different kinds" of inquiry—into language and into air—were, moreover, intimately linked. His *Rudiments of English Grammar* (1761) even held that a "*Grammar* may be compared to a treatise of *Natural Philosophy*; the one consisting of observations on the various changes, combinations, and mutual affections of words; and the other on those of the parts of nature."[18] As for *Experiments and Observations*, its great contribution to the systematic study of "nature" was the parsing of the "common air" we breathe into the "parts" that compose it.

These "parts" are what we now call gases, and they include both carbon dioxide and oxygen, as well as twelve others (from ammonia to nitrogen) that Priestley's writing opened to view. Priestley himself, we should note, chose not to call them gases, though the word had been available ever since the thickly accented Belgian chemist Jan Baptista van Helmont had speculated about "chaos"—a word corrupted in its passage through the air so that his English-speaking listeners heard it as "gas."[19] For his part, Priestley does not seem to have heard it at all; in his book, gases are "factitious airs," so called to advertise the fact that they had to be made (*facere*), often by artificial means, in order to be known. As he manufactured his "factitious" airs by means of siphons and pneumatic vats, sequestered them, and then observed, manipulated, and traced their behavior, Priestley lent his shoulder to shift British scientific understanding of "common air" away from Boyle's "physico-mechanical" model of "air" as a singular object of quantitative analysis toward a sensed theater of ambient "airs." In this project he was aided by the volatile intellectual atmosphere of his own times—by what, as his recent biographer Steven Johnson puts it, was simply "in the air" that had not been there a century before.[20]

At the same time, Priestley's work opened up the possibility that the "factitious airs" that compose common air might best be known not via elimination in the vacuum of Boyle's celebrated air pump but rather through internalization and immersion in them. *Experiments and Observations* may thus be said to have given rise to everything from the "balloonomania" of the 1780s to Thomas Beddoes's founding of Bristol's Pneumatic Medical Institution in 1799. In conversation with Priestley, his American friend Benjamin Franklin's pursued his own adventures with kite and key, thereby establishing that electricity is ambient and ambiance electric.[21] "Experiments" such as Benjamin Wilson's, designed from within elite science to rebuke many of the conclusions that Franklin drew, soon followed.

The reflexive, volatile, heterogeneous, and all-encompassing atmosphere that Priestley opened to view escaped into the liveliest sociopolitical metaphors of his era, even as those metaphors saturated it.[22] Priestley's preface to *Experiments and Observations* notoriously declared that "the English hierarchy (if there be anything unsound in its constitution) has equal reason to tremble even at an air pump, or an electrical machine."[23] Defenders of that hierarchy borrowed the terms that arose from Priestley's pneumatic science to figure a perceived threat to what Priestley himself, genially playing into the game, was wont to depict as a Gothic "system which had stood for ages" now "vanish[ing] like an enchanted castle in a romance"[24] "The wild *gas*, the fixed air is plainly broke loose," an alarmed Burke wrote in 1790 of the "Revolution in France." William Pitt also marked "the effervescence of the public mind" at that time, though Burke, naturally, was the one to recommend that we "suspend our judgment until the effervescency is a little subsided, till the liquor is clearer, and until we see something deeper than the agitation of the troubled and frothy surface."[25]

The trouble with Priestley was precisely his commitment to troubled and frothy surfaces. *Experiments and Observations* preceded the American Revolution by one year and the French by fifteen, and there are numerous analogies between Priestley's chemistry and the way those unsettling events had unfolded—analogies reinforced by his own liberal, unto radical, politics. *Experiments and Observations* opens, innocently enough, with the author strolling past a Leeds brewery and noticing the bubbles that seethe in its vats. Intrigued, he acquires an air pump and proceeds to induce effervescence artificially, thereby testing the character of the "fixed air"—carbon dioxide—that turns out to be one potential constituent of "common air." Though he was the first to simulate it in a wide range of media—most notably water—"fixed air" itself was not Priestley's discovery: the Middlesex clergyman Stephen Hales had long since derived it from roasted cucumbers, and a chemically minded Scot, Joseph Black, had paid close attention to it too.[26] One of Priestley's friends, the eccentric Henry Cavendish, conducted his own inquiries. But in contrast to all of these players in the so-called chemical revolution, what Priestley most wanted was simply to find out what fixed air is like. How did it behave, interact with other entities, appear, disappear, and respond under pressure? Priestley was, so to speak, interested in the "air" of air. So he sought no calmer, clearer, or "deeper" substance underpinning fixed air's effect of "effervescency." Instead, he assumed—and, to the mind's eye, displayed—the integrity of effervescency, hence of "fixed air" itself.

Generally speaking, air was anything but fixed in Priestley's revolutionary chemistry. His work on air "unsettled every thing, and settled nothing" because it presented air as something far other than the elastic but unitary substance that Boyle had assumed and on which he had settled the physico-mechanical properties of spring and weight. Priestley instead approached "common air" as a lively composition of nonhierarchical and equally "factitious airs." "There are," he declared in the second volume of *Experiments and Observations*, "very few maxims in philosophy that have laid firmer hold on the mind, than that air, meaning atmospherical air (free from various foreign matters, which were always supposed to be dissolved, and intermixed with it) is a simple elementary substance, indestructible and inalterable." Revolting against this maxim, he pronounced himself "satisfied that atmospherical air is not an inalterable thing" (*Experiments*, 2:104), even in its purest forms. In a suite of narrative pieces that "immerged" the common reader in his own strangely solitary encounters with such pneumatic effects as inflammation, effervescence, putrescence, and evaporation, Priestley's book established for good that what we call "atmospherical" or "common air" is nothing but a composition—and a perpetually unsettled one at that.

More important, Priestley took this composition to *be* a good, exulting outright in the transient appearances and antics of the "factitious airs." His method was to "put the substances which [one] examine[s] into various new situations, and observe the result of circumstances" (*Experiments*, 1:viii). As they were plunged into new predicaments (or at least new "phials"), each of Priestley's "factitious airs" proved, in its purest form, fatal to animal life. Once that was over, most of them naturally sped the decomposition of organic bodies into the "airs" that in turn composed them. What this meant, if not "chaos," was that there was no longer any such thing as "the" air, even as it also appeared, now, that all air is composed of the same elements, in differing combinations.

What communicative form could begin to sustain, or even convey, such paradoxes? Priestley's account of how and why he first became engaged with "effervescency" hints at one. "It was in consequence of living for some time in the neighbourhood of a public brewery, a little after Midsummer in 1767," he wrote at the outset of *Experiments and Observations*, "that I was induced to make experiments on fixed air, of which there is always a large body, ready formed, on the surface of the fermenting liquor" (*Experiments*, 1:43). In these words, it is hard not to hear echoes of Horace Walpole's *The Castle of Otranto* (1764). Walpole's then-recent "romance" was structured entirely around the parlous state of a very "large body" whose reassembly into ready form triggers the collapse of an entire castle, if not patriarchy itself. Priestley's words, in turn, work within the premises of a "large body, ready formed" to unsettle it. The key ones are "ready formed" and "surface"—terms that even without Walpole's example would identify scientific "progress" with what Priestley's contemporary Clara Reeve termed "the progress of romance."[27] That would be the apparent progress that occurs only within fixed boundaries and only by decomposing and rearranging what was there to begin with.

In his intuitive preference for any communicative form that might "gain readiest admission to the mind," Priestley was already on record as preferring the "writer of romances" to the "faithfull historian." The latter, he declared in his *Course of Lectures on Oratory and Criticism*, can engage only "the disposition" of events. The former can at least hope to gain "some access to the springs of the human passions."[28] Notwithstanding an early "aversion to plays and romances, so that I never read any works of this kind except *Robinson Crusoe*," Priestley grew up to become a professed admirer of "*Clarissa*, George Barnwell, Eloise, and any other well-contrived fictions," making no important distinction among them. But in order to provide "access" to otherwise hidden "springs," fictions did have to make their own fictionality—their formal "disposition," as opposed to the spontaneous unfolding of historical "events"—apparent. "The thought of its being a fiction enables us to make but a feeble and ineffectual effort to repress our feelings," Priestley observed, and those who "have acquired an aversion to all works of fiction" are incapable of the "unprejudiced attention which this experiment requires."[29]

Thus Priestley the literary theorist. But what Priestley the pneumatic chemist was asking his reader to do was to "experiment" with him imaginatively on the "factitious airs" themselves and thereby virtually to enter the contingent and composite forms that they assumed at any given moment and in any given place. Priestley's open, friendly style guaranteed just about any English reader admission to the world that these forms composed. All that reader needed to be able to do was accept the ever-shifting alliances of differing airs as the precondition of what we experience as the surrounding uniformity of "common air."

Experiments and Observations entertains the possibility that engagement with the fixed forms of writing themselves can complement and extend experiment, restaging it in the curved notional space that seems to arise from them. Priestley offers a vivid, eventful, and often extremely beautiful depiction of the "factitious airs," one that does nothing if not exercise the "unprejudiced attention." In one experiment, "some delicate flowers" are "suspend[ed] within the region of fixed air over . . . fermenting vats" (*Experiments*, 1:285), and "contact" with two different "kinds of air" gives rise to a "beautiful white cloud" (*Experiments*, 2:315). The delight of the spectator lapses into that of the taster and breather. "Drawing [pure oxygen] through a glass siphon," Priestley reported, "I fancied my head felt peculiarly easy and light for some time afterward" (*Experiments*, 2:162). Fizzy water was so delicious that he thought he might go ahead and market it.[30] Of his fluor acid experiments, Priestley confesses that "few philosophic experiments exhibit a more pleasing appearance than this" (*Experiments*, 1:343); later, watching bubbles form, rise, and froth in a basin of water, he declares that he has "met with few persons who are soon weary of looking at it." Indeed, "some could sit by it a whole hour and be agreeably amused all the time" (*Experiments*, 2:345). Having crystallized "nitrous air" in a vial, the chemist returns to find the container "almost filled with the most beautiful crystallizations imaginable," and his pen renders the feathery form they take in the most delicate of strokes: "A more beautiful appearance can hardly be imagined," he concludes, "and I am afraid I shall never see the like again" (*Experiments*, 3:154).

Euphoric experiments such as these discovered openings within a closed system. But they did not necessarily liberate the investigator. To the contrary, they could also give rise to unanticipated effects: fear, as here, or even a heightened moral sense. When his nitrous feathers "see[m] to decay," reduced to "one compact mass, without anything of the beautiful appearance they had made before" (*Experiments*, 3:159), Priestley hurries to "repair the injury." Ever on the lookout for an air of "superior goodness," he is fascinated by the fact that, though oxygen—"vital air"—is the reason whatever it is we breathe gives life, we cannot thrive unless that air is adulterated. "We might, as may be said, live out too fast and the animal powers be too soon exhausted in this pure kind of air," Priestley mused. "A moralist, at least, may say, that the air which Nature has provided for us is as good as we deserve" (*Experiments*, 2:169). Though he could not pretend to be this "moralist," he could "at least" take on the air of one, nurtured by the freedom created by what "may be said" and by what a certain kind of philosopher "may say." Such fictional equivocations allow the language of air to synthesize immaterial and material forms of reality, revealing passageways between them that would not have been evident otherwise. So Priestley discovered that the conceits of good and bad air could even be drafted into the service of historical fantasy. Whether air "will be in future time, better or worse than it is at present, is a curious speculation," he conceded, "but I have no theory to enable me to throw any light upon it" (*Experiments*, 2:170).

Experiments and Observations determinedly avoided "theory," even hypothesis, in order to simulate the experience of air as a common medium that consisted in endlessly recombinant forms. Despite chemistry's subsequent—if ambivalent—claims to it, this wonderful book thus might be best classified as a set of unclassifiable "speculations." These are deliberately composed in and through the very aeriform medium that gave rise to them. Indeed, with respect to "the aerial form of substances," Priestley—unfixing everything indeed—found that "there is no substance in Nature but what is capable of assuming that form" (*Experiments*, 2:403). Why couldn't the matter of his own writing count as one such substance? *Experiments and Observations* certainly is reflexively conducted from within the arbitrary system of difference that is language. It discriminates among the "different kinds of air" only in order to suggest the contingency and transience of the differences that we think we see among the larger forms that they compose—between, say, a body and its environment; between different bodies; between different environments; between a body living and a body dead.

Consequently, what Priestley called his "doctrine" of the air (*Experiments*, 2:44) could not have been less doctrinaire. For all that it's been implicated in eighteenth-century chemistry's drift toward ever more rigid classification, his work headed in exactly the opposite direction precisely because of its commitment to the "doctrine" of air as one of freely and infinitely recombinant form.[31] Embrace of this "doctrine," in turn, owes less to single-minded belief than it does to a state of mind that anticipates the "willing suspension of disbelief" that Priestley's friend and admirer Coleridge found to be the primary act of the poetic imagination. As a chemist, Priestley was, of course, inclined to take "suspension" literally, even as he consciously practiced it in language. This is to say that he insisted on "using the term air as expressive of the mere form in which a substance is exhibited, without any consideration of the elements in which it consists" (*Experiments*, 1:10).[32] Because Priestley treated the word "air" as a purely formal term—which is to say as a term that looked toward compound forms rather than toward any necessary substance—it was potentially knowable through the forms of language in a way that it had not been before. Even in a branch of natural philosophy unusually attuned to verbal nuance, Priestley's (English) language-based experiments and observations gave adjectives and prepositions—typically regarded as the most incidental, transient, and secondary parts of speech—unprecedented weight, spring, and authority.

Cases in point are the names of Priestley's many airs. In the manner of the rounded characters then beginning to turn up in the English novel, each possessed a first and a last name, though the last—"air"—was always and everywhere the same. So in *Experiments and Observations* we meet "mephitic air," "fixed air," "vital air," "dephlogisticated air," and "nitreous air." Because they were all airs, the "different kinds of air" enacted otherwise mutually exclusive principles of diversity, universality, and commonality. As they hovered somewhere between adjectives and nouns, place markers and thing markers, their very names kept them—both conceptually and in terms of how they were used socially—up *in* the air. Likewise, the verbs that Priestley selected to describe the behaviors of his airs tended not to be objective, abstract, or determinate ones such as "compound" or "combine." They were instead relatively subjective and often explicitly figurative. So Priestley's airs were "imbibed," or themselves "impregnated"; they became "depraved," got "injured," or became "infected"; they were "diminished" and "vitiated" (*Experiments*, 2:229, 247, 252). They even proved their "goodness" (*Experiments*, 1:28).

And yet the author and disposer of these suspended airs pronounced himself to have "been under a necessity" when he so disposed them. "No person was ever more temperate, or more cautious, than I have been in the introduction of new terms," Priestley claimed, and "it was with great hesitation, though compelled by necessity, that I did it at all, . . . and always adopting such as were analogous to others in established use" (*Experiments*, 1:9). Here we are once more brought to mind of the "ready formed." The linked freedoms of conception and expression prove contingent on what inhibits them—repetition, "hesitation," and the pressure of "necessity." But the marks of constraint that they bear also guarantee that Priestley's airs, whenever they are invoked, will reveal the historical, social, physical, and conceptual conditions under which they are known and collectively acknowledged, and therefore believed, to exist.

Excuses have been made for Priestley's peculiar vocabulary.[33] One is simply his resistance to hypothesis and cautious preference for empirical description over theory.[34] In fact, he appears to have aimed to keep air up "in the air." He thus made air the conceptual—as it is the real—estate of everyone. "Those who chuse to apply air to a *substance* and not to a *form* are certainly at full liberty to do so if they please," Priestley therefore assured his reader, "and provided we understand one another, no inconvenience will result from our use of a different language" (*Experiments*, 1:10). As for himself, his own freedom lay, if paradoxically, in the fact that "the language that I adapt . . . applies no attachment to any hypothesis whatsoever, and may still be used though I should change my opinion on the subject" (*Experiments*, 1:10).

Priestley's suspenseful and democratic lexicon of the air(s) also shows his alienation from natural philosophy's discursive conventions as these were consolidating both in England and in Europe at large. In a meticulous analysis of the chemical revolution in eighteenth-century France, Wilda Anderson identifies the ascent of Priestley's competitor Antoine-Laurent de Lavoisier with the "reform of chemical nomenclature." In Anderson's view, that reform took place in the context of an array of "rhetorical and literary maneuvers" that had been orchestrated by earlier chemists such as Pierre Joseph Macquer and Étienne-François Geoffroy. As early as 1717, Geoffroy's *Table des Rapports* had begun to map out chemistry's conceptual field in graphic space.[35] Priestley's approach to air through writing appears to have aligned him with the French chemical community before Lavoisier; in terms of the English chemical establishment, however, Priestley grew up in Yorkshire of all places, received no elite education to speak of, taught in a dissenting academy, very far from either Oxford or Cambridge, and pursued his experiments off center, in Birmingham and Bath. After his sympathies with the French and American revolutionaries got his laboratory and library torched in the Birmingham riots of 1791, he took his "phials and retorts" all the way to the banks of the Susquehanna, there (as he predicted in *Experiments and Observations*) "ceasing to breathe when I have just begun to know, what it is I breathe" (*Experiments*, 1:5).

Priestley's nomenclature explicitly and deliberately works within a culturally cohesive set of verbal conventions. In this manner, his sentences limit knowledge of the objective world even as they open knowledge of the symbolic structures in which that knowledge is wrapped and exchanged. However frustrating, this may also be seen as a form of enlightenment in its own right. Samuel Johnson, for example, complained that Priestley "borrowed from those that had been borrowers themselves, and did not know that the mistakes he adopted had been answered by others" (Boswell, *Life of Johnson*, 1386). But this is only to say that Priestley's writing about the air marked, within literary space, the social and conventional dimensions of our knowledge of it. His preferences for terms such as "factitious airs," "fixed air," and the like was a preference for making apparent the man-made, communicative dimension of knowledge—the part of it that is, in fact, the most affined with air itself. In this manner, Priestley's words drew attention to their own work of mediation, and so to their effectual identity with the medium from which they most differ, air itself.[36] Within this context, however, these same words retained the freedom to differ from themselves, just as "common air" necessarily differs from itself, depending for its being and perception on the airs that compose it at any given moment. Priestley's words thus revealed air's composition in and through their own. At the same time, Priestley's naming of the airs was also driven by something more personal: to avoid "the" air not just as an improperly unified concept but also as a material entity. If we are to see how this is so, we need to look more closely at the voice that "speaks" in *Experiments and Observations.*

Horror of Impediment

Of course, we cannot really look at speech. But the pretense that we can mattered very much to Priestley—who, like Boyle, had trouble talking. Indeed, for Priestley, writing itself was an exercise in avoidance. Priestley's memoirs draw attention to the "*impediment in my speech*, which I inherited from my family, and which still attends me. Sometimes I absolutely stammered," he recalls, "and my anxiety about it was the cause of much distress to me" (*Memoirs*, 78). Though he grew up to become a Unitarian minister, the "*impediment in [his] speech*" plagued him "so much as to make preaching very painful." Against it Priestley sought the aid of one Mr. Angier, who had perfected a method "to cure all defects of speech." Whatever this method entailed, from it Priestley "received some temporary benefit; but soon relapsed, and spoke worse than ever" (*Memoirs*, 80). Meanwhile, Priestley's "defect of speech" conspired with "the weakness of my constitution" until, he reported decades later, "I felt occasionally such distress of mind as it is not in my power to describe, and which I still look back upon with horror" (*Memoirs*, 71).

Priestley's indescribably horrifying "defect of speech" spurred an early and deep investment in the written word as a form of regulation and compensation. He found himself helped somewhat by reading slowly aloud and helped very much by writing instead of speaking. Before long, Priestley's pen was automatically transcribing both what he heard *and* what he would otherwise have said. It became his childhood "custom . . . to recollect as much as I could of the sermons I heard, and to commit it to writing. . . . [W]riting at home almost as much as I had heard, I insensibly acquired a habit of composing with great readiness; and from this practice I believe I have derived great advantage through life, composition seldom employing so much time as would be necessary to write in longhand anything I have published" (*Memoirs*, 74). In adulthood, writing remained both a form of material management and an amelioration of troubled utterance, even as it meshed seamlessly with personal interaction. Happy to be "writing on any subject by the parlour fire," Priestley never felt "the necessity of retiring from company in order to compose anything" (*Memoirs*, 91).

Even Priestley's scientific pursuits were as likely to spring from the reflexive social forms of writing as from, say, watching things bubble and seethe. One friend, the poet Anna Barbauld, depicted his laboratory as "a strange mixed olio" of "papers and books" where

> A mass of heterogeneous matter,
> A chaos dark, nor land nor water;
> New books, like new-born infants stand,
> Waiting the printer's clothing hand.[37]

Having mentioned to another friend, Benjamin Franklin, "an idea that had occurred to me of writing a history of discoveries in electricity," Priestley took up his pen "without having the least idea of doing anything more than writing a distinct and methodical account of all that had been done by others." It was only from and through writing that he was "led . . . by degrees into a large field of original experiments."[38] These compose the second half of his 1767 *History and Present State of Electricity*; the first half recapitulates the large body, ready formed, of the writing of other "electricians," thereby at once marking and mediating Priestley's place in the community of ideas that suffused and integrated a far-flung community of persons.

The frisson of seduction, if not compulsion, betrays more than a little discomfiture between the otherwise redemptive lines of literary composition. Committed to "methodical arrangement," for example, Priestley relied on various "mechanical expedients to secure and arrange my thoughts, which have been of greatest use to me in the composition of large and complex works" (*Memoirs*, 89–90). This security system was motivated by a "defect in point of recollection"; despite the high "degree of composure" (*Memoirs*, 158) and "freedom from anxiety" (*Memoirs*, 93) that Priestley claimed to possess, the poverty of his memory was unnerving. His own *Memoirs* remember (though how perfectly would by definition be hard to say) one incident "when I was composing" the preface to what would become his *Harmony of the Gospels* (1777). As usual, Priestley first read several writers and then "digest[ed] the result in the compass of a few paragraphs, which I wrote in shorthand," only to "misla[y] the paper." But "in the space of a fortnight I did the same thing over again, and should never have discovered that I had done it twice, if, after the second paper was transcribed for the press, I had not accidentally found the former, which I viewed with a degree of terror" (*Memoirs*, 89).

This "degree of terror" revives "the horror" with which Priestley had "look[ed] back upon" his speech impediment. In more hopeful and composed moods, Priestley could suppose that while "a defect in point of recollection . . . may be owing to a want of sufficient coherence in the association of ideas formerly impressed," it might also "arise from a mental constitution more favorable to new associations" (*Memoirs*, 96). Still, the shadow of "want," incoherence, and involuntary repetition menaced not just Priestley's "expedients" for composition, with their bright promise of unity and integrity, but even "that degree of composure which," Priestley insisted, "I have enjoyed through life" (*Memoirs*, 88). An air at least facetious clings to his claim that "all seasons have been equal to me" (*Memoirs*, 86), that while "many things must have occurred to mortify and discompose me, nothing has ever depressed my mind beyond a very short period" (*Memoirs*, 87).

So does the ghost of the unspeakable and uncontrollable, the alienated and aggrieved, haunt Priestley's literary compositions. Even the most carefully composed of his written sentences tripped over the speech they labored to compose. "As I may say" is a common evasion throughout *Experiments and Observations*, and certain words are so often repeated that apologies are in order. "I wish my reader not quite tired with the frequent repetition of the word *surprise*, and others of similar import," Priestley offers at one point, "but I must go on in that style a while longer" (*Experiments*, 2:113–14). Impediment, dissolution, and ungovernable recombination attend air's transformation into the visible and potentially transcendent forms of writing. Those forms indirectly illuminate these features of the air even as they labor to redeem them into unity.

Grammars of the Air

Priestley's interest in the relationship between the language we see and the language we don't was one that he shared with other members of the so-called Lunar Society of Birmingham, a group of fourteen provincial scientists, industrialists, and educators whose numbers included Erasmus Darwin, Richard Lovell Edgeworth, James Watt, and Josiah Wedgwood. Over the late 1770s and the early 1780s, the Lunar Men supplemented their viva voce meetings—held monthly by the atmospheric light of the full moon—with the incessant exchange of letters. Darwin's posthumous poem *The Temple of Nature* (1803) described an object of special interest to the group: a "diabolicall" wooden mouth that with the help of some bellows behind it could form such words as *mama*, *papa*, and *map*.[39]

Priestley's own "large body" of writing about language reveals his personal fascination with the ways in which the supposed second-order medium of writing is implicated in the supposed primary medium of air. Two of his three books on language are supposedly transcribed from "a course of lectures," and he was confident that "all ideas analogous to those of sounds and intervals, admit of a natural expression by words," and hence "words may not be mere *arbitrary* signs of such ideas, but bear a real *resemblance* to them" (*Lectures on Oratory*, 289). Words, ideas, and air thus naturally express one another; they are all aspects of one whole. So "by means of its elasticity (whereby it is capable of transmitting every possible variety of vibrations) the air affords to all animals that live in it a most convenient medium of communication by sounds." But *human* articulation "consists not only in varying the aperture of the mouth . . . but chiefly, in checking and stopping it in a great variety of ways."[40] Even as obstruction ("checking and stopping") unbinds other and more integrated forms of expression, a "universal" phenomenon—speech—can be conceived only through "varying" and "variety." Here, in other words, is a supplement to Priestley's scientific discovery that "common air" is nothing more or less than the "different kinds" in which it consists.

All of this confirmed Priestley's impression that speech, musically, "consists of sounds divided by a great variety of intervals." Differences in "checking and stopping" air were meanwhile responsible for grosser differences among languages. They even explained differences within languages both as these existed synchronically ("the principal Northern tongues," for example, "did not give the air a free passage out of their mouths, but made great use of their tongues and lips in retarding and modulating it") *and* as they changed diachronically.[41] True to form, Priestley mined his impressions from a rich lode of linguistic examples, albeit ones he was less likely to have heard than to have seen transcribed in writing. He looked at tongues from Arabic to "Brazilian" and "Huron," locating their commonality in the limitation and incompleteness, the forced yet variable stops and starts, that typified all of them.

As his 1762 *Theory of Language and Universal Grammar* would attest, Priestley was philosophically (and, as a Unitarian, religiously) committed to the idea that all human languages, their differences notwithstanding, shared one origin and were moving toward a single destiny.[42] But the unity at both ends of the history of language was finally better approximated through visual language than through its spoken counterpart, "connect[ing], as it were, the living, the dead, and the unborn" (*Universal Grammar*, 22). Writing seemed marvelous because it could (as romance did those of fiction) make language's current and intermediate state, and the principles of composition that applied to it, visible to everyone. It thus preserved the vitally "compositional" or composite character of human language. These Priestley saw to be so "liable to many intermediate fluctuations" that "no internal constitution [could] preserve them either from the general revolutions or the particular accidents" (*Universal Grammar*, 169). Writing's vivacity and autonomy came from its ability to establish limiting factors such as relationship and position as constituents of meaning and from its unique potential to display an "internal" and self-correcting principle reminiscent (or anticipatory) of Adam Smith's invisible hand.

Here English commanded special respect. If Priestley considered the English language itself not to have been "fixed till about the reign of queen *Ann*," this was in the sense that air could be "fixed" (*Universal Grammar*, 179). Fixity was but one state among many and, by definition, always transforming. *Rudiments of English Grammar* often noted his native tongue's preference for composition over inflection. This preference was especially apparent in the future tense, which in English is usually conveyed through the use of auxiliaries such as "shall," "will," and "must" rather than through some mutation of the main verb. As far as Priestley was concerned, this was just as it should be: "We have no more business with a future tense in our language, than we have with a whole system of Latin moods and tenses," he maintained. Furthermore, "the true symmetry of the grammar is lost" whenever "inflections are confounded with grammatical uses of the combinations of words, of the order in which they are placed, or of the words which express relations" (*Rudiments of Grammar*, viii). Here, as in his aerial philosophy, Priestley liked a compositional structure better than a "simple elementary substance, indestructible and inalterable." In manifesting particularity, composition also expressed relationships. The spaces between verbal elements not only separated and limited those elements but also displayed their unique, if temporary, forms and the pressure that other such forms put on them.[43]

Equally significant, it is in grammar's technical lexicon that the words that have become synonymous with literary atmosphere—"mood" and "voice"—assume a literal meaning, albeit one entirely detached from personality or subjectivity, and relevant merely to the way that words are said to behave in relation to other words. Priestley saw mood and voice—unlike, say, auxiliary verbs—as acquired characteristics of English, genetic material that made that tongue inherently rather than accidentally other than itself. The "optative mood" for example, and the "perfect and pluperfect tenses of the passive voice" had been "absurdly transferred from the Greek language into the Latin, where there were no modifications of verbs to correspond to them," and thence into English (*Rudiments of Grammar*, viii).

Priestley's own hope was to eliminate the absurd transferences still borne in the moods and voices of English. He would thereby "introduce uniform simplicity" into his native tongue. This ambition has caused his language theory to be classified as prescriptive and uncharacteristically autocratic rather than simply descriptive.[44] At the same time, though, Priestley's desire to expel all forced "analogies" between Latin and English because they were "contrary to our modes of speaking" also led him to spot conformity between common speech and elite letters and to promote "analogies of other languages more like our own" (*Rudiments of Grammar*, v). As for English itself, what Priestley called its "paucity" turned out be its highest recommendation, a virtually moral value that allowed the language to reveal both the equality and consistency of its constituent parts and the possibility of their eventual confluence into a unified whole. So Priestley was most interested in the times "when two words are used to compose one, in order to make one name of a thing." Under such circumstances, the two words "often coalesce into one word, and are written close together," although at other times they might "remain in a kind of middle state" (*Rudiments of Grammar*, 140). The compound words that arise ("unheard-of," remarkably, was a favorite example) are the heroes of Priestley's English grammar. They are, like certain of the factitious airs, "easily understood, because their meaning out of composition is retained when they are compounded" (*Rudiments of Grammar*, 141). So chemical epithets such as Priestley's remark in *Experiments and Observations* that "the circumstances in which any kind of air is decomposed tend to discover its composition" also encompass (if they do not directly reference) the English in which they are after all composed.

A language philosopher of an earlier era might have found English a laudable tongue when and because it appeared to deliver a superior transparency and immediacy of reference to things. But for Priestley the advantage of English's conspicuously composed "nature" was that it gave rise to a formal integrity within the entire field of language, one that paradoxically promised to unify all of its constituent parts and thence the speakers thereof. To demonstrate English's virtue, Priestley ruthlessly culled linguistic examples without regard for their content. English translations served this end especially well, and a favorite source was Tobias Smollett's recent translation of Voltaire. From a pneumatic perspective, Smollett is probably loveliest in light of his novel *Humphry Clinker* (1771). Here Scotland, "like Wales, cursed with a weeping climate," is nonetheless extolled for its "healthy" air, even as a "high exalted essence of mingled odours" and "a thousand frowzy steams" are listed in a sardonic analysis of "the fragrant aether we breathe in the polite assemblies of Bath."[45] Smollett's fictive airs are the same because they differ, and they differ because they are the same. As a linguist, Priestley singled out Smollett's translation of Voltaire (1761–65) because it automatically foregrounded the virtues of English composition at the point of transfer into, or out of, a different language. Ultimately, Priestley was not interested in the promotion of English above other languages. He looked instead for a more profound linguistic unity that would eventually subsume English air. In time, even English would turn out to be but an intermediate composition, blessed only because it so transparently (yet so beautifully!) displayed its own work of mediation.

Picture Writing

Priestley's English was, at its best, written English: air apparent. But for all his love of airs and English letters, Priestley found that phonetic alphabets, grounded as they were in differences of utterance, finally lacked the universal intelligibility of "*picture-writing*, with that contraction of it into *Hieroglyphics*" (*Universal Grammar*, 32). And like the flowers and feathers that the factitious airs presented to the eye, "*picture-writing*" fascinated, especially when Priestley imagined that "by degrees [the picture writers] would learn to contract [their] pictures, and only to draw so much of their first outline as was sufficient to distinguish one expression from another" (*Universal Grammar*, 33). Here, the differences lay *within* the formal system rather among the sounds—the aerial phenomena—outside (or at least around) it. This difference gave picture writing its own analogy to air not because it represented the way sounds were made in the air but because it shared air's natural synthesis of different parts all equally internal to it. Egyptian hieroglyphs and Chinese ideograms marked further contractions of "this picture or analogical writing" (*Universal Grammar*, 34). But they were also limited in that they could not be combined to make new ideas and only represented those that already existed; indeed, some "combinations of ideas" were impossible in certain written languages.

A further advantage of the graphic medium, especially if it were pictorial, had to do with the like-mindedness that it presumed and also, at least in Priestley's view, produced. Priestley even idealized geometry, praising Euclid for having developed the most effective "form in which evidence is presented." In his *Course of Lectures on Oratory and Criticism*, Priestley presented geometry as a form of "synthetic demonstration" that "gains the readiest and most irresistible admission into the mind" because its structures conform to those of the mind itself (*Lectures on Oratory*, 45). Geometry's defining principle was self-evidence. Its strings of axioms, lemmas, and propositions allowed temporal representation to approach synopsis—to verge, in effect, on a "present state" that encompassed and transcended history without sacrificing its own expository potential. But such a state could only be attained if the media that articulated its various aspects and the transitions among them were as conspicuous as what they attempted to make so.

Musical composition, by contrast, perfectly condensed the "medium" and the message. Notes set down on a staff made apparent the very air they appeared to divide. Heard melodies were a trickier story that Priestley did not undertake to tell. In writing, however, "the form of the sentence" had to "truly express the state of mind of him that use[d] it" even as it conformed to "the texture of the brain" of its reader. Here "figurative speech" took special priority because it could best convey and address states of mind "not by means of the words it consists of, considered as signs of separate ideas . . . but as circumstances naturally attending those feelings which compose any state of mind." On the strength of its intimacy with attending circumstances, figurative language could be "viewed in the same light as attitudes, gestures, and looks"—as, in a word, airs (*Lectures on Oratory*, 77).

And this is what Priestley's theory of language in all its forms really evoked: airs. These could be airs of the kind that a person puts on, stylistic arrangements whose content is their form. A figure whose politics opposed Priestley's, Edmund Burke, would still share this sense of air, maintaining in his *First Letter on a Regicide Peace* (1796) that "manners are what vex or sooth, corrupt or purify, exalt or debase, barbarize or refine us, by a constant, steady, uniform, insensible operation, like that of the air we breathe in."[46] Burke here seems anxious not to penetrate air's occult power, and still less that of manners. Like his aerial philosophy, Priestley's language theory offered a vision of mediating form—of manner—that could remain bound to, and thus reveal, the nature of the material world. It did so in part by registering its own difficult implication in that world. Never mind Burke. This was an ideal that bent the story of English air back to John Wilkins and John Milton. To the former, Priestley's linguistic romance would have seemed to move toward Wilkins's mercurial dream of instantaneous transmission through writing. To the latter, it would have evoked the prelapsarian equipoise in which angelic communication with the human had been possible. Convinced that he lived near the end of time (including the time of air), the Unitarian Priestley was at least as anxious as the Puritan poet to discover what the fallen medium of writing might do amid the natural degeneracy of air and toward its rehabilitation. Especially apt to restore air to its native unity was the newborn "art of printing, by which [we] communicate [our] several feelings, and thereby bring [our] tastes nearer to a perfect similarity" (*Lectures on Oratory*, 135).

Like his fascination with "picture-writing," Priestley's preference for the "perfect similarity" that print composition can bring about stemmed not just from its potential universality but from its clarity of outline. Outline both makes an underlying order apparent and reminds us that that order can only be known through mediation. When Priestley published his *Course of Lectures on Oratory and Criticism*, he thus apologized that "these lectures are much too short and too concisely written." But, he explained, "it was my custom to write only the outlines of what I delivered in the class" (*Lectures on Oratory*, v). Outline registered the constraints of "custom" as fully as it did the possibility of immediacy and unity.

Made explicit in the coordinated outlines of the typeset alphabet, this effect may also be visualized in at least one version of the pottery manufactured by Priestley's friend and fellow "Lunatick" Josiah Wedgwood. Priestley's chemistry enabled many of the techniques of firing and coloration that distinguished Wedgwood's ceramics, and Wedgwood in turn supplied Priestley with his tubes and retorts. In light of the close reciprocity between the two men, it is fitting that Wedgwood's signature china pattern—white silhouettes against a sky blue background, or perhaps blue shapes against white ones—should at no cost to their specificity abstract the diverse elements of form into a single impression. Wedgwood's designs thus seem to express one of Priestley's most cherished symbolic ideals: the integration of visibly composed difference within a single impression. Obviously factitious, this impression is for that very reason both revelatory and liberating.

Priestley was no stranger to the visual arts. His *Familiar Introduction to the Theory and Practice of Perspective* (1770) opened by reiterating his preference for visual composition: "Those who are desirous of receiving information find, that it is, in many cases, conveyed with unspeakably more ease or certainty, by the eye than by the ear."[47] "Every verbal description of any object must be reduced to a lineal one in the imagination, before any distinct idea can be conceived of it, and this is often a difficult and painful operation" (*Perspective*, iv). Priestley sought a communicative mode that could shorten the distance between "verbal description" and "lineal representations," promising that, if his book were properly read, "the whole of this art" could be "mastered in a week." Indeed, certain aspects of it were so easily communicated that "the whole mystery of it might be learned in a few seconds" (*Perspective*, x).

Theory and Practice of Perspective sought immediacy through, not in spite of, the necessity of communication through the constraints of space and time. Indeed, Priestley defined drawing itself not as a means of gaining unobstructed access to real things but rather as "the art of delineating objects as they would appear upon an upright plane, interposed between them and the eye; for instance as they appear upon a pane of glass when they are seen through a window" (*Perspective*, 7). This medium-based view of perspective is itself explicitly transferred through the written instruction manual. So "a window" also figures the paper on which Priestley is describing it: "The surface of this paper is designed to represent the pane on which the objects are to be drawn," he explained. His words referenced the hypothetical drawing paper, of course, but indirectly they made apparent the page on which his own words were written. There, "I draw, or suppose to be drawn . . . a line AB" (*Perspective*, 8). Also "suppose[d] [was] a Y plane, transparent, so that objects situated on the plane X could be seen through it" (*Perspective*, 12). And supposition was vital; Priestley, typically democratic, even promised that while his first rules of perspective composition "suppose a previous knowledge of the real distances, magnitudes, and relative positions of all the objects that are introduced in the picture, . . . in many cases a person may be so situated that this knowledge cannot be obtained" (*Perspective*, 57). Here protractors and other instruments could take a hand; this was the difference between what Priestley called "stereographic perspective," in which the eye is at a fixed distance from the perspective plane, and what he called "orthographic" perspective, "in allusion to the method of projecting the sphere orthographically" (*Perspective*, 63).

This projected orthographic sphere is often virtually indistinguishable from Priestley's own orthographic projection of *it*; his instructions manage to turn the temporally experienced elements of writing—lines and alphabetical characters—into spatially oriented directions in their own right. "Draw the point Fg," he thus instructs, pointing out "where the shadow enters the oblique surface, at *e*, and where it leaves at *d*" and predicting that "the lines QD and SD will meet at D on the perspective plane, the representations of RD, TD, and all other lines parallel . . . meet in the same point at sight D" (*Perspective*, 116). A work whose own verbal composition compresses representing and seeing into a single experience of transference takes a natural interest in shadows, even if their value is mainly "ornamental" (*Perspective*, 75). To produce the effect of shadows, Priestley recommended drawing lines from a light source, with the point of their intersection yielding "the out-lines of the shadow of the whole field." The objective? To find where the light would have fallen if no solid had been present and thence to "proceed with it as if nothing solid had been in the case" (*Perspective*, 76). The "as-if" procedure of "drawing through . . . interposed objects, as if they had not been in the way" (*Perspective*, 83) finally limned a conditional order of reality that had a shape and persistence of its own. Adumbrating a gothic stereotype—themselves composed within an explicit fiction—Priestley's shadows both suggested and *were* this order. Consequently, it didn't matter whether their sources were natural or man-made. The thick shadows that fell only in lamplight were of a piece with the "imperfect, obscure kind of shadow" we see "in the light of a cloudy day, or in a room where the sun does not shine" (*Perspective*, 87).

Priestley's perspective manual premised "how to" on "what if." Partly because it speaks to the genre of literary instruction to which it belongs, *Theory and Practice of Perspective* easily doubles as a guide to the making of any kind of atmosphere through the uniquely compromised medium of the prefigured page. Beyond awareness of the inevitability of mediation, little else is required: good news for less privileged pursuers of this art. Priestley's own early drawings of the "electrical machine" in *The History and Present State of Electricity*, he pointed out, had turned out just fine even though they had been "made when I had very little knowledge of the theory or practice of perspective" (*Perspective*, 91).

Perspective drawing contends with objects in space. As for those that exist in time, Priestley published his truly new *New Chart of Biography* in 1765 and a *New Chart of History* four years later. Here again Priestley leaned heavily on the compositional medium of the page, for both "charts" sought to exploit lines, colors, and the contoured spaces between words and letters to achieve a sense of coherent immediacy. Not for Priestley the narrative progression through a series of described events. He instead contrived pages wherein "if the reader carries his eye vertically he will see the contemporary state of all empires," and if the reader "cast[s] his eye a little to the side of the vertical line, he will see what empires had lately gone off the stage, which were about to come on." Priestley's charts were distinctive also in that the "thin and void places" in them were, "in fact, no less instructive than the most crowded, in giving us an idea of the great interruptions of science, and the intervals at which it hath flourished."[48]

Daniel Rosenberg points out that this peculiar "device of reference" allows every individual reader to "construct a historical context for every individual biography."[49] At the same time, however, Priestley's charts were also devices of transference, composed so as to "represen[t] the situation of [a] life, and every part of it, in universal time, and the proportion it bears to the whole period which the chart comprises." Throughout this book, there are nothing but "lines . . . which suggest the ideas, and this they do immediately, without the intervention of words." Thus, it is "the black line under each name which is to be attended to," with the names added only because "there was no other method of signifying what lives the lines stand for" (*New Chart of Biography*, 9). Rosenberg suggests that Priestley's aim was the representation of time itself, and his ideal here seems to have been the transcendence of names. But what his charts themselves project is the impossibility of leaving these behind. Because it must be communicated through them, the ideal of immediacy is inevitably bound within man-made forms. These both are and are represented by any "name."

Such projections take us back (if also chronologically forward) to Priestley's pneumatic chemistry. *Experiments and Observations* is full of temporal markers that somehow manage to be both imprecise and compulsive. "Having at one time"; "at another time"; "in the mean time"; "some time after": such iterations mark time not as duration—how long, for example, it takes for a mouse to suffocate or a mint sprig to wilt or for beeswax, sugar, and charcoal to absorb "acidic air"—but as a symbolic environment. Once its limits are admitted, such an environment institutes its own flexible, yet individuated genres of experience and perception. In this spirit, Priestley's temporal markers possess no particular quantitative or referential value. They are instead projections within a medium of recombinant elemental forms, one that they both explicate and participate in. This is exactly what bound Priestley's scientific research to—or unbound it into—an aesthetic model of the sublime.

Sublime Composition

No less than his aerial philosophy, his grammatology, and his writing on perspective and history writing, Priestley's *Course of Lectures on Oratory and Criticism* emphasized the "disposition of . . . materials" and expressed a moral and aesthetic preference for explicitly intermediary forms over both originary and ultimate ones. In the chemical process of sublimation, substances are transformed from solid into gas without passing through the intermediate state of liquid. Priestley's own sublime, however, countered this process by foregrounding just that elided intermediate state—the state of composition—and working through it. "That close attention to a subject which composition requires, unavoidably warms the imagination," he wrote. "Then ideas crowd upon us, the mind hastens, as it were, in the midst of things" (*Lectures on Oratory*, 31). This movement of "the mind" into "the midst of things" was possible only if literary composition had first explicitly occurred. The mind's necessary surrounding by its own visible objects, its sense of being among them, was not just an ideal but a goal that could really be achieved in the context of explicit composition. So Priestley deplored the parataxis he saw to be characteristic of the modern ode; here too often one found "no traces of [the] medium of transition" so that the reader could "perceive no connection" (*Lectures on Oratory*, 67) among different parts.

Edmund Burke's far more famous sublime depended on obscurity for its integrity and specificity. Priestley's insisted that compositional forms are the condition of our experience of what resolves them into a unified whole.[50] Hence, this loftiest of aesthetic values could repose in the humble media of communication themselves, "in geometry and even in algebra." In fact, "theorems may also be sublime by their relating to great objects." But when "they fill and charm the soul," it is not the objects themselves but "the causes, the adjuncts, or the effects of these things that are contemplated" (*Lectures on Oratory*, 157). It is because the sublime can be transferred to what transfers some sense of *it* that "a reader of a lively turn of mind though of good discernment, may enter into the scene" of sublimity (*Lectures on Oratory*, 90).

So when we return to *Experiments and Observations* and its opening claim that "the English hierarchy (if there be anything unsound in its constitution) has equal reason to tremble even at an air pump, or an electrical machine," we find such technologies to be sublime—productive of fear and trembling anyhow—because they openly and visibly disperse knowledge "in all directions." Here knowledge is transmuted from a form into a common medium, a kind of "atmospherical air" in its own right. Although we ourselves might see it as a precursor of the information age, this dispersal was, for Priestley, a "Prospect truly *sublime* and glorious," and to his mind the best metaphor for it was Pope's poetic "description of traveling among the Alps, with this difference, that there is not only a *succession*, but an *increase* of new objects and new difficulties" (*Experiments*, 1:xix).

Priestley's version of the sublime accommodates the same "difficulties" he acknowledges in his naming of the airs. In turn, the open acknowledgment of impediment holds Priestley's exhilarated science to a unique epistemological and ethical practice. At its center huddle limitation, disappointment, and even subjection. Indeed, few of Priestley's experiments were free of these elements. Even the most successful of his "researches" often had "at first quite another object" (*Experiments*, 3:349). "I was deceived at the beginning of the process," he would report (*Experiments*, 3:518). "New experiments unhinged whatever I had thought the best established" (*Experiments*, 3:534). "I could not immediately upon seeing this phenomenon, form an idea of the nature of this air." "I kept viewing this production of air with a considerable degree of surprise and anxious expectation" (*Experiments*, 3:112).

Such gothic formulations of life in the lab have looked disingenuous to many historians of science. But precisely as such they float an invisible hypothesis: that simply to perceive that one might be inside a scheme is to experience, critically, a legitimate form of reality itself. From this point of view, it is only information—never knowledge—that we can possess. And even that, like common air, is always rearranging itself into new (yet uncannily familiar) forms as different elements happen to enter and exit it for reasons known only to themselves. It is, after all, Priestley insists, only because he happened to live beside a brewery in Leeds that he even noticed his pet gas, "fixed air," in the first place. Likewise, "if . . . I had not happened, for some other purpose, to have a lighted candle in front of me, I should probably have never made the trial" that led him to the discovery of oxygen, or "vital air" (*Experiments*, 2:114).

Priestley, however, never quite made that discovery, never knew or explicitly identified as oxygen what, after many guttering candle flames, he himself comes across in *Experiments and Observations*. To a Frenchman, Antoine Lavoisier, fell the prestige of having properly discovered that kind of air and conferred its "real" name on it. Priestley, by contrast, retreated to a finally alchemical—indeed, a historically gothic—convention of naming and conceptualization. He preferred to call this particular kind of air "dephlogisticated air," holding that, for "common air" to support breath, the fiery element of phlogiston—in alchemy the mysterious fifth element, long imagined to be the secret of air's known combustibility—had to be absent from it. In this view, burning bodies released phlogiston into the air, ultimately depriving it of its life-sustaining powers; only when common air was empty (of phlogiston) did it become its better self. This exactly backward morality tale Priestley maintained to the bafflement and irritation of historians of science to this day and to the mortification of the English pneumatic chemistry that he otherwise helped to transform and revitalize.[51]

But what demands that we see it that way?[52] Priestley had established that the "air" we breathe and assume depends for its meaning on customary, "inherited" form—on what is, metaphorically, in the air at the time. In a sense that matters, it is a fiction that dissolves into the conventional linguistic forms that we use to talk about it. We need it to. Priestley's aerial philosophy confronts this need and actively tests the limits of the symbolic conventions it sponsors. It's thus very natural for him to have kept the secret of oxygen, even from himself. Keeping that secret opened up to his readers a very different kind of knowledge: knowledge of themselves as subjects of language, enthralled by the materially immaterial environments—the large bodies, ready made—that it creates.

The scenes in *Experiments and Observations* where the chemist fails to discover "vital air," even as he learns everything about it, will strike today's reader as the most gothic in his book. Throughout these pages, blood changes color, flames gutter, and mint leaves shrivel, at the eleventh hour restored to greenish life. Enlightenment arrives without appearing to, or it appears to and does not. As their imagery suggests, these are scenes that a "real" gothic novelist, Ann Radcliffe, would rewrite in her popular fictions of the 1790s. In many ways—and not least because she was also a writer, compelled to work within fixed conventions—Radcliffe was the reader that Priestley seems to imply throughout his work: remote, even isolated, from the exotic objects that compel her imagination and yet preternaturally close to the media that at once simulate and convey information about them.

Radcliffe's most atmospheric romance, *The Mysteries of Udolpho*, is also the one of all her novels that borrows its constitutive elements from a then-consolidating British literary canon, as if to investigate that large body, ready made, by decomposing it from within. As citations of the "aerial forms" of

Thomson, Milton, Collins, and Gray pile up like cloud banks, the investigation can feel automatic if not mindless, the story hard to tell from the syllabus. In an early scene that dramatizes the affinity between Radcliffe's method and Priestley's in *Experiments and Observations*, Radcliffe's heroine, Emily St Aubert, finds herself in her dead father's library. He was, in point of fact, a naturalist, and this might as well be a laboratory because Emily is supposed to be burning some letters of his. The catch is that she has been directed not to read a word of them; because she is a good girl, she manages halfway. But she can't help deciphering a few words as the flame is devouring the paper. They tell her nothing. But they do give rise to dark moods and spectral visions, and as such they offer an instance of—even a model for—the kind of atmospheric reading that the novel they are in also demands. Media of knowledge the burning letters are not. But they do allow Radcliffe's reader to catch sight of the process of sublimation that allows her to experience this fiction not as a story, and still less as a set of referential signs, but as an enveloping and informing experience. In its affinity with them, Radcliffe's fiction thus reveals both the limits of Priestley's experimental writings on the "different kinds of air" and what rendered them, in his own sense, "sublime."

9

Priestley's Radcliffe and the Experimental Gothic

> Of her works, we will now speak at some length, for they are getting less known daily. The generation that read them in youth is passing away, and unless they are speedily reprinted we must look to remote provinces and old-fashioned libraries for those once celebrated romances. From the great literary centre they have all but utterly vanished.
>
> JULIA KAVANAUGH, *English Women of Letters: Biographical Sketches* (1862)

Julia Kavanaugh knew that Ann Radcliffe belonged among her *English Women of Letters.* Still, she did fear that the gothic novelist might be disappearing into thin air. By 1863 Radcliffe's "once celebrated romances" had dispersed to the cultural periphery, and soon they might be gone forever.[1] It was a fate figured in those "romances" themselves, their scenes ever dissipating into the "persistent haze over [the] landscape," that Walter Scott had observed.[2] And yet today, as in her own time, Radcliffe is read more often than Priestley: maybe not in the "great literary centre," but there is no longer one of those anyway. There is barely any literature. Yet this makes it easier, somehow, to see Radcliffe's writing as a despondent version of Priestley's contemporary experiments on, and in, the "different kinds of air."

Radcliffe's Victorian predicament was a feature of Priestley's pneumatic chemistry from the start. In fact, he and his radical cohort of aerial philosophers never generated a consolidated body of knowledge on the order of the French *Encyclopédie.*[3] Their personal associations remained diffuse, sustained across vast distances by letters. These could, and did, converge temporarily in compendia such as Priestley's friend Benjamin Franklin's epistolary *Experiments and Observations on Electricity* (1751). But they cohered only to dissolve and scatter yet again, taking up new forms and new homes in new books all the time. In this way, works of pneumatic chemistry expressed the sublime atmospheric phenomena of electricity, putrefaction, and combustion that they analyzed. Priestley's own atmospheric writing shares its euphoric, expansive "air" with this wider sphere of writing.

Such effervescent mobility is hard to detect in Radcliffe's writing. Two generations on, Kavanaugh could praise her work's "dim gorgeousness," but there has never been any denying that it *is* dim or any shirking its melancholy weight (Kavanaugh, *English Women*, 1;264). Radcliffe's gloom is nowhere deeper than in her fourth and longest novel, *The Mysteries of Udolpho* (1794). Nor is the air anywhere more apparent. This coincidence attends the arrival of literary atmosphere as we now automatically think of it, even as it predicts atmosphere's persistence as a realizing device in nineteenth-century novels from *Jane Eyre* (1847) to *Jude the Obscure* (1895). Unsurprisingly, perhaps, given atmosphere's opacity ever since—not to mention our inability to account for it—it is also in the lugubrious airs of Udolpho that we find literary fiction estranged from the "other kinds" of air study in which it once had played a part.

Unfixed Air

In the history of science, Priestley's romance of atmosphere mediated between the romantic organicism of the future and the corpuscular and mechanistic paradigms of the past. No character in it did so with more agility than the gas that made Priestley's name: fixed air. Carbon dioxide (as we call it) was, as such, a fixed entity, yet it was also a variable and thus gloriously *un*fixed element of the larger whole we call "air." Once Priestley figured out how to isolate fixed air, he could go on to see what other kinds of "factitious air" might make up "common air." He blithely extinguished candles and studied the fanciful shapes that the resulting smoke might take around him. Later, when he stumbled across "vital air"—this being the oxygen that he so notoriously failed to name—he inhaled it so that what had previously contained him he now himself contained.

Such reversals were possible only because Priestley, with what jubilation we have seen, had stumbled on a world of traceable differences that nonetheless lacked determinate boundaries and definite shape. As the portal to this world, "fixed air" made his name. One reason was that bubbles of fixed air were relatively visible, and they were themselves *un*fixed in space; once people knew what to look for, fixed air started showing up all over the place. "Being in Germany in the summer of the year 1774," Priestley himself recalled, he "happened to pass by the famous spring of seltzer water," and here he observed "a bubbling of air, exactly similar to that in the Bath waters."[4]

Fixed air comes up in Radcliffe's writing too. The popular travelogue of her 1794 tour of the German states devotes pages to a side trip to "Selters, . . . which under the name of Seltzer, is so celebrated throughout Europe, for its medicinal water." When Radcliffe mixed "Seltzer" water with Rhenish wine and sugar, she found that "the acid of the wine, expelling the fixed air of the other ingredients, occasions an effervescence, like that of Champagne." The novelist supposed that "the danger of drinking [carbonated water] is that the acid may be too powerful for some constitutions."[5] What made this "acid" so threatening to "some constitutions"? One thing is that it reveals all constitutions to be just that: constituted of elements that are, in Priestley's phrase, "exactly similar" from constitution to constitution.[6] What, for example, *is* the difference between a German cocktail and, say, French champagne? Between "Selters" and "Seltzer"? Between the "bubbling of air" in a German spring and the same thing in a vat in Bath? Between bubbl*ing* and bubb*les*? Between the experiments and observations of a respected chemist and a travelogue by a best-selling novelist? Both Radcliffe and Priestley raised these suddenly unanswerable questions within the interchangeable elements of the English print medium. But at the level of those typographical elements, how English was that freedom, necessarily?

For Radcliffe the traveler, such problems of discrimination compounded at the German frontier town of Heilbrunnen, whose springs contained "so much fixed air, as to effervesce slightly when mixed with wine" (*Journey*, 322). This was just what had supposedly distinguished Selters (or Seltzer). Now it scarcely could be said to. Again and again, Radcliffe's *Journey* less resolves the difficulty of where to draw the lines between entities "exactly similar" in their elements than discovers that difficulty in the visible elements of language. When she traveled, thus, Radcliffe encountered both the internal fixity of specific linguistic markers and the hypnotic uniformity of the larger bodies into which they constantly dissolved. This is exactly the paradox that, for Priestley, attended (and legitimated) the sorting of airs and words through the medium of literary composition.

Referentially speaking, there appear to be two "kinds of airs" in Radcliffe's *Journey*: the sensible but invisible ones that presumably distinguish individual places, and the feigned ones that pertain to people and buildings. In the first instance, the traveler finds the air of England "thin," while in the Scottish fells "the air . . . was boisterous, intensely cold, and difficult to be inspired"; the "sereneness of the air" in Eltenberg counters a "freer air" but a few miles away (*Journey*, 459, 88, 91). Then there is a second, more obviously factitious kind of air. Dutch girls display an "air of decorous modesty" (*Journey*, 72). Rotterdam's buildings have "too much the air of prettinesses," a statue an "air more striking and grand than can be described" (*Journey*, 16, 56). Both "kinds of air," natural and affected, appear not as entities or essences but rather as formal containers. Even common natural air is "loaded with the seculence of putrid kernels" or "filled with scents from the woods," and these "persuad[e] to such tranquil rapture" as the English poet William Collins once expressed in his "Ode to Evening" (*Journey*, 103, 88). Likewise, to Radcliffe's eyes, the airs that people and buildings assume convey such intangible qualities as "decorous modesty."

All of Radcliffe's airs thus share the structure of communicative form. This gives them what Priestley the grammarian would have called a "strict analogy" to Radcliffe's own language. But that is all they can be said to possess in any conventional sense: otherwise they *are* possessed, and ever only temporarily. Radcliffe's "blue and purple tints of air" convey mutability and perpetual mobility. Ever ephemeral, air states are in turn never the permanent or inherent property of any specific place. Here, Radcliffe's *Journey* registers a contemporary shift in the physical and conceptual habitation of the air. By the early 1790s, a new chemistry of transport was everywhere apparent. Britons experimented with the inebriating effects of certain factitious airs—taking, as we'd say, the kind of "trips" where nobody goes anywhere. And they embarked on the first air travel, an activity that both presumed and revealed that air was not intrinsically different from place to place, only differently composed. Coincident with such literary works as the anonymous *Adventures of a Hot Air Balloon* (heatedly narrated by one and in its fifth edition in 1780), the hot air balloon flights of the 1780s, for example, arose from the recent isolation of helium. They made possible not just movement from place to place through the common element of the air but the probing of the upper atmosphere in terms of its continuity with the lower one.[7] Likewise, weather systems no longer seemed indigenous to particular places but were for the first time perceived to move from place to place, like soldiers ("fronts") native to none of them and common to all. As Mark Monmonier has shown, this perception arose in part from fin-de-siècle innovations of the graphic medium of the weather map, including novel applications of color, to coordinate changes in time with changes in location.[8] Just so, climates once thought to be fixed and specific to discrete places were now understood as differing compositions of elements that could be found anywhere.[9]

Radcliffe's travel writing gives this paradox of difference within sameness a specifically lexical grounding. Her own *Journey* does not pass through "Holland" or "England," let alone present-day "Germany," so much as through particular pieces of these projected national entities. Radcliffe's account remixes the airs that compose each town or village while in her pages dark horizontal lines divide each stop from those that appear before and after it. Radcliffe's *Journey* obsessively likens every vacation spot to her native England, and the effect is both to mark proper distinctions among discrete places and, strangely, to erase them, or at least to foreground the arbitrariness of their arrangement into larger wholes. So to drift down the Rhine is to float through a "stream" of changeful names:

> The stream, which bends to the west, takes the name of the Waal; that which flows in the general direction of the river, retains its name, for a few miles, when another stream issues to the northward, and takes that of the Yffel. The old river is still recognized . . . and the town of Rhenen takes its name from it, but about a mile longer, and yields to the denomination of the Lech, which, like that of the Waal, does not long enjoy its usurped distinction. The Waal, or Wahl, being joined by the Maas, is to the Dutch, or the Neuse, as the French call it, near Bommel, takes the name of that river, and, soon after, the Leck merges in their united stream, which carries the title of the Maas . . . into the German ocean. (*Journey*, 348)

The travelers themselves see through so many different "denomination[s]" not because they, possessively, stand outside them but because they are, as Radcliffe liked to say, "immerg[ed]" in them: "We did not yield to this artificial distinction so far as to think ourselves taking leave of the Rhine" (*Journey*, 348). In turn, a stream of man-made "title[s]" pilots Radcliffe's own reader through the maze of a single river. Diversity is revealed as singularity *through*—not despite—the factitious flow of arbitrary names and perpetually "usurped distinction[s]."

Back in England, Radcliffe encounters point after point at which English loses itself in German. She stops in the village of Kendal, "white smoking in the dark vale," the "outlines of its ruinous castle . . . just distinguishable through the gloom" (*Journey*, 386). Here she enters a church whose features, she notes, have already been recorded by Thomas Gray. He failed, however, to notice a virtually obliterated inscription to one "Dame Thomasin Thornburgh Wiffe of Sir William Thornburgh Knyght . . . the ellventhe of August On Thousand fyue hundreth eightie too." Without referring to the dead "Wiffe," Radcliffe finds material artifacts of linguistic difference to appear only at the moment of their transition into the "gloom" of uniformity. She thus supposes that "the Saxon has been so strongly engrafted on our language, that, in reading old inscriptions, especially those, which are likely to be spelt, according to the pronunciation, one is frequently reminded by antient English words of the modern German synonyms. A German of the present day would say for eleven, eilf, pronounce long like eilve, and for five, funf, pronounced like fuynf" (*Journey*, 383).

Quite as Priestley had in his *Rudiments of English Grammar*, the female traveler of the present age isolates phonetic writing ("spelt" elements) as a point where graphic signs exhibit and succumb to the pressures of their own circumstances—"smoking" air, for example. This process cannot be distinguished from the dissolution of both English and German into their common Saxon. Indeed, when "one is frequently reminded by antient English words of the modern German synonyms," "antient" and "modern" cannot be told apart either. It's even possible, all of a sudden, to be "reminded" by something—the "antient"—of which, properly speaking, one should only *be* reminded.

At this point, Radcliffe could easily be standing in the middle of one of Priestley's charts of history or biography. So entirely immersed is she in the elements of composition that they have become something other than themselves; they have become their own almost mystically atemporal medium or, to borrow the imagery of the Rhine (or the Waal, or the Maas), their own "united stream." Elsewhere, a brass plate marking where "lyeth the bodye of Alan Bellingham," father of "eight daughters" ("dyed ye 7 of Maye A-dm 1577"), brings to mind nothing to do with Bellingham or his daughters except, apparently, insofar as any "Bellingham" is an autonomous lexical form that eventually confesses its impurity. So Radcliffe remarks only on the "correctness of inserting unpronounced consonants in the words Eight and Daughters." This yields "proof of the universality of the Saxon mode of spelling, with great abundance and even waste of letters; a mode, which is so incorporated with our language, that those, who are for dispensing with it in some instances, as in the final k in 'publick' and other words, should consider what a general change they have to effect, or what partial incongruities they must submit to" (*Journey*, 388–89).

Here again Radcliffe discovers a universal "mode . . . so incorporated with our language" that its externality to that language makes no difference, and freedom from "incongruity" depends on "submit[ting]" to it. Though shadowed by consciousness of the social and sexual conventions of submission, Radcliffe's is a grammatical analysis from within the strictures, and structures, of English. The forms of freedom that she discovers are ones that Priestley had figured in his pneumatic lexicon of "common" yet endlessly differentiating airs. In this manner, Radcliffe's travel writing literalized, without however naturalizing, the political and historical processes through which female names are lost. The literary aesthetic that arises from this mode of analysis is very naturally conducted in and through the grammar of air.

Radcliffe's own journey took her to only three countries, and one of them was her own. All were distinguished from the rest of Europe not by any features of their landscape but rather by the transient appearances of the air that condenses in them. Radcliffe saw this phenomenon figured both in the similarity of English, German, and Dutch to one another and in their common difference from, say, French. And yet, disembarking at Dover, she was momentarily transfixed by the distant "scene of pre-eminent combination" that is Calais (*Journey*, 369). There were also French troops in Germany; one

Austrian solider even displayed an "air [that] might have announced the French courtier, if his rank had been unknown" (*Journey*, 264). The composite effect of these similarities is to reveal a common organizing and animating medium, an "air." And the same infinitely differentiating and yet universal "air" that made an Austrian almost "announc[e] the French courtier" composes the physical atmosphere in such a way that it can only be presented as a shared sphere of transient appearances. In the border town of Cologne (or is it Köln?), even the substantive bodies of the mountains thus "appea[r] upon the sky like supernatural transparencies" (*Journey*, 124). Contemplating "the appearance of the northern Fells" of England, Radcliffe notes that "in a bright day their appearance is beautiful." Under that very circumstance, "their grand outlines are distinctly drawn upon the sky" to make them "a vision of the Alps" (*Journey*, 380).

Radcliffe identifies the dissolution of differences into spectral community with the ubiquity of the print medium: "It is not in this age of communication and intelligence, that any person will be credulously eager to suppose the inhabitants of one part of the island [of Great Britain] considerably or generally distinguished in their characters from those of another" (*Journey*, 397). Traces of that medium surface in the landscape: at Ullswater the traveler thus watches the lake itself assume "the form of a letter S," while "blue lines," "dark lines," "broken lines," and "outlines" mark every terrain, at once constituting and conveying it (*Journey*, 410, 350, 488, 435, 408). So does Radcliffe's travel writing experiment perpetually with the relationship between the graphic medium and the world that it at once makes and mediates.[10]

Elastic Airs and the Romance of the Forest

Radcliffe's *Journey* discovers that different airs can simultaneously compose and elucidate a common medium, one continuous with—and legible through—Radcliffe's literary one. Her book thus materializes through a dynamic analogy between the lexical points and lines that literally compose it and the figurative ones through which its narrative passes. In this way Radcliffe's *Journey* registers the first and only time that our author laid eyes on the continent that provided the setting for almost all of her fiction.

But "setting" may be too flat a term, just as "continent" may take us a bit too far afield should we want to know just where it is that Radcliffe's romances are located. The fact that Radcliffe had only ever read about the places in her fiction helps to explain why, from the start, her representations of those places tended to foreground states of the air in them, and why mists and vapors constantly intervene in her descriptions of natural features of the landscape: mountains, trees, and water. Even Radcliffe's relatively sketchy early work transpires in nominally real and yet purely imaginary locales such as Scotland, Sicily, and the lower reaches of the Italian peninsula. When they are not thinly disguised copies of the canvases of Nicolas Poussin and Salvator Rosa, such places look to have been cut and pasted from the pages of English pastoral romance, Shakespearean or otherwise. *The Castles of Athlin and Dunbayne* (1789) and *A Sicilian Romance* (1790) may thus be said to take place less in the places of their titles than in transferred descriptions of them. By the time we arrive at the 1790s and her three longer novels of that decade, Radcliffe's settings had grown no less stylized, no less obviously prefabricated. But they were much more minutely discriminated both internally and from one another, and Radcliffe was proportionately more committed to tracking the presence or absence of air in them.

It was through exactly such tracking, however, that Radcliffe came to animate her set scenes analytically from within. The Neapolitan "breeze" that lifts Ellena di Rosalba's veil in the first pages of *The Italian* (1797), the "balmy freshness of the air" that just as independently inspires Adeline de Montalt in *The Romance of the Forest* (1791), and, above all, the "blue tinge of air" that suffuses the first paragraph of *The Mysteries of Udolpho* all introduce an active and reflexive syntax into an otherwise inanimate topography.[11] In *The Romance of the Forest*, for example—or at least in the overwrought French Arden of the title—"the air was still." But when a "dewy freshness was diffused upon the air," a mechanical medium of description seems to inflate and quicken; attention is drawn to its nature and efficacy *as* a medium (*Romance*, 83). Meanwhile, through Radcliffe's almost awkward grammatical composition ("diffused" by what?), a manner of reference most immediately registers as just that: a manner. Two sentences later, Adeline—a reluctant sojourner in this "woody tract" and more bewildered by it than one would expect of a compulsive reader of the English literary tracts from whence it springs—is "repeat[ing] . . . stanzas" fraught with "fervid air" (*Romance*, 83). These lines' reproduction through the medium of the '"real" air makes the latter an instrument for investigating sylvan tracts not as flat, penciled redactions of painted tableaux but instead as dynamic environments.[12]

"Assisted by a fine constitution," Adeline uses verbal repetition to absorb, comprehend, manage, and at last distinguish herself from the overly conditioned air of her surroundings But she's soon spirited away from "the Forest" by a designing libertine whose rococo chateau floats on an "extensive garden resembling more an English pleasure ground, than a series of French parterres." Perhaps because the garden is of a piece with Radcliffe's modern English depiction of renaissance France, "not a breath waved the foliage, and not a sound stole on the air" (*Romance*, 165). If anything moves, it is only when musical "notes floated on the air in soft undulations" (*Romance*, 163). Just so, the content of this poetical air (its first line is "In the sightless air I dwell") combines two factitious media into a single efficacious one. Adeline sobs, the spell is broken, and she is inspired, if not compelled, to escape.

Ah, she flees, but only to find herself in much the same place. True, it is the Alps this time, but here is a mountain "lake, unruffled by the lightest air" (*Romance*, 341). Not only does LaLuc, the elderly virtuoso who takes in the wandering Adeline, inhabit this hermetically sealed outdoors; his name transposes the letters that compose the lake. Small wonder that he should maintain a bookless library "fitted up with chymical apparatus, astronomical instruments, and other implements of science" (*Romance*, 248). As a pneumatic chemist, LaLuc murders to dissect something that has an air of never having been alive in the first place. But his daughter Clara has a different relationship to the air and so discovers something vital in it: *she* employs a lute to "repeat her favorite airs" (*Romance*, 252). Moving the air she breathes through an act of composition, Clara turns it into an encompassing, essential, and dynamic element—into atmosphere. LaLuc, we should note, discourages Clara from playing her lute—far too much pleasure in that. But there were other pneumatic chemists, including Priestley, who actually used lutes in their pneumatic experiments. Priestley himself was an avid, if not especially competent, student of the flute as well. In any case, his experimental romance validates Clara's route to atmospheric knowledge through immersion in her own recompositions of affixed "airs."

Radcliffe's Adeline isn't much of a musician, but she is a reader and a poet. Though French, she rearranges the English writers whose work she seems to breathe, "repeat[ing]" them in English poetic airs that turn the romance "of the forest" inside out by dissolving the boundaries between the inside of Radcliffe's own narration and its epigraphic frame.[13] Indeed, Adeline's nondemonstrable Frenchness keeps the English medium that conveys her conspicuous. "Elastic" Adeline herself (*Romance*, 9), ever stepping with an "airy lightness," is as likely to register the "churlish chiding of [Shakespeare's] winter's wind" that appears outside a chapter as she is the haunted "solitary air of [any] place" ostensibly located in that same chapter (*Romance*, 33). So does Adeline realize Radcliffe's own literary medium through her physical continuity with it. In return, that medium seems to realize *her*—but not as a person, exactly. Adeline will never be a Jane Eyre, or even a Catherine Morland, for all that the latter would be modeled on her (or would so model herself) through acts of reading that replicate Adeline's own anxious encounters with rolls of manuscript secreted in moldering chests. No, her airs cannot be said to realize Adeline "as if" she were a person. But they do realize the air of as-ifness itself. That turns Adeline into an embedded model for how one might inhabit Radcliffe's own aeriform medium and even make answerable inquiries of it.

The Airs of Udolpho

The case of Emily St Aubert, the heroine of *The Mysteries of Udolpho*, is not so different from Adeline's, or at least her air is not: it can be hard to distinguish between two figures who share such "airy lightness," not to mention the same "light veil" (*Romance*, 29, 7; *Udolpho*, 155), and blue eyes, the same "mind[s]" almost preternaturally responsive to the "fresh breeze of the morning" (*Romance*, 9; *Udolpho*, 619). Adeline's spirits, however, refuse to "los[e] by long oppression [their] elastic energy" (*Romance*, 9). The same cannot be said of Emily, who in her defense has many more atmospheres to contend with (almost twice as many pages' worth) and is more inescapably identified with them. Her name, no less than those of Priestley's factitious airs, ends in air, after all.

As Emily St Aubert herself will do, the romance that seems to contain her drifts from airspace to airspace. The "partial vapours" of Emily's native Gascony yield to the "serene air" of Languedoc, where her ailing father is sent for—what else?—"the air." After he expires and Emily is turned over to his vain, ineffectual sister and her scheming new husband, Montoni, she is transported to the "cool breeze" (*Udolpho*, 190) of Venice (where the musical airs of the gondoliers seem to transform water vapor into air itself) and thence to the dank and draughty castle of Radcliffe's title. Here there is always a "passing air" to "threate[n] to extinguish" the tremulously lifted lamp (*Udolpho*, 452). Two spectacular lightning storms later, we are back in Languedoc. New "voices [rise] upon the air" (*Udolpho*, 484), and new "breeze[s] . . . trembl[e] upon the light foliage" (*Udolpho*, 473). But, and alas, Emily is no closer than she ever was to figuring out the source of the mysterious music that has followed her everywhere, haunting whatever air she happens to be breathing at the time.

There are plenty of other "mysteries" too. The most famous concerns the ontological status of the apparition that Emily believes she sees at Udolpho. But there are also the contents of letters her father has ordered her to burn without reading them and the identity of the face in a miniature over which she is sure she has seen him weep. The face turns out to belong to St Aubert's murdered sister, through whom Emily eventually regains the "real" estate that he had lost through a series of mistakes too tedious to recount. In the end, she retires to her reclaimed native home with her amazingly vapid first love, Valancourt. All these mysteries are underwritten by the more pressing one of the air that somehow informs all the persons, places, and things involved in them.

Air carries everything in *Udolpho*, from "long groans that see[m] to pass on the air" at Udolpho itself (*Udolpho*, 464) to lightning to untraceable but ubiquitous musical "airs."[14] If technically invisible, air nonetheless appears on every page. And as it does, it mixes with so much that—as both Scott and Hazlitt were to note—it threatens to erase the distinctions among plots, characters, places, and properties. Each looks to be nothing but a temporary composition of the same elements that compose everything else. So here the air with Emily herself "float[s] . . . lightly among the leaves" (*Udolpho*, 587); there it builds into fractious gales. But what is it in itself? Is there anything there at all? In Priestley, strictly speaking, there was not: there were many "factitious airs" but no "air," except in social and conceptual formulations of it through language. But this mystery galvanized Priestley's own literary medium to give it uncommon affinity with "the" air itself. Is this anything like what happens to Radcliffe's air?

One difference is the sociopolitical resonance of that air, which is obsessively linked to matters of property and propriety. This is true of at least three of the "different kinds of air" we find in *Udolpho*. First there is the air that Emily seems to breathe and that Radcliffe so often describes as if we could see it. Emily opens her window to "inspir[e] the pure air," only to be confronted with a flood of "sweet and *picturesque* sounds, if such an expression may be allowed" (*Udolpho*, 75). In no time, "her ideas [have] arranged themselves in . . . lines" of poetry, itself full of "genial air and "fresh . . . breeze" (*Udolpho*, 76). Emily's lines seem to be her own, but more often various airs have been prearranged into writing by English poets. Moving in the direction of Gascony, for example, Emily and her expiring father pause in "the serene air of [a] summit" and "ope[n] a basket of provisions while

> by breezy murmurs cool'd,
> Broad o'er *their* heads the verdant cedars wave,
> . . . *they* draw
> Ethereal soul, there drink reviving gales
> Profusely breathing from the piney groves.
>
> (*Udolpho*, 57)

Here, the St Auberts literally breathe literary air: that of James Thomson's *Spring*. These "reviving gales" belong, properly, neither to a posited natural world nor to a manifestly artificial one, neither to Radcliffe nor to Thomson, neither to her France or his England. Radcliffe's italics further compromise

Thomson's pronouns to underscore the indeterminacy of possession. A spectralized typeface even realizes this transfer of property from a material body to an immaterial one, incorporating the ghostly "gales" into the page where they appear.

Later, Emily pauses to listen to one of the "popular airs of her native province," presumably Gascony. The lyrics—"Pleasant as the gale of spring, that sighs on the hunter's ear"—make air the theme, main metaphor, *and* content of this "popular ai[r]" as well as its form (*Udolpho*, 388) At the same time, because Gascony is in France, and this air is English, it feels, well, factitious: in the very moment of its consolidation, it disperses into what can only be called thin air. Radcliffe's own footnote to this passage reads simply "Ossian," thereby doubly alienating the air that Emily hears, not just from the heroine's "native province" but from Radcliffe's own fiction, to which it clearly does not belong. But then again, "Ossian" was nobody but a medieval Scots bard infamously invented by the latter-day forger James Macpherson. "Ossian" is but a sibilant marker in a lexical medium, albeit one meant to conjure the ghosts of an airborne culture from that same medium. So this "air" cannot be said to belong to "Ossian" either.

But if a particular air cannot be owned, it can be realized, temporarily, through the function of mediation that is left to it. Interestingly, this is a function that this "popular air" shares with Emily, who in listening to it becomes its mediator. She is the means through which, for Radcliffe's reader, it assumes a set of coherent properties, a discernible and meaningful form. In turn, it's through her interactions with elusive air—through her attempts to absorb, track, inhabit, and differentiate it—that Emily literally interacts with Radcliffe's unpossessable writing. She thereby realizes its properties for *Udolpho*'s own reader, if only in the fleeting moment of literary experience. No wonder this aeriform book was itself so "popular." Continuous with air itself, it belonged to anybody, everybody, and nobody—to the Emilys of the world.

But what of the obviously factitious airs—the kind, for example, that a person might put on? When Emily tries to read other characters, she (like us) has only their airs to attend to. Some of these characters, usually "with a dejected air" (*Udolpho*, 588), step straight out of the twilit air itself; others display a pesky "resemblance to [one another] in air and figure" (*Udolpho*, 125) that mocks the impression that such appearances lead much of anywhere at all. Still others are recognizable only by the airs that they affect. Madame Montoni, for one, "seem[s] to assume the airs of a princess" (*Udolpho*, 179), and a dying nun mysteriously projects an "air of high rank" (*Udolpho*, 580). And yet it turns out that this nun, heavily implicated in the long-ago murder of Emily's aunt, is indeed "of high rank," and the point of Madame Montoni's lofty airs is that you can see right through them. So, in a version of what was true of the "gale of spring" itself, we are left with airs that, for all their guilty association with the realm of pure appearance, do not actually lie. Through the logic of negation that caused Priestley to see oxygen as "dephlogisticated air," they produce possessable experience *because* they are factitious.

This paradox pertains to yet a third "kind of air": apparitions. Since Wollstonecraft at least, much has been made of Radcliffe's determination to track them to purely natural causes. When M. St Aubert's dead face floats in midair, his daughter dismisses it as a mere phantasm of living grief. Later, the specter of a livid corpse turns out to be nothing but a waxwork memento mori. Haunting melodies follow Emily everywhere, only to reveal themselves as compositions of the living, if depressed. The many explained ghosts of *Udolpho* have been brilliantly, even definitively, explained by Terry Castle as proleptic artifacts of an emergent psychology grounded in melancholy and narcissism.[15] But Radcliffe also retains something of the residual grammar of romance that would place them outside the psychological mind. From this perspective, her apparitions are best identified with a transpersonal aesthetic, one that Radcliffe always grounds in the literary medium. Because, at the narrative level, that medium is identified with "air," Radcliffe's explained supernatural is a red herring in its own right—not because it is part of an ordinary mentality but because, strictly speaking, there *are* no mentalities here beyond those formed in the act of transference. The charge of Radcliffe's fiction is therefore not to explain the supernatural in terms of what seems to be the natural. It is all air, after all. A different aim instead appears: to model and cultivate in Radcliffe's reader an epistemological practice—ultimately a belief state—that can find both autonomy *and* some purchase on what passes for reality.

In the only theoretical writing that Radcliffe seems to have produced, she defined (and defended) "the supernatural in poetry" very much on these grounds. Her discussion of the matter is conducted through two interlocutors. "With an air of discovery," one of them holds that "the only real witch" is "the witch of a poet." But that doesn't mean he will not "thrill with delighted awe" at her appearance every time.[16] In the less rarefied practical environment of Radcliffe's fiction, however, we have only a dispossessed young girl to show us how to keep things up in the air. Though she occasionally gains the "upper region of the air," Emily barely has a chance to inhale up there before she is once more pressed with the problem of the "lower airs" and the unhappy task of figuring out just how to take them—of reading the book that she happens to be in (*Udolpho*, 418).

This isn't the same as sorting out what is and is not true in itself. As with

Priestley, Emily's predicament is one because all individuals share a single communicative medium. That medium can become transparent only when its opacity is acknowledged. It seems coherent only when its multiplicity is admitted. Still, surviving intelligently in this element is a sad and lonely proposition. Emily's parents both are dead before *Udolpho*'s first volume is half over. Valancourt is detained for most of the novel in Paris, that dissolute metropolis where not only had Priestley's rival Lavoisier just managed to name oxygen but (in a no doubt related development) "affectation so often distort[s] the air" (*Udolpho*, 549).

No wonder Emily is always struggling to "suppres[s] a heavy sigh" (*Udolpho*, 94)! She has been brought up in the pastoral air of La Valée. *The Mysteries of Udolpho* itself begins, if not *in medias res*, then certainly *in medias aer*, with a description of this property. But, properly speaking, it is not a description at all. While an initial series of prepositional phrases locates Emily's girlhood home, "the chateau of Monsieur St. Aubert," "on the pleasant banks of the Garonne," "in the province of Gascony," " in the year 1584," the rest of Radcliffe's first paragraph shows (or doesn't) "the majestic Pyrenees, whose summits, veiled in clouds, or exhibiting awful forms, seen, and lost again, as the partial vapors rolled along, were sometimes barren, and gleamed through the blue tinge of air" (*Udolpho*, 3). The effect of this verbal cloud bank is less to blur the subject of the sentence and the conditions that modify it than to diffuse that subject through those conventions. Consequently, the subject of the sentence belongs to no one and everyone. What, then, are we supposed to see in Radcliffe's chorography of La Valée?

The answer is, likely nothing. What we are supposed to do is see *it*. In so doing we see through it. Radcliffe's grammatical structure here expresses its *rapport* with Priestley's aerial grammars, to borrow a term from French pneumatic chemist Pierre Macquer. If anything, its "blue tinge" alone makes Radcliffe's air visible as a medium of communication indistinguishable from the matter communicated. Just so, these "partial vapours" might very well enhance the Burkean sublimity of a landscape half-obscured. They might even pave the way to William Hazlitt's perception that for all that Radcliffe's "descriptions of scenery, indeed, are vague and wordy to the last degree" and "her story comes to nothing," her "indefiniteness is an advantage." For Hazlitt, it was an advantage because it compels a reader to "add expression to the inanimate outline, and fill up the blank with all that is amiable, interesting, and romantic."[17] As Hazlitt hints, Gascony's "partial vapours" supply a paradoxical medium of enlightenment. They place Radcliffe's reader within an environment composed of mediating forms and secondhand impressions. But they also guarantee that reader, whoever she may be, freedom to compose and recompose the larger forms and impressions that result. Mounting amid such vapors, a "mystery" strikes us less as a secret that needs to be revealed than as an inscrutable but essential medium that has to be managed from within.

No one is more familiar with this distinction than Emily St Aubert. The airs of La Vallée are hers—literally so, because the airy elements that compose *it* turn out to compose *her* as well. For turn the page a time or two, and here is Emily: "In person," we find, she "resembled her mother; having the same elegant symmetry of form, the same delicacy of features, and the same blue eyes. But, lovely as was her person, it was the varied expression of her countenance . . . that threw such a captivating grace around her" (*Udolpho*, 7). At the referential level of Radcliffe's plot, of course, what matters is that Emily "resembled her mother," who in turn resembles Emily's traduced aunt; there is even a mysterious miniature portrait floating around to pretend to fix these resemblances in visual space. But as Radcliffe's references to "expression" suggest, more pressing are the interactions among the media that at once make up Emily and convey something of her: the "symmetry of form," the "delicacy of features," the "blue eyes" that are "the same" not only as her mother's but as the "tinge of air" that has appeared but a page before.

Emily's mother is alive at the beginning of *Udolpho* but immediately catches a fever from her husband. "Oppressed" by "languor and dejection" (*Udolpho*, 20), she has expired by the end of the first chapter and been "interred" by that of the second. The tears that her husband and daughters shed over an open prayer book, not to mention the entirely literary conception of the grave that Radcliffe was to develop in her *Journey*, hint at a burial in letters worthy of Defoe's H. F. In any case, and sad to say, Emily does not appear to give her mother a second thought. But perhaps she does not need to; Mme St Aubert's sky-blue eyes and transient formality put this maternal figure in the air itself. Mother and air are both registered, glancingly and indirectly, through secondary properties and oblique effects. In turn, Emily's unshakable languor and dejection, so often expressive of the atmosphere about her, also mark her of one being with a mother who was never quite there—and thus was always everywhere—to begin with.

Emily's mother is literally compounded with what we might call *Udolpho*'s ambience. It was in Radcliffe's lifetime that the word "ambience" began to travel indiscriminately across English, Italian, and French.[18] In literary terms, ambience may be understood as the synesthetic effect we try to capture when we speak of a literary work's voice or mood. It is possible to identify this effect with the maternal: Gabriele Scwab, for example, analyzes "literary moods," or *Stimmung*, as oblique and undifferentiated effects that replicate "the earliest experience of voice" in the body of complex cultural objects. Poetry, music, and other works of art thus resonate with a preverbal "ecology" that, because it unfolds in infantile exchange with the mother, is intrinsically interactive and communicative. Consequently, what can seem to be merely "unformulated inner spaces" accessible at best through the "unthought known" are part of a process of transference.[19] In Radcliffe's writing, such spaces also exist outside the logic of property. Schwab's concern is with poetry and its essentially musical qualities of rhythm, melody, polyphony, and recursivity. Radcliffe's fiction, however, evokes these qualities in visual, visibly factitious, distanced forms that retain only the *air* of immediacy.

So it's probably a good thing that her father is so determined to communicate to Emily the habit of "cool examination" (*Udolpho*, 5). Although in principle this habit should detach Emily from her surroundings and give her some analytic purchase on it, in effect it often does the opposite, giving her an active and entitling role in their production. For Emily St Aubert is *too* cool; she is colder than Lennox's Arabella ever was, and she is about to shiver her way down to the Apennines, then back the way she came. She is warm only when the air is warm, and that—for all that we're never out of southern Europe—is hardly ever. She thus is perceptibly at one with the partial vapors that surround her—not in a romantic way, or even in the thermometric way that Castle ascribes to the eighteenth-century heroine of sensibility, though (Castle, *Female Thermometer*, 21–43). Rather, it's through Emily that the conventions that compose this novels' environment are themselves coldly analyzed *as* compositions that arbitrarily and autonomously arrange themselves into convention.

These are in the most immediate sense linguistic compositions and conventions. Hence, Emily's best bet if she wants to keep breathing is not to hold on to the property that, plotwise anyway, threatens to slip through her fingers. It is, as it is for Radcliffe at the compositional level of writing, to learn how to compose second-order *properties*. Fortunately, as Priestley had shown, the properties of air are all there is to it, making it at once the most and least "real" estate around. And here is where M. St Aubert, his own "benevolence unchilled" (*Udolpho*, 1), comes in. He happens to be a scientist—a botanist avant la lettre, this being 1584, after all.

In 1794, however, botany was a branch of natural philosophy that had recently expanded its own authority through two fundamentally symbolic innovations: Linnaean taxonomy and the introduction of color to botanical drawing.[20] The latter development is obviously related to advances in the chemistry of pigmentation—a chemistry indebted to Priestley's analysis of the factitious airs. As for Linnaeus, while traveling in the German "frontiers,"

Radcliffe was to mark where "choice shrubs, marked with Linnean distinctions, composed the thickets" (*Journey*, 263). Her words make no distinction between the "shrubs" themselves and the "Linnean distinctions" that "mar[k]" them, and in this way they ironically measure the extent to which a verbal frame of knowledge can be taken for its object.

With respect to *Udolpho*, it matters that the paternal legacy so conspicuously transferred to Emily should be "the mind of the naturalist" who often "passed the day in pursuits of his favourite science." There's not much difference between that and reading a novel. The lesson Emily's father might impart, however, is the composure not of vegetable forms through "Linnean distinctions" and the happy application of watercolor but of the person. Having noted his daughter's "uncommon delicacy of mind" and "degree of susceptibility too exquisite to admit of lasting peace," M. St Aubert "endeavour[s], therefore to strengthen her mind, to enure her to habits of self-command; to teach her to reject the first impulse of her feelings, and to look, with cool examination, upon the disappointments he sometimes threw in her way" (*Udolpho*, 5).

He can hardly hope for success in this Edgeworthian scheme, one that Claudia Johnson exposes as both an attempt to mask St Aubert's own appropriation of feminine inefficacy and a girding of Emily in traditionally masculine virtues that can do her, a woman, no good.[21] Or can they? It is Emily's *failures* of composure that keep her from figuring things out: from reading key letters, from looking twice at objects that distress her. And her capacity *for* composure helps her to figure things themselves and thereby to reinforce her own continuity with what Hazlitt called "the flimsy texture of a modern romance" (Hazlitt, "Heroes of Romance," 130–31). The apple might well fall a little way from the tree insofar as, instead of botany, Emily specializes in Priestleyan pneumatic chemistry. Still, her father's lessons in composure condition her to know the contingent, evanescent, and endlessly recombinant forms through which reality itself is transmitted and through which it must be perceived.

Unlike botany, which could afford to finesse the matter, pneumatic chemistry had, since Boyle, openly embraced its own figural dimension. It was, as we've so often seen, forced to do so by both the invisibility and the inescapable immediacy of what that chemistry attempted to analyze. Like Priestley, Emily knows the airs that surround her by composing them from within: she reads, writes, sings, and draws. Like Priestley's "experiments and observations," these are airings of personal curiosity that mark Emily's participation in—even her oneness with—the aeriform media that convey her. Thus aligned with more than a century of British insular writing about the air, Emily emerges as an autonomous investigator of her immediate environment. If that environment seems to have little choice but to seem, by turns, despotic and depleted, this is no fault of Emily's. Nor does Radcliffe fail to display either its workings or, ultimately, its factitiousness.

This is a project we have seen undertaken in various ways by Milton's Eve, Pope's Belinda, Defoe's Crusoe, Fielding's Partridge, and Lennox's Arabella. Radcliffe's air, however, is a far more oppressive medium than any encountered by these forebears. Emily's reward for her investment in the air is qualified at best. If she learns about her own environment by subjecting herself to it, and if she analyzes it only by composing and recomposing it, her only freedom is at last to disappear into it. Likewise, if Radcliffe's fiction makes air visible as a medium of oral culture, music, and life-giving breezes—so these forms count as sources of "true" knowledge—that knowledge has nowhere to go. On the affirmative side, that is, Radcliffe's literary composition actively tests the air, much as Priestley did, to see what it will do, how much it can bear, what forms it takes, and what styles of belief it can support. But none of the answers to any of these questions proves translatable outside the literary medium. Radcliffe's fiction, subtle and reflexive as it is, thus captures English air in the phial of fictional form and, refusing to release it, has to let it die there.

Just so, in Radcliffe's plot, air kills at least as often as it refreshes and revives. Emily's mother succumbs to an airborne fever. When her replacement, Madame Montoni, breathes her last, it is in the midst of a "tremendous storm that shook the air," and Emily cannot "doubt but that the violent change in the air, which the tempest produced, had effected [the] fatal one, on [her] exhausted frame" (*Udolpho*, 377). And yet all of Emily's time in Udolpho is spent seeking an open window, and it is only by tracking the air that anyone inside that labyrinthine structure can tell how near an exit might be. But then again, the fault might not lie in the air itself or even with Emily's final powerlessness to master its meanings even while knowing its grammar. What, in other words, about Udolpho itself?

The Mystery of Udolpho

One of the "mysteries of Udolpho" is that of what "Udolpho" itself might be. Only a part of Radcliffe's narrative—albeit the middle one—actually takes place in the decomposing Italian castle, "silent, lonely and sublime," of her title (*Udolpho*, 227). Literally speaking, "Udolpho" is a series of sounds, all but one of which seethe gently on the verge of dissolution in the mind's ear. "Silent," indeed. The castle itself always tends that way, and in so doing makes the silence perceptible as a form. Air thus moves conspicuously through all its many passages, its cargo but the abstraction of coldness and sounds that are ever dying away. Just so does Udolpho itself, even at first glance, appear to be disappearing into thin air, an autocritical figure of Radcliffe's own writing.

In terms of her plot, meanwhile, Udolpho once belonged to Emily's murdered aunt. Montoni, the husband of her surviving aunt, has an inkling that it might now rightfully belong to Emily herself. So he brings her there in hopes of either selling her off or through her forced signature gaining the property for himself. The problem is that Udolpho is less a property than a set of secondary properties and transient effects. Approaching it, Emily thus "anxiously survey[s] [its] edifice," only to find that "the gloom, that overspread it, allowed her to distinguish little more than a part of its outline." Just so, Udolpho's "mouldering stones . . . seemed to sigh, as the breeze rolled past"; its "arches opened to a lofty vault" (*Udolpho*, 229–30). The truth, which Radcliffe tells, is that this vapid structure isn't there at all.

Ambiguities such as these are usually attributed to Udolpho's romantic power to give subjectivity external form. It's in this spirit that Robert Harbison associates Udolpho with the "nonprogressive fiction of pure setting." This association gives Udolpho an intimate psychological relevance: it is, just as Harbison says, impossible to determine "who is from inside and who is from outside the castle."[22] It's equally difficult to determine who is inside and who is outside any individual mind. Melancholy the Radcliffean heroine most certainly is. She is anxious, frightened, capable even of tranquility and rapture. But even so nimble a critic as Johnson is forced to look for signs of coherent interiority in the passing air of indirect free speech, while Castle turns to the larger cultural phenomenon of the haunted mind to account for the images through which Emily, for example, presents her own thoughts and feelings to herself (Johnson, *Equivocal Beings*, 76; Castle, "Spectralization," 231). What Udolpho's confusion of what lies inside a form and what lies outside it *can* be shown to express, however, is a linguistic phenomenon, not a psychological one. "The cold winds, that blow through the castle in winter" (*Udolpho*, 231) have the immediate purpose of confirming a fundamental identity between "Udolpho" the composition of letters and "Udolpho" the decomposing place to which that composition only appears to refer.

It was on the strength of this identification that Scott considered Udolpho an internal trope for Radclifffe's aesthetic program as a whole. We're now free to return to his assessment of that program. "There is," he wrote, "as it were, a haze over her landscapes, softening indeed the whole, and adding interest and dignity to particular parts, and thereby producing every effect which the author desired, but without communicating any absolutely precise or individual image to the reader. The beautiful description of the Castle of Udolpho, upon Emmeline's first approach to it, is of this character" (Scott, "Mrs. Ann Radcliffe," 1:379). If Radcliffe's Emily here fades into the heroine of a different novel altogether (presumably Charlotte Smith's gothic heroine Emmeline), that is but an epiphenomenon of Radcliffe's obstinate refusal to "communicat[e] any absolutely precise or individual image to the reader." And Scott's observation captures something more. All the effects he notes are, explicitly, the effects of letters, which is to say of the medium that Radcliffe had most immediately to hand—"the weak hand," as Radcliffe's last sentence puts it, "that has recorded this tale" (*Udolpho*, 679). Millions of typographic "character[s]" compose this medium, of course, and one paradox is that their sheer multiplicity nonetheless yields no stable overarching form—no "large body, ready formed," as Priestley would have said (Priestley, *Experiments*, 1:43).

Hence even the kindly disposed reader's trouble recollecting any part of it. Even Radcliffe's "heroes have no character at all," complained Hazlitt. "Theodore, Valancourt,—what delightful names! And there is nothing else to distinguish them by." These same qualities, however, also let us see what and how acts of mind produce engagement. As Hazlitt put it, "A sounding name, a graceful form, are all that is necessary to suspend the whole train of tears, sighs, and the softest emotions upon" (Hazlitt, "Heroes of Romance," 130). This is to bring forward the shaping power of the medium itself—the name and graceful form—as the condition of suspense. And it is what Scott registered. For—rearranging the letters of Emily's name, subtracting one and tossing a few more in—what he says is perfectly true: "Emmeline" *is* approaching not "the Castle of Udolpho" itself but rather "the beautiful description of it."

The "as it were" at the beginning of Scott's discussion of Radcliffe's "landscapes" proves uncannily accurate. It even "immerg[es]" Scott's own reader quite literally into the stream of Radcliffe's writing. And isn't "immerging" exactly what the despondent Emily is doing as she involuntarily approaches the beautiful description of Udolpho? "From the deep solitudes, into which she was immerging, and from the gloomy castle, of which she had heard some mysterious things, her sick heart recoiled in despair, and she experienced, that, though her mind was already occupied by peculiar distress, it was still alive to the influence of new and local circumstance; why else did she shudder at the idea of this desolate castle?" (*Udolpho*, 227). Why else indeed? Emily's immerging mixes an "idea" *and* "some mysterious things" that were first borne on the air into an environment that can be "experienced." The awkward mechanics of Radcliffe's hyperperiodic sentence ("she experienced, that") only make the discrete and fundamentally communicative features of this experienced environment apparent. And it's to this environment that we must look if we want to know what Udolpho is, what relationship it has to the book that bears its name. No wonder Emily fears she's suffocating in "dismal galleries and halls . . . fit for nothing but ghosts to live in" (*Udolpho*, 234). We can, with her, only "shudder at the idea."

But how, exactly, is it that one can shudder at an idea? How, that is, does an abstraction elicit a bodily response? This new mystery of *Udolpho* turns on the obscure relationship between what is invisible and the visible forms that it assumes, and more immediately on the question of how airs become letters. In life, Radcliffe appears to have been keenly interested in the transactions between aeriform language and lexical formation. Both Scott and her 1824 biographer, Thomas Noon Talfourd, underscored the pleasure she took in "listening to any good verbal sounds"—a pleasure seemingly grounded in the experience of transference itself. She "desire[d]," anyway, "to hear passages repeated from the Greek and Latin classics, requiring, at intervals, the most literal translations that could be given."[23]

"Literal translations" between air and image appear everywhere in *The Mysteries of Udolpho*. Airs heard *on* the air fade into figures that emerge from it. Emily seems to pluck her own compositions out of thin air, and the visual form they take on Radcliffe's page is steeped in the aerial transactions to which they often continue to refer. Hence her airs' eerie interchangeability—with their settings, with each other, with Radcliffe's own poetry, and with the epigraphs from the English poets that Radcliffe places at the head of every chapter. The effects of such literary air can be downright exhausting. They are indiscriminate, monotonous. An American reader complained that Radcliffe's "style is a constant ringing of bells, & mists, & glens, & portcullises, & moats, &c. &c. the miserable and tattered coverings of a feeble imagination & a languid & barren understanding."[24] Even her "admirers found that her richly coloured representations" could easily degenerate into "high sounding and ambiguous expressions" that could "impede the distant presentiment of material forms." In short, Radcliffe's compulsive "repetition of the same images of rock, wood, and water, and the same epithets of grant, vast, and sublime, which necessarily occur, must appear tautologous" (Talfourd, "Memoir of the Life," 118). But such tautologies and repetitions are also modes of transparency. At the very least, they realize Radcliffe's writing *as* writing: "It is a very faint outline . . . that can be exhibited by writing," one reviewer observed, "and the ideas we form [from it] are not so much real images of those objects as confused combinations and analogies drawn from others that had previously come within the sphere of our own perceptions."[25] Radcliffe's "elements" are, with Priestley's factitious airs, reshuffled borrowings that nonetheless identify ideation with the repetition of forms through convention.

One effect of such monotony is the blurring of the line we think divides the outside of a book from the inside of it; only the line itself proves real. An Emily St Aubert can profit from this phenomenon, for it brings her closer to reality. Hence Virginia Woolf was not only inclined to see all gothic writing as transpiring in the "ordinary library" but, with respect to Radcliffe, let the critic Edith Birkhead speak for (and through) her: "'Mrs. Radcliffe's heroines,' said Miss Birkhead, 'resemble nothing more than a composite photograph in which all distinctive traits are merged into an expressionless type.'"[26] If this assessment is strangely anachronistic (the "heroines" in question can hardly be said to resemble photographs that did not yet exist) stranger yet is the implication that all of Radcliffe's heroines together resemble one photograph, itself a "composite." In this blending of the many with the one, the charge of "expressionless type" kicks us from the photograph back to the printed page—the only medium that can conceivably "merg[e]" disparate points in time and space or compose the outlines of "distinctive traits" into a uniform field of visible speech.

That is, of course, the world that Radcliffe builds up in *The Mysteries of Udolpho*. It is a world compulsively ordered through puns that, by definition, record the arbitrariness, if also the inescapability, of fixed systems of naming. Emily St Aubert is bred in a valley that *is* its name—La Vallée—and that valley abuts "the plains of Languedoc," whose name historically differentiates it according to a dialect or variant form of a single language ("langue") in which "oui" is pronounced "oc" (*Udolpho*, 5). Radcliffe's sentences, for their part, treat such place names as locations within themselves: "We now return to Languedoc and to the mention of the Count de Villefort" (*Udolpho*, 466), the narrator characteristically announces. "Udolpho" is full of thugs whose own names tend either to spin nonsensically off from its own (Ugo) or simply to run together: Bertrand, Giovanni, Roberto. "It is all one for that," Bertrand snaps when Emily points out that he himself keeps mixing them up (*Udolpho*, 405). The last setting of the novel, Chateau le Blanc, is the page-like background for the "bluish line[s]" (*Udolpho*, 458) and "dark lines" that typify the Languedoc landscape, and Radcliffe even installs a Blanche in Chateau le Blanc to reinforce the typographical point. Oars can only "imprint the water" (*Udolpho*, 484). There are many "volumes of vapour," not to mention more lines everywhere—a "line of separation" as waters fade into the sky, "a blueish line that appeared upon the horizon" (*Udolpho*, 621, 413, 458). At one juncture, Emily even arranges the "line" of the horizon *into* "following lines" (*Udolpho*, 414) of her own.

What, then, turns these lines into kinds? What makes them discernible not just as hermetically sealed signs but as forms and conditions of experience? What allows "airs" to rise from them so that the cold, abstracted, yet suffocating elements of this medium will take us in? When Emily opens that window and seems to breathe "sweet and picturesque sounds, if such an expression may be allowed" (*Udolpho*, 73), a clue surfaces as to how, in effect, atmosphere is made. Emily seems to know the airs of Radcliffe's fiction for what they are, and through her Radcliffe encodes a set of directions for both consciously and skeptically—for, indeed, ironically—negotiating it as a reality. Emily is nothing if not an atmospheric reader, and if this way of reading falls to her because of her de facto exclusion from other forms of knowledge, this is not to degrade her experiments.

Emily first tries atmospheric reading in her father's library. Instructed to burn his letters without reading them, we have seen—but having made out a few "words, which had roused equally her curiosity and terror"—she wrestles her own "inflamed . . . imagination" until at last she persuades herself to "consign[n] the papers to the flames." Her "eyes watched them as they slowly consumed," and she "shuddered at the recollection of the sentence she had just seen, and at the certainty, that the only opportunity of explaining it was then passing away for ever" (*Udolpho*, 104–6).

Surely what Emily is reading (or not) is gothic fiction. At the same time, as we saw at the end of the last chapter, this scene evokes Priestley discovering—and yet not discovering—oxygen as he burned red calx in his laboratory. Priestley was bound by the forms through which his own practice had been transmitted. Radcliffe narrates Emily's literary experiment as a "triumph of integrity" over imagination, but less pyrrhic is one kind of reading's conquest of another (*Udolpho*, 105). The kind of reading of which Emily deprives herself is one in which graphic signs are explained instantaneously by other graphic signs next to them. The kind to which she consigns herself is one in which graphic signs are explained over time and through their interaction with a layered, complex, polyvalent environment in which they also participate. Figured in the absorption of the paper into the flame, this interaction demands new styles of knowledge. The next six hundred pages of Radcliffe's novel—the "following lines," as she herself might say—are devoted to exploring these styles in themselves as forms of experience. In other words, Emily may not discover what her father's letters refer to. But she does discover atmosphere.

This is, perhaps, inevitable, because the many books scattered throughout *The Mysteries of Udolpho* actively participate in the composition and decomposition of the air. Everybody turns out to be a reader, from the skeptical

Count de Villefort, with his taste for Tacitus, to M. St Aubert, the whole "west side" of whose chateau is "occupied" by a library. Emily and Valancourt, their ethereal passion esoterically mediated by Petrarch, differ little from the chattering, superstitious servants who steal books and slip them to one another. In its ubiquity and universal availability as a medium of communication, writing might as well *be* air. It shares air's chemical properties, is traded as freely, and just as entirely constitutes environments. And because it is an environment rather than an object, writing demands exactly the kinds of apprehension that Emily—her eyes ever likely to "wande[r] from the page to the landscape" (*Udolpho*, 418)—cultivates with respect to it. When, for example, Emily reads "lines written with a pencil on a part of [a] wainscot," she finds that "these lines were not inscribed to any person" and that she "therefore could not apply them to herself." Nonreferential language—which Radcliffe reproduces on the next page—leaves Emily "compelled to rest in uncertainty" (*Udolpho*, 7).

Room after room takes shape in and through combinations and recombinations of letters. Emily fingers a tapestry at Chateau le Blanc, only to "discove[r], by verses in the Provençal tongue, wrought underneath each scene, that it exhibited stories from some of the most celebrated ancient romances" (*Udolpho*, 534). Such examinations are not reading in any familiar or conventional sense, and the inscriptions under examination compel us both to experience and to imagine writing, especially as it transfers air into picture, as something very different from diacritical marks disjoined from their referents. Books themselves are always turning up in strange settings—Emily has "put two or three . . . into the carriage" that takes her from La Vallée, but she doesn't *read* them, really. Instead, "she sought for one, in which Valancourt had been reading the day before, and hoped for the pleasure of re-tracing a page, over which the eyes of a beloved friend had lately passed, of dwelling on the passages, which he had admired, and of permitting them to speak to her in the language of his own mind, and to bring himself to her presence" (*Udolpho*, 60).

Here, not only is composing continuous with reading, but Valancourt might well have been visible "in" the book he was reading "in," for all that Radcliffe's grammar tells us. In any case, the book in question seems to have vanished into thin air:

> On searching for [it] [Emily] could find it no where, but in its stead perceived a volume of Petrarch's poems, that had belonged to Valancourt, whose name was written in it, and from which he had frequently read passages to her, with

> all the pathetic expression, that characterized the feelings of the author. She hesitated in believing, what would have been sufficiently apparent to almost any other person, that he had purposely left this book, instead of the one she had lost, and that love had prompted the exchange; but, having opened it with impatient pleasure, and observed the lines of his pencil drawn along the various passages that he had read aloud, . . . the conviction came, at length, to her mind. (*Udolpho*, 60)

Here, reading has little to do with deciphering marks on a page to arrive at an abstract or absent referent. Instead, it involves absorbing a mixture of forms: graphic ("lines of his pencil drawn along the various passages"), aural ("various passages he had read aloud"), and lexical ("whose name was written in it"). Emily's response to the synesthesia that "is" atmosphere? She "hesitated in believing" and awaits the autonomous arrival of "conviction."

Atmospheric reading—suspended, compelling, and strangely self-aware—is also contagious. One chapter begins with a citation of Hamlet's "airs of heaven" and then delivers a servant eager to test floating "belief" about a haunted room. He spends a night in that "gloomy apartment" armed only with a book that "had been lent to him" by the housekeeper, Dorothée. She herself has retrieved it from "an obscure corner in the Marquis's library," the air there having given the servant "some excuse for detaining it from its proper station" (*Udolpho*, 553). This same air has also given even a female servant at least temporary possession of it. "The damp corner into which it had fallen, had caused the cover to be disfigured and mouldy, and the leaves to be so discoloured with spots, that it was not without difficulty the letters could be traced" (*Udolpho*, 553). And what might this book be about? Nothing, really. In it, "the appearance of a stranger of noble air" turns out to be a ghost who leads a baron to his own corpse, which "form gradually vanished from his astonished sense." Strongly affected by these visions, its plebeian reader, Ludovico, soon imagines that "he saw a man's face, looking over the high back of his arm-chair"; shortly hereafter, he disappears, his fate unknown for more than a hundred pages (*Udolpho*, 559).

Because Ludovico turns out to have been kidnapped by brigands hiding near the chateau, his story certainly looks like an instance of the explained supernatural, of one kind of air exposed as factitious airs, and so as nothing more than air. But Radcliffe does not really seem to have felt pressed to choose between natural and supernatural models of the air—between air as ghost and air as physico-mechanical body—or even between innocent air and air depraved by fiction. These distinctions, like the others in her novel, are finally meaningless. What matters more is the transaction between the air, what*ever* it might be, and the literary medium that does not merely describe but actively composes it and in so doing gives rise to new modes of intelligence and belief. Within this medium, there is no meaningful distinction between what floats in the air and the air itself. Like the supernatural and its explanation, both are equally factitious.

This perspective owes something to Priestley's *Experiments and Observations*, if not to the decades of English aerography behind that book. Priestley had drawn his reader into a world of recombinant forms whose equality was, if disturbing, liberating too. But in Radcliffe, liberation and autonomy do not necessarily accompany a new form of air that can be made and unmade through composition. This is because the literary medium is finally the sole available register of experience. Hence the last chapter of *Udolpho* has *only* the "air" of a happy ending, albeit one replete with two conventional signatures of happiness: Emily's reunion with Valancourt, and the restoration of the lost property of La Vallée to her, as well as of Udolpho to the female line from which it was deflected by her aunt's murder. But these are only signatures, and their effect is to trap Emily, and the way of reading she models, in the purely fictional medium—the "nothing" that both Hazlitt and Wollstonecraft remarked on—that Priestley's comparable chemistry was able to escape.

When Valancourt and Emily, so "alike in [their] tempers" (*Udolpho*, 636), are at last reunited in a twilit garden, it is while Emily is playing "her favourite air." The music seems to call Valancourt out of thin air, though his words put Emily there instead: "'Ah Emily!' said Valancourt, 'that air, those words—alas!'" Yes. Alas. As the air fades away, Valancourt seems to convince Emily that his "conduct has been misrepresented." Nonetheless, "the deep twilight would not allow [him] to distinguish the astonishment and doubting joy, that fixed her features" (*Udolpho*, 670). Astonishment and doubting joy are, of course, exactly the combined affective and epistemological manners conducive to the experience of fiction, as opposed to the polite scanning of it. But in experiencing these mixed modes of feeling—thus reading her medium as she has been designed to—Emily herself fades into "that air." While Valancourt professes "a conviction that overpowers me," his bride-to-be remains stranded in the gloom of the medium that conveys her (*Udolpho*, 670). And that medium here loses its own discrimination, its capacity for self-reference. Drained of its animate reflexivity, it is no longer inclined to announce its own agency or particularity. The marriage that takes place in the next chapter might as well unfold in an "enchanted palace"—which is exactly what Emily's servant Annette supposes, having "read the fairy tales" (*Udolpho*, 673).

Pneumatic Institutions

The first edition of *The Mysteries of Udolpho* appeared in print in 1794. The next year saw major publications by two of Priestley's fellow "Lunar Men": Thomas Beddoes's *Letters on Asthma, Consumption, Fever and Other Diseases* and the first of Beddoes's and James Watt's five-part *Considerations on the Medicinal Use and on the Production of Factitious Airs.* Such "considerations" materialized in advance of the physical structure of Beddoes's Pneumatic Medical Institution in Bristol. As Mike Jay has recently explored in iridescent detail, the Pneumatic Medical Institution opened its doors (in 1799) because Beddoes "supposed that Factitious Airs might be advantageously introduced into the Practice of Medicine."[27] *Pace* Priestley, Beddoes divided "the inferior region of atmosphere" into "two kinds of air"—"vital, dephlogisticated, or oxygene [*sic*] air" and "azotic, phlogisticated, foul, or bad air."[28] Retaining, even doubling, the suspense that Priestley had built into atmosphere's nomenclature—and retaining too his awareness of how small a part of the atmosphere ("little more than a quarter") is any good at all—Beddoes supposed that "oxygene air," isolated and inhaled, could counter the diseases, such as consumption, then thought the result of "bad air" in general and of British insular air in particular (Beddoes, *Considerations*, 2:10). Before long, though, Beddoes's attention drifted to nitrous oxide. The "nitrous" part of this compound is the bad "kind of air," instigator of corruption and decay.[29] Ameliorated as an oxide, however, Priestley's "dephlogisticated nitrous air"—also known as "sweet air"—became for a time Bristol's Pneumatic Medical Institution's star gas.

What can its brief ascendancy have meant? Jay emphasizes that Beddoes himself was not convinced of what he said in public about the therapeutic potential of any of the factitious airs, confessing to his protégé, Humphry Davy, that "I do not believe in my own theories."[30] Robert Schofield likewise notes that, at the cultural level, "belief in the medical powers of gases [had] beg[u]n to wane" even as "conditions became right for [the Pneumatic Medical Institution's] opening"; indeed, the Pneumatic Medical Institution lasted less than ten years in its original configuration, soon dissolving into the Medical Institution for the Benefit of the Sick and Drooping Poor (Schofield, *Lunar Society*, 376). Its anesthetic potential largely overlooked, nitrous oxide itself proved significant mainly for its hallucinatory properties. Impervious to the caricature and contempt that their pneumatic revelries provoked, Beddoes, along with Coleridge, Lovell Edgeworth, Watt, Davy, and assorted others, giddily demonstrated sweet air's disposition to take those who inhaled it anywhere and nowhere.[31]

Built on disbelief and devoted to the suspense of ordinary sensation, Beddoes's Pneumatic Medical Institution looms as a fantastic analog not just to the architectural forms that dominated contemporary gothic fictions but to those fictions themselves, especially insofar as they instituted the thrill of transaction with factitious airs. Many of the "pneumatic revelers" who joined Beddoes and Davy in Bristol wrote delirious internal travelogues of their experiences with the "laughing gas," so-called because one of nitrous oxide's effects was the spontaneous maniacal laughter that turns up as a stock feature of gothic novelistic atmospheres. Stimulated by nothing but what was in the air—or rather by the airs that were now inside those who had inhaled them—pneumatic laughter mimics the mixed response that gothic tropes solicit. Once he has drawn this particular "air" out of a bag, Beddoes wrote, through a siphon, and into his own nose, the typical inhaler will "exhibi[t] a ludicrous struggle between the propensity to laugh undoubtedly produced by the air, and an eager desire to continue the inhalation." One such inhaler—Robert Southey, as it happens—reported that "his first definite sensations were a fullness and dizziness in the head, such as to induce fear of falling. This was succeeded by a laugh, which was involuntary, but highly pleasurable, accompanied by a peculiar thrilling in the extremities."[32]

In this apparent parody of gothic reading protocols, different kinds of sensation (pleasure, pain, or "fear") and contradictory proclivities (to giggle, to inhale) are synthesized within an experience that seems to originate outside the human mind but in fact arises quite literally within it. The "extremities" that typify that experience are no less those of the body that undergoes it. Jay notes the heightening of sensation in which the inhaler (in this case Davy) "'lost all connection with external things' and, paradoxically, entered a self-enveloping interior world of the senses" (Jay, *Atmosphere of Heaven*, 198). Sense experience here is linked to subjective pleasure, not to the sensus communis that had long validated both social and natural knowledge and, indeed, once shuttled between them. The essence of the nitrous oxide experience is dissociation, not association. At the same time, however, to take in this factitious air was also to disport in it notionally, and in practice it was still to do so with like-minded others and then to realize the experience in the sociable sphere of letters. Written accounts of pneumatic revelry thus rendered the factitious airs a vehicle of transport between autonomy and compulsion, isolation and community. Such transport occurred, if anywhere, within a shared sphere of provisional, if not pretended, belief in the possibility of an experience that was as concentrated and intense as it was abstracted from the outside world.

Radcliffe's fiction, for its part, makes no direct mention of nitrous oxide, carbon dioxide, or oxygen. Her "ether" is not the organic compound of growing interest to pneumatic chemists but the rarefied "blue" element of the "upper regions of the air." Chemical ether's absence from the gothic pages on which the word "ether" appears makes sense insofar as ether, finally, is not a factitious air that has to be synthesized. It obscures the role that human manufacture plays in its production and, unlike Radcliffe's factitious airs, is not visibly involved with physical and aesthetic processes of corruption and vivacity. As if to demonstrate gothic romance's continuing commitment, rather, to ether's others—the factitious airs—even Matthew Lewis's lurid *Ambrosio; or, the Monk* (1796) unfolds in Spanish crypts as likely to house a "laboratory" as a tomb. The "greening liquors," putrescent, grotesquely revivifying bodies, and "sulphureous flames" that stock Lewis's noxious catacombs amplify Radcliffe's airs even as they satirize recent experiments with the "mephitic" air of prisons and other enclosed spaces. Meanwhile, Lewis's theatrical Ambrosio wears both an "air of conscious superiority" and a "complexion . . . of a deep but clear brown" that encodes the sultry climate of his Jamaican origins, not to mention those of the Lewis family fortune. Both renditions of air make visible the engrossing properties of Lewis's romance.[33] Later, Lewis's *Journal of a West India Proprietor* (1834) would contemplate the Caribbean climate's "seasoning" of African slaves, even as it complained that "as to air, we have not enough to bless ourselves with."[34] Lewis's earlier gothic fiction rehearses this complaint as it systematically replaces natural air with factitious ones upheld by superstition and convention.

Lewis had set about writing *The Monk* so as to mock Radcliffe. But he ended by resynthesizing the factitious airs that built her fictions—airs that retain aesthetic coherence and sensational form even as their analytic grip on the moral, political, and physical world slips. Indeed, *The Mysteries of Udolpho* institutionalizes exactly this aspect of gothic fiction, if not of "fiction" now broadly understood. Instead of legitimating literary fiction's claim to articulate and investigate actual states of the air, "atmosphere" ceased to be invoked explicitly except in order to signify a malevolent, compelling integrity. For example, in *Things as They Are*; or, *The Adeventures of Caleb Williams*, the political novel he published the same year as *Udolpho*, William Godwin sought "a unity of spirit and interest" that would "giv[e] it a powerful hold on the reader." Holding of this sort, identical with grip, required that Godwin himself be held by "an atmosphere of romance." Thus of his novel's chief antagonist—the seductive Falkland, made possible by eighteenth-century social forms—Godwin later confessed that "it was necessary to make him, so to speak, the tenant of an atmosphere of romance, so that every reader should feel compelled almost to worship him for his high qualities."[35] In Godwin's novel, Falkland seems to control the very air; his supposed victim, Caleb Williams, blames him for a "grey and cloudy atmosphere," likens his sphere of influence to the Roman empire's "spread from climate to climate," and records that, with respect to England, Scotland, and Wales, Falkland has forbidden him to "go out of these climates," thereby essentially rendering them one single, oppressive entity (Godwin, *Caleb Williams*, 145, 289, 288). Falkland's "atmosphere of romance" is not Priestley's or Radcliffe's—or even Lewis's—conspicuously composed "air": it is a tendentious Jacobin nightmare about elite sociability that ultimately purports to give us, as Godwin's title has it, "things as they are."

Detached from overt rhetorics of political justice and unconvinced that fiction's business is things—"as they are" or otherwise—Radcliffe's factitious airs analyze apparent unities from within. And yet, for this very reason, her fictions make their man-made particularities apparent in a way that Godwin could not. His atmospheric novel's "hold on" us thus differs radically from Radcliffe's hold *of* us. And it was Radcliffe's subtler atmospheres, not Godwin's, that would persist in Edward Bulwer Lytton's "dark and stormy night"; in the mood-setting meteorologies of Collins, Hardy, and the Brontës; and in the "rain ever falling drip drip drip" that realizes Dickens's fantastic moral worlds.[36] As is not the case with Radcliffe, however, the expressive atmospheric effects so interwoven with anglophone writing after the end of the eighteenth century are ones that we take entirely for granted.[37] They lack the analytic power that Radcliffean air still carried, thanks to its commerce with the airs so apparent to earlier English writers. In a telling iteration of Marx's "all that is solid melts into air," air's fantastic properties were to disappear into the English novel's realist program.[38] This means that the conditions of fiction's realization—and of the impression of immediacy that it creates—were themselves obscured, where once they would have been elucidated through atmospheric signs. We are, for instance, unlikely to think twice about the last name of Charlotte Brontë's Jane Eyre; instead of emerging as a subjected expression of the "rain, wind, and darkness that filled the air," Jane seems to stand out from and against the very elements that in reality convey her to Brontë's hailed "Reader."[39]

In nineteenth-century comic fiction, however, this appears to be a little less true. "To save the space usually devoted to explanations of the state of the weather in books of this kind," the first copy-text of Mark Twain's *Pudd'nhead Wilson* (1894) deleted a sardonic "system of weather signs." That series of ideograms would have signaled conditions ranging from "sunny" to "pitch dark" and from "moonlight" to "fog" (the icon of an umbrella would indicate when it was "rainy").[40] As Twain's example suggests, the kind of atmosphere that eighteenth-century writers had in mind would linger in the ironists who allegedly repudiated Radcliffe's gothic style. Jane Austen's famous parody of Radcliffean romance, *Northanger Abbey* (1818), is a case in point. As everybody knows, Austen's heroine in training, Catherine Morland, has "read all Mrs. Radcliffe's works, and most of them with great pleasure." As for *The Mysteries of Udolpho*, she "remember[s] finishing it in two days—[her] hair standing on end the whole time."[41]

Catherine's infantile attachment to Radcliffe's writing is usually interpreted as a symptom of her seduction by the gothic tropology that had come to frame female knowledge of the social world. But what *of* that hair standing on end? Far less explicit—and thus far more tenacious and pervasive—is the Radcliffean atmosphere that Austen's own writing appears to have absorbed. How much may be measured on the day that Catherine is hoping for a walk through Priestley's (and Joseph Glanvill's) Bath with her new friends, the Tilneys. Weather intervenes in the form of ominously gathering clouds. But with it arrives a splendid opportunity for projection: "A bright morning so early in the year, [Catherine] allowed would generally turn to rain, but a cloudy one foretold improvement as the day advanced" (Austen, *Northanger Abbey*, 61). Catherine hopes for affirmation of her optimism, but her LaLuc-like guardian, "not having his own skies and barometer about him, declined giving any absolute promise of sunshine," and by late morning "a few specks of small rain" are striking the window. "The clock struck twelve, and it still rained." Catherine wishes "that we had such weather here as they had at Udolpho, or at least in Tuscany and the South of France!—the night that poor St. Aubin died!—such beautiful weather!" (Austen, *Northanger Abbey*, 61).

Like Scott, Catherine gets the names wrong. It was St Aubert, Emily's father, who died. But she also gets the atmosphere wrong. "Such weather . . . as they had at Udolpho" is exactly the weather they are having here at Bath. Catherine's missing what never was yields only superficial difference between her "common" world—the world of literary realism—and the "factitious" sphere of Radcliffean romance. The difference is marked only by the fact that no one in Radcliffe's novel is equipped with an umbrella, whereas in *Northanger Abbey* "there are four umbrellas up already!" (Austen, *Northanger Abbey*, 61). Catherine's querulous reaction to this fact ("How I hate the sight of an umbrella!") retains the realizing, yet self-reflexive potential that Radcliffe transferred from earlier English aerographies into her atmospheric fiction. So does Austen's comedy of manners—of airs apparent—at once leash and unleash a long century's unsettling conviction that there may be something in the air besides the atmosphere—or there may be nothing there at all.

Notes

Introduction

1. The latest proponent of this view is Steven Johnson, *The Invention of Air: A Story of Science, Faith, Revolution, and the Birth of America* (New York: Riverhead, 2008). Its most influential proponents are Steven Shapin and Simon Schaffer in *Leviathan and the Air Pump: Hobbes, Boyle, and the Experimental Life* (Princeton, NJ: Princeton University Press, 1985).

2. Deidre Lynch offers a witty summary of this view in "Gothic Fiction," in *The Cambridge Companion to British Romantic Fiction*, ed. Katie Trumpener and Richard Maxwell (Cambridge: Cambridge University Press, 2008), 47–65.

3. Gabriele Schwab engages the elusively atmospheric aspects of poetry in terms of Christopher Boas's theory of an "unthought known" rooted in the mother-infant bond and in "Words and Moods: The Transference of Literary Knowledge," *SubStance* 26 (1997): 107–27. A related concept is the German one of *Stimme*, or "voice-feeling," whose sense overlaps with the notion of *Stimmung*, meaning "temper," "mood," or, indeed, "atmosphere." See Geoffrey Hartman, *Saving the Text: Literature, Derrida, Philosophy* (Baltimore: Johns Hopkins University Press, 1981), and *Stimme(n) im Text: Narratologie und Positionsbestimmungen*, ed. Andreas Blödorn, Daniele Langer, and Michael Scheffel (Berlin: W. de Gruyter, 2006). It is, unsurprisingly, critics of Henry James who have been most drawn to the link between literal depictions of states of the air and the more subtle aspects of "atmosphere," as witness Thomas J. Otten, *A Superficial Reading of Henry James: Preoccupations with the Material World* (Columbus: Ohio State University Press, 2006). See also Ronald Schleifer's "Material Voices: Tourette Syndrome, Neurobiology, and the Affect of Poetry" in *Intangible Materialism: The Body, Scientific Knowledge, and the Power of Language* (Minneapolis: University of Minnesota Press, 2009). In his critique of a "strictly mechanist conception of matter" (Schleifer, 36) that relies on an organized system for its coherence, Schleifer proposes reassembling the parts of such a system to produce new orders of experience that are not reducible to their material constituents.

4. William Empson, *Seven Types of Ambiguity* ([1930] Harmondsworth, UK: Penguin, 1961), 17.

5. Karl Marx, *Capital: A Critique of Political Economy*, ed. Frederick Engels, trans. Samuel Moore and Edward Aveling, 2 vols. ([1867–94] New York: International Publishers, 1967), 1:74.

6. John Wilkins, *A Discovery of a New World, or, A Discourse Tending to Prove, that 'tis Probable there may be another Habitable World in the Moon* ([1638] London, 1684), 92.

7. John Wilkins, *Mercury; or the Secret and Swift Messenger* (London, 1641), 152.

8. Arthur Stanley Eddington, *New Pathways in Science* ([1935] Ann Arbor: University of Michigan Press, 1956), 8.

9. Boyle was established as a literary technologist in Shapin and Schaffer's *Leviathan and the Air Pump*, and this status has been reaffirmed by that book's many disciples. Notable here are Rose-Mary Sargent, *The Diffident Naturalist: Robert Boyle and the Philosophy of Experiment* (Chicago: University of Chicago Press, 1995), and Jan Golinski, "Robert Boyle: Skepticism and Authority in Seventeenth-Century Chemistry," in *The Figural and the Literal: Problems of Language in the History of Science, 1630–1800*, ed. Andrew Benjamin, Geoffrey Cantor, and John R. R. Christie (Manchester, UK: Manchester University Press, 1987), 58–82.

10. Early modern natural philosophy's reliance on specifically literary tropologies has been heavily investigated, as witness James Bono's discussion of scientific "bookishness" in *The Word of God and the Languages of Man: Interpreting Nature in Early Modern Science and Medicine* (Madison: University of Wisconsin Press, 1995). See also Bono, "Science, Discourse, and Literature: The Role/Rule of Metaphor in Science," in *Literature and Science: Theory and Practice*, ed. Stuart Peterfreund (Boston: Northeastern University Press, 1990), 59–90; Jan Golinski, "The Literature of the New Science," in *The Cambridge History of Romantic Literature*, ed. James Chandler (Cambridge: Cambridge University Press, 2008), 527–52; numerous essays in *The Literary Structure of Scientific Argument: Historical Studies*, ed. Peter Dear (Philadelphia: University of Pennsylvania Press, 1991), but especially Lissa Roberts, "Setting the Table: The Disciplinary Development of Eighteenth-Century Chemistry as Read through the Changing Structure of Its Tables," 99–132; and Wilda Anderson, *Between the Library and the Laboratory: The Language of Chemistry in Eighteenth-Century France* (Baltimore: Johns Hopkins University Press, 1984).

11. John Arbuthnot, *An Essay concerning the Effect of Air on Human Bodies* (London, 1733), 11.

12. A paradigmatic treatment of questions of eighteenth-century climate and ecology from a cultural studies perspective may be found in the many fine essays gathered under the rubric "Ecohistoricisms" in the *Journal for Early Modern Cultural Studies* 8 (2008).

13. Samuel Johnson, in *The Works of Samuel Johnson*, vol. 9, *A Journey to the Western Islands of Scotland*, ed. Mary Lascelles (New Haven, CT: Yale University Press, 1971), 115.

14. The standard study of the fixation on bardic culture as a form of empire building remains Katie Trumpener's *Bardic Nationalism: The Romantic Novel and the British Empire* (Princeton, NJ: Princeton University Press, 1997), though the specific role that Johnson's writing played in this process is taken up by Deidre Lynch in "'Beating the Track of the Alphabet': Topography, Lexicography, and Johnson's Visions of the Ideal," *ELH* 57 (1990): 357–405. Leigh Eric Schmidt demonstrates "the eye's clear eclipse of the ear [and] the decline of listening in the face of the ascendant power of vision" (Schmidt, 15) in the eighteenth century and traces the persistence of auricular communicative forms in *Hearing Things: Religion, Illusion, and the American Enlightenment* (Cambridge, MA: Harvard University Press, 2000). See also Constance Classen, *Worlds of Sense: Exploring the Senses in History and across Cultures* (London: Routledge, 1993). The growing legitimacy of the sense of smell at the end of the eighteenth century is taken up by Alain Corbin in *The Foul and the Fragrant: Odor and the French Social Imagination* (Cambridge, MA: Harvard University Press, 1986). See also Constance Classen and David Howes, *Aroma: The Cultural History of Smell* (London: Routledge, 1994).

15. Bruno Latour, *Science in Action: How to Follow Scientists and Engineers through Society* (Cambridge, MA: Harvard University Press, 1987). For further commentary on this idea and its relevance to the imbrications of "natural" and cultural knowledge, see Jan Golinski, *Making*

Natural Knowledge: Constructivism and the History of Science (Cambridge: Cambridge University Press, 1998), 37–39.

16. Bruno Latour, *We Have Never Been Modern*, trans. Catherine Porter (Cambridge, MA: Harvard University Press, 1993).

17. The locus classicus of this view is, of course, Keith Thomas's magisterial *Religion and the Decline of Magic: Studies in Popular Beliefs in Sixteenth and Seventeenth Century England* (Oxford: Oxford University Press, 1971). Thomas's work follows the still valuable paradigm established by W. E. H. Lecky in *History of the Rise and Influence of the Spirit of Rationalism in Europe* ([1878] New York: Braziller, 1955). The secularization of magic via new technologies of vision and its persistence as a shaping force in modern culture is the thesis of Simon During's *Modern Enchantments: The Cultural Power of Secular Magic* (Cambridge, MA: Harvard University Press, 2002). Terry Castle subtly traces the transposition of the uncanny into atmospheric devices such as the thermometer and the barometer—and their migration thence into cultural metaphor and the language of the everyday—in *The Female Thermometer: Eighteenth-Century Culture and the Invention of the Uncanny* (Oxford: Oxford University Press, 1995). On the ascendance of phantasmagoria as pure entertainment via the camera obscura and other precursors to photography and cinema, see Marina Warner's wide-ranging *Phantasmagoria: Spirit Visions, Metaphors, and Media into the Twenty-First Century* (Oxford: Oxford University Press, 2006). Still the most useful and concrete documentation of the persistence of straightforward investment in the supernatural is Michael Hunter's introduction to *The Occult Laboratory: Magic, Science and Second Sight in Late Seventeenth-Century Scotland* (Rochester, NY: Woodbridge, 2001), 1–31. See also Montagu Summers, *The Geography of Witchcraft* (Evanston, IL: Northwestern University Press, 1958), and Herbert Leventhal, *In the Shadow of the Enlightenment: Occultism and Renaissance Science in Eighteenth-Century America* (New York: New York University Press, 1976).

18. Here one can only cite the many studies of the novel that have grappled with Ian Watt's thesis in *The Rise of the Novel* (Berkeley: University of California Press, 1957). Especially relevant, however, are Michael McKeon, *The Origins of the English Novel, 1600–1740* (Baltimore: Johns Hopkins University Press, 1987); Nancy Armstrong, *Desire and Domestic Fiction* (Oxford: Oxford University Press, 1987); Homer Brown, *Institutions of the English Novel* (Philadelphia: University of Pennsylvania Press, 1994); and William B. Warner's pivotal "The Elevation of the Novel in England: Hegemony and Literary History," *ELH* 59 (1992): 577–96.

19. On character: Deidre Lynch, *The Economy of Character: Novels, Market Culture, and the Business of Inner Meaning* (Chicago: University of Chicago Press, 1994). On plot: Peter Brooks, *Reading for the Plot: Design and Intention in Narrative* (Cambridge, MA: Harvard University Press, 1984). On style: D. A. Miller, *Jane Austen, or, the Secret of Style* (Princeton, NJ: Princeton University Press, 2003).

20. Thing theory—and the tradition of reading from this perspective—begins with Bill Brown, *A Sense of Things: The Object Matter of American Literature* (Chicago: University of Chicago Press, 2003), and has been extended into eighteenth-century studies by Cynthia Wall's luminous *The Prose of Things: Transformations of Description in the Eighteenth Century* (Chicago: University of Chicago Press, 2006); Christopher Flint, "Speaking Objects: The Circulation of Stories in Eighteenth-Century Prose Fiction," *PMLA* 113 (1998): 212–22; and Mark Blackwell, ed., *The Secret Life of Things: Animals, Objects, and It-Narratives in Eighteenth-Century England* (Lewisburg, PA: Bucknell University Press, 2007).

21. Hubert Damisch, *A Theory of /Cloud /: Toward a Cultural History of Painting*, trans. Janet Lloyd (Stanford, CA: Stanford University Press, 2002), 35, 17.

22. Mark Dorrian, "Clouds of Architecture," *Radical Philosophy* 144 (July 2007): 26.

23. On the priority of intelligibility within the mechanistic science of the period, as opposed to the instrumentality that "displaced" it, see Peter Dear, *The Intelligibility of Nature: How Science Makes Sense of the World* (Chicago: University of Chicago Press, 2006). In Dear's reading, the overriding value of intelligibility selected specific textual praxes.

24. Mary Favret, *War at a Distance: Romanticism and the Making of Modern Wartime* (Princeton, NJ: Princeton University Press, 2010).

25. Jean-Louis Barrault, *The Theatre of Jean-Louis Barrault*, trans. Joseph Chiari ([1959] New York: Hill and Wang, 1961), 61.

26. Mary Wollstonecraft, *A Short Residence in Sweden, Norway and Denmark*, ed. Richard Holmes ([1796] London: Penguin, 2007), 63.

27. Kenneth R. Olwig, "All That Is Landscape Is Melted into Air: The 'Aerography' of Ethereal Space," *Environment and Planning D: Society and Space* (2011): 525.

28. Joseph Glanvill, *Saducismus Triumphatus; or, Full and Plain Evidence concerning Witches and Apparitions* (1681), in *Collected Works of Joseph Glanvill*, ed. Bernhard Fabian, 9 vols. (Hildesheim, Germany: Georg Olms Verlag, 1978), 9:115.

29. Jane Austen, *Northanger Abbey*, ed. John Davie and Claudia L. Johnson ([1817] Oxford: Oxford University Press, 2003), 82.

Chapter One

1. Samuel Johnson, *Idler* 11 (June 24, 1758), in *The Works of Samuel Johnson*, vol. 2, *"The Idler" and "The Adventurer,"* ed. W. J. Bate, John M. Bullitt, and L. F. Powell (New Haven, CT: Yale University Press, 1963), 36.

2. These and other weather events at the end of the Little Ice Age are depicted in H. H. Lamb, *Climate, History and the Modern World* (London: Methuen, 1982), 232, 208. For a summary of the instability that attended the end of the Little Ice Age, see Robert Markley, "'Casualties and Disasters': Defoe and the Interpretation of Climate Instability," *Journal for Early Modern Cultural Studies* 8 (2008): 102–3.

3. Samuel Johnson, *A Dictionary of the English Language*, 2 vols. (London, 1755), 1:103.

4. Samuel Johnson, preface to *A Dictionary of the English Language*, in *Works of Samuel Johnson*, vol. 18, *Johnson on the English Language*, ed. Gwin J. Kolb and Robert DeMaria (New Haven, CT: Yale University Press, 2005), 113. In *Johnson's Dictionary and the Language of Learning* (Chapel Hill: University of North Carolina Press, 1986), Robert DeMaria considers Johnson's *Dictionary* as a "book of books" (DeMaria, 4) in a culture devoted to making all knowledge bibliographic. For the *Dictionary* as a response to the large-scale objectification of fact through print and the loss of "aura" through the repression of orality, with a discussion of Johnson's own sense of defeat, it is best to consult Alvin Kernan, *Samuel Johnson and the Impact of Print* (Princeton, NJ: Princeton University Press, 1987), 181–203.

5. George Cheyne, *The English Malady; or, A Treatise of Nervous Diseases of all Kinds, as Spleen, Vapours, and Lowness of Spirits* (London, 1733), 1.

6. Raymond Williams, *Keywords: A Vocabulary of Culture and Society* ([1976] New York: Oxford University Press, 1983), 18; Freya Johnston, "Accumulation in Johnson's *Dictionary*," *Essays in Criticism* 57 (2007): 306. Johnston also comments on Williams's literal understanding of Johnson's air of "frigid tranquility." On melancholy and climate, see Eric Gidal, "'O Happy Earth! Reality of Heaven!': Melancholy and Utopia in Romantic Climatology," *Journal for Early Modern Studies* 8 (2008): 74–101.

7. Isaac Watts, *Logick, or the Right Use of Reason in the Enquiry after Truth* (London, 1725), 130. In an inadvertent demonstration of the airy inconsistency of print, Johnson actually misquotes Watts as follows: "If I were to tell what I mean by the word *air*, I may say, it is that fine matter which we breathe in and breathe out continually; or it is that thin fluid body, in which the birds fly, a little above the earth; or it is that invisible matter, which fills all places near the earth, or which immediately encompasses the globe of earth and water." Johnson, *Dictionary*, 1:103.

8. Karl Marx, *The German Ideology*, trans. C. J. Arthur ([1845] New York: International Publishers, 1939), 51.

9. Aristotle, *Meteorologica*, trans. H. D. P. Lee (Cambridge, MA: Harvard University Press, 1952), 33.

10. Arden Reed, *Romantic Weather: The Climates of Coleridge and Baudelaire* (Hanover, NH: University Press of New England, 1983), 38–39.

11. John Mills, *An Essay on the Weather* (London, 1770), 89.

12. On the natural complicity of lyric and narrative forms in eighteenth-century writing despite their recent segregation in the criticism, see G. Gabrielle Starr, *Lyric Generations: Poetry and the Novel in the Long Eighteenth Century* (Baltimore: Johns Hopkins University Press, 2004).

13. Lennard Davis traces the emergence of a special category of fiction as a kind of *vraisemblance* detached from laws of reference out of an undifferentiated fact/fiction matrix in *Factual Fictions: The Origins of the English Novel* (New York: Columbia University Press, 1983).

14. *A Wonderful History of all the Storms . . . that Have Happened in England* (1704), 65–66.

15. Virginia Woolf, *On Being Ill* ([1926] Ashfield, MA: Paris Press, 2006), 26.

16. Mark Twain, "The Weather in This Book," in *The American Claimant*, ed. Shelley Fisher Fishkin ([1892] Oxford: Oxford University Press, 2010), 11. For his part, Twain vows that "no weather will be found in this book. This is an attempt to pull a book through without weather." He ultimately relents and decides "to gather such weather as is necessary for the book from qualified and recognized experts. . . . This weather will be found over in the back part of the book" (Twain, 11).

17. Thomas H. Uzzell, *Narrative Technique: A Practical Course in Literary Psychology* (New York: Harcourt Brace, 1923), 363–65. Thanks to Jonathan Grossman for this reference.

18. John Wilkins, *A Discourse concerning a New World, and Another Planet* ([1638] London, 1640), 113.

19. Joseph Glanvill, *The Easie, True, and Genuine Notion and Consistent Explication of a Spirit* (1683), in *Collected Works of Joseph Glanvill*, ed. Bernhard Fabian, 9 vols. (Hildesheim, Germany: Georg Olms Verlag, 1978), 9:99.

20. Marshall Brown, *The Gothic Text* (Stanford, CA: Stanford University Press, 2005).

21. Steven Shapin and Simon Schaffer, *Leviathan and the Air Pump: Hobbes, Boyle, and the Experimental Life* (Princeton, NJ: Princeton University Press, 1985), 344.

22. Steven Johnson, *The Invention of Air: A Story of Science, Faith, Revolution, and the Birth of America* (New York: Riverhead, 2008), 62. For a recent and wider-ranging popular study of air that complements Johnson's, but appeared after this book was complete, see Steven Connor, *The Matter of Air: Art and Science of the Ethereal* (London: Reaktion, 2010).

23. Vladimir Janković, *Reading the Skies: A Cultural History of the English Weather, 1650–1820* (Chicago: University of Chicago Press, 2000); Mark Monmonier, *Air Apparent: How Meteorologists Learned to Map, Predict, and Dramatize the Weather* (Chicago: University of Chicago Press, 1999); and Jan Golinski, *British Weather and the Climate of Enlightenment* (Chicago: University of Chicago Press, 2007). (The Windy City would seem to be a natural publishing venue for books about the weather.) A very different perspective on early nineteenth-century sky study

is offered in Richard Hamblyn's study of Howard's romantic-era *On the Modifications of Clouds* (1804) in *The Invention of Clouds: How an Amateur Meteorologist Forged the Language of the Skies* (New York: Picador, 2001). On the systematization of meteorology over the eighteenth century, see Theodore S. Feldman, "Late Enlightenment Meteorology," in *The Quantifying Spirit in the Eighteenth Century* (Berkeley: University of California Press, 1990), 143–78. The conversion of weather prediction into elite science in the nineteenth century is taken up by Katherine Anderson, *Predicting the Weather: Victorians and the Science of Meteorology* (Chicago: University of Chicago Press, 2005).

24. Francis Bacon, *Works of Francis Bacon*, vol. 5, *The History of the Winds*, ed. James Spedding, Robert Leslie Ellis, and Douglas Denon Heath (London, 1889), 143, 200.

25. Edmund Bohun, *A Discourse concerning the Origine and Properties of Wind* (London, 1671), 1, 4.

26. Thomas Short, *A General Chronological History of the Air, Weather, Seasons, Meteors &c in Sundry Places and at Different Times* (London, 1749), vi–vii.

27. Edward Barlow, *Meteorological Essays* (London, 1715), 42.

28. John Pointer, *A Rational Account of the Weather* (London, 1738), 164.

29. John Claridge, *The Shepherd of Banbury's Rules to Judge of the Changes of the Weather, Grounded on Forty Years Experience* (London, 1744), viii.

30. John Mills, *An Essay on the Weather; with remarks on The Shepherd of Banbury's rules* ([1770] London, 1773), 2.

31. *A Calendar of the Weather for the Year 1781* (London, 1781), 9. Golinski sorts through the shepherd's complex literary genealogy and also notes that his ties to the classical pastoral gave him additional status within elite culture. See Golinski, *British Weather*, 72–73.

32. Terry Castle, *The Female Thermometer: Eighteenth-Century Culture and the Invention of the Uncanny* (Oxford: Oxford University Press, 1995), 25.

33. Recent summaries of climate theory in the context of modern forms of identity, including racial, national, and personal, may be found in Dror Wahrman, *The Making of the Modern Self: Identity and Culture in Eighteenth-Century England* (New Haven, CT: Yale University Press, 2004), 83–126, and Felicity A. Nussbaum, *Torrid Zones: Maternity, Sexuality, and Empire in Eighteenth-Century English Narratives* (Baltimore: Johns Hopkins University Press, 1995).

34. John Arbuthnot, *An Essay concerning the Effect of Air on Human Bodies* (London, 1733), 151–52.

35. Hippocrates's scheme also informed new assumptions about the relationship between psychology and environment, such as David Hartley's notion that "internal Feelings of the Mind" were forged by some "subtle elastic Fluid" that was "the Means for their rise and free Propagation"; this material was "so subtle and elastic a Fluid as the Aether described by Sir Isaac Newton" and could "reasonably be supposed to generate the Pores of the medullary Substance . . . in the same Manner as Air penetrates grosser Cavities and Pores, and, like Air, both be itself agitated by Vibrations from a Variety of Causes, and also communicate these to the medullary Particles." See Hartley, *Observations on Man* (London, 1749), 1:25.

36. Mark Harrison's *Climates and Constitutions: Health, Race, Environment and British Imperialism in India, 1600–1850* (Oxford: Oxford University Press, 1999) begins with an extraordinarily detailed and fluent survey of the role of particle theory—which "conceive[s] of particles as participants in a regular, dynamic system whose harmonies could be expressed in mathematical terms" (Harrison, 41)—in dismantling Hippocrates's deterministic model from within by guaranteeing "the *indeterminacy* of the air's action upon human bodies" (Harrison, 65). See also

Rajani Sudan, "Chilling Allahabad: Climate Control and the Production of Anglicized Weather in Early Modern India," *Journal for Early Modern Cultural Studies* 8 (2008): 56–73.

37. David Hume, "Of National Characters," in *Political Essays*, ed. Knud Haakonssen (Cambridge: Cambridge University Press, 1994), 85. Hume's treatment of climate theory is treated as bankrupt political theory in Lucian Boia, *The Weather in the Imagination*, trans. Roger Leverdier (King's Lynn, UK: Reaktion, 2005), 50–54.

38. David Hume, *A Treatise of Human Nature*, vol. 1, *Texts*, ed. David Tate Norton and Mary J. Norton ([1739–40] Oxford: Oxford University Press, 2005), 23.

39. "The same phenomenon occurs very often with the south wind, when the atmosphere is especially heavy and thick." Seneca, *Naturales Quaestiones*, trans. Thomas H. Corcoran, 2 vols. (Cambridge, MA: Harvard University Press, 1872), 1:25.

40. See Daniel Tiffany, *Toy Medium: Materialism and Modern Lyric* (Berkeley: University of California Press, 2000), 201–2, and Michel Serres, *La Naissance de la physique dans le texte de Lucrèce: Fleuves et Turbulences* (Paris: Minuit, 1977).

41. This is the view painstakingly developed by Richard Kroll in *The Material Word: Literate Culture in the Restoration and Early Eighteenth Century* (Baltimore: Johns Hopkins University Press, 1991).

42. Joseph Addison, *Spectator* 110 (July 6, 1711), in *The Spectator*, ed. Donald F. Bond, 5 vols. (Oxford: Clarendon, 1965), 1:455.

43. Henry More, "The Immortality of the Soul" (1662), in *A Collection of Several Philosophical Writings of the Late Henry More* (London, 1712), 169. The 1712 edition is the fourth in which More's Neoplatonic treatise appeared.

44. On the history of theological views of Satan as a tenant of the air, as opposed to Hell, see Henry Ansgar Kelly, *Satan: A Biography* (Cambridge: Cambridge University Press, 2006), 174–81.

45. This is the thesis of E. J. Clery's enlightening and perceptive *The Rise of Supernatural Fiction, 1762–1800* (Cambridge: Cambridge University Press, 1995), and also the way that Michael McKeon accounts for the fate of apparition narrative in *The Origins of the English Novel, 1600–1740* (Baltimore: Johns Hopkins University Press, 1987), 83–89.

46. Steven Shapin, *The Scientific Revolution* (Berkeley: University of California Press, 1996), 50.

47. See, among her many contributions to the topic, Katharine Mary Briggs, *The Anatomy of Puck: An Examination of Fairy Beliefs among Shakespeare's Contemporaries and Successors* (London: Routledge, 1954), and *The Vanishing People: A Study of Traditional Fairy Beliefs* (New York: Pantheon 1978).

48. Robert Kirk, *The Secret Commonwealth; or A Treatise Displaying the Chief Curiosities among the People of Scotland as They are in Use to this Day* (1692), reprinted in Michael Hunter, ed., *The Occult Laboratory: Magic, Science and Second Sight in Late Seventeenth-Century Scotland* (Rochester, NY: Woodbridge, 2001), 78–79.

49. Robert Plot, *A Natural History of Staffordshire* (London, 1686), 10.

50. Dorothea von Mücke, *The Seduction of the Occult and the Rise of the Fantastic Tale* (Stanford, CA: Stanford University Press, 2003). Mediated through a careful reading of Ludwick Tieck's "Über Shakespeares Behandlung des Wunderbaren" (1793), von Mücke's analysis of the "mixed semiotic universe" that blends "markers of the real" and "markers of the imaginary" (von Mücke, 6–7) is unusually persuasive.

51. Benjamin Franklin, *Experiments and Observations on Electricity at Philadelphia in America* ([1751] London, 1769), 11.

52. Ludwig Wittgenstein, *Philosophical Investigations*, trans. G. E. M. Anscombe ([1953] New York: Macmillan, 1983), 607.

53. *Compact Edition of the Oxford English Dictionary*, 2 vols. (Oxford: Oxford University Press, 1984), 1:135.

54. Joseph Priestley, *A Familiar Introduction to the Theory and Practice of Perspective* (London, 1770), 63.

55. Leo Spitzer, "Milieu and Ambiance," in *Essays in Historical Semantics* (New York: Russell and Russell, 1948), 179–80.

56. Kenneth R. Olwig, "Performance, *Aetherial* Space, and the Practice of Landscape / Architecture in the Case of the Missing Mask," *Social and Cultural Geography* 12 (2011): 305–18.

Chapter Two

1. All references to *Paradise Lost* are taken from *Poems of John Milton*, ed. John Carey and Alistair Fowler (London: Longman, 1968) and are cited parenthetically in the text by book and line number.

2. Early but influential interpretations of Eve as the first reader include Christine Froula, "When Eve Reads Milton: Undoing the Canonical Economy," *Critical Inquiry* 10 (1983): 321–42, and Stanley Fish, *Surprised by Sin: The Reader in Paradise Lost* (Berkeley: University of California Press, 1971).

3. Such an accident would have aligned him with the eldest of his three daughters, whose "difficult utterance of speech," according to Milton's nephew Edward Phillips, was "to say truth I doubt the principal cause of excluding her" from the task of reading to him. See Edward Phillips, "The Life of Milton," in Milton, *Complete Poems and Prose*, ed. Merritt Y. Hughes (Indianapolis: Hackett, 2003), 1037.

4. John Milton, *Areopagitica*, in *Complete Prose Works of John Milton*, vol. 2, *1643–1648*, ed. Ernest Sirluck (New Haven, CT: Yale University Press, 1959), 524. In a valuable contribution, Arden Reed discusses Milton's engagements with the language of weather throughout his *Romantic Weather: The Climates of Coleridge and Baudelaire* (Hanover, NH: University Press of New England, 1983), 35–37. Abbe Blum links Milton's sense of speech in *Areopagitica* to classical rhetorical modes and modern political exigency in "The Author's Authority: *Areopagitica* and the Labor of Licensing." in *Re-Membering Milton*, ed. Mary Nyquist and Margaret Ferguson (New York: Methuen, 1988), 74–96.

5. Milton's angels are treated as artifacts of declining belief and in relation to the turbulent contemporary debate on the supernatural that lured scientists such as Boyle in Robert H. West, *Milton and the Angels* (Athens: University of Georgia Press, 1955).

6. John Wilkins, *Mercury; or the Secret and Swift Messenger* (London, 1641), 118–19. Katherine Ellison considers Wilkins's cryptography as a protofictional reading manual in "Cryptogrammatophoria: The Romance and Novelty of Losing Readers in Code," *Eighteenth-Century Fiction* 20 (2008): 281–306.

7. George Hammond, *Angelographia* (London, 1690), 120.

8. John Evelyn, *Fumifugium; of the Inconveniencie of the AER of London* (London, 1661), 47. Ken Hiltner links Satan with contemporary air pollution in "'Belch'd Fire and Rowling Smoke': Air Pollution in Paradise Lost," in *Milton, Rights and Liberties*, ed. Christophe Tourner and Neil Forragh (Bern, Switzerland: Peter Lang, 2007), 293–302.

9. Milton's tussles with the contemporary notion that "northern climates had a deleterious

effect on the human mind" (Fink, 67) is taken up in Z. S. Fink, "Milton and the Theory of Climatic Influence," *Modern Language Quarterly* 2 (1941): 67–80.

10. Alvin Snider, "Hard Frost, 1684," *Journal for Early Modern Cultural Studies* 8 (2008): 8.

11. Robert Burton, *Anatomy of Melancholy*, ed. William A. Gass (New York: New York Review Books, 2001), 61, 148.

12. Carolyn Merchant considers Eve as the first scientist in *The Death of Nature: Women, Ecology, and the Scientific Revolution* (Ithaca, NY: Cornell University Press, 1976).

13. Karen L. Edwards, *Milton and the Natural World: Science and Poetry in Paradise Lost* (Cambridge: Cambridge University Press, 1999); Joanna Picciotto, "Reforming the Garden: The Experimentalist Eden of Paradise Lost," *ELH* 72 (2005): 23–78.

14. Samuel Taylor Coleridge, "Notebook 10," cited in *Coleridge on the Seventeenth Century*, ed. R. F. Brinkley (Durham, NC: Duke University Press, 1955), 579.

15. Our most comprehensive studies of Milton in relation to contemporary science somehow ignore his relationship to air studies as developed by Boyle or anyone else. Kester Svendsen considers the incipient disciplines of astronomy, geology, botany, zoology, and anatomy; John Rogers contemporary vitalism; and Karen Edwards and Desiree Hellegers botany and zoology. See Edwards, *Milton and the Natural World*; Desiree Hellegers, *Handmaid to Divinity: Natural Philosophy, Poetry, and Gender in Seventeenth-Century England* (Norman: University of Oklahoma Press, 2000); John Rogers, *The Matter of Revolution: Science, Poetry, and Politics in the Age of Milton* (Ithaca, NY: Cornell University Press, 1996); and in a now infamous disavowal of any real connection between Milton and contemporary science, Kester Svendsen, *Milton and Science* (Cambridge, MA: Harvard University Press, 1956). Stephen Fallon offers a more generalized analysis of Milton's relationship to materialism in *Milton among the Philosophers: Poetry and Materialism in Seventeenth-Century England* (Ithaca, NY: Cornell University Press, 1991). Intellectual historians interested in the relationship between science and theology in the late seventeenth century are more likely at least to imply links between *Paradise Lost* and the study of the air. See West, *Milton and the Angels*, 98, and Georgia Christopher, *Milton and the Science of the Saints* (Princeton, NJ: Princeton University Press, 1982).

16. Steven Shapin and Simon Schaffer, *Leviathan and the Air Pump: Hobbes, Boyle, and the Experimental Life* (Princeton, NJ: Princeton University Press, 1985), 31–32. Robert Markley also explores the ways that passages describing the air pump and its operations were "designed to mimic the immediacy of pictorial representation." See Markley, *Fallen Languages: Crises of Representation in Newtonian England, 1660–1740* (Ithaca, NY: Cornell University Press, 1993), 224.

17. Samuel Johnson, *A Dictionary of the English Language* (London, 1755), 1:104. On Boyle and the leaky pump that opened him to the Hobbesian charge of pure rhetoricity, see Shapin and Schaffer, *Leviathan and the Air Pump*, 276–81.

18. Gilbert Burnet, "Memorandum," reprinted in Michael Hunter, ed., *Robert Boyle by Himself and His Friends* (Cambridge: Cambridge University Press, 1994), 27.

19. Picciotto also links Boyle to Milton but very much by way of contrast, depicting Boyle as a hopeful eviscerator of the material world, though also one forced to admit its capacity to enchant him. See Picciotto, "Reforming the Garden," 23–24.

20. See Shapin and Schaffer, *Leviathan and the Air Pump*, 290.

21. Robert Boyle, *New Experiments, Physico-Mechanical, touching the Spring of the Air* (1660), in Thomas Birch, ed., *The Works of the Honourable Robert Boyle* (hereafter cited as *Works*), 6 vols. (London, 1772), 1:11. Future references to Boyle's writing will be to the Birch edition,

still preferable in terms of user friendliness to the online *Works of Robert Boyle*, ed. M. Hunter, A. Clericuzio, and L. Principe (1999–2000).

22. Boyle often references his writing as "composure," particularly in the theological instance. See, for example, *Seraphick Love* (1660) in Robert Boyle, *Some Motives and Incentives to the Love of God* (London, 1708), 41.

23. Boyle's attraction to the occult, through the alchemical tradition from which his skeptical chemistry nonetheless sprang, through his theological considerations, and through his independent interest in psychic phenomena, is explored in a number of pieces by Michael Hunter, though see especially "Alchemy, Magic, and Moralism in the Thought of Robert Boyle," *British Journal for the History of Science* 23 (1880): 387–40, and "The Discovery of Second Sight in Late Seventeenth-Century Scotland," *History Today* 51 (2001): 48–53. Also pertinent are Keith Hutchison, "Supernaturalism and the Mechanical Philosophy," *History of Science* 21 (1983): 297–333, and especially Lawrence Principe, *The Aspiring Adept: Robert Boyle and His Alchemical Quest* (Princeton, NJ: Princeton University Press, 1998).

24. Robert Boyle, *General History of the Air*, in *Works*, 5:720.

25. This thesis is elaborated most fully by J. Paul Hunter in "Robert Boyle and the Epistemology of the Novel," *Eighteenth-Century Fiction* 2 (1990): 275–91.

26. On both of these matters, see Michael Hunter, *Robert Boyle (1627–91): Scrupulosity and Science* (Woodbridge, UK: Boydell, 2000). See also Peter Dear, "*Totius in Verba*: Rhetoric and Authority in the Early Royal Society," *Isis* 76 (1985): 145–61.

27. Peter Shaw, *The Philosophical Works of the Honourable Robert Boyle, Esq; Abridged, Methodized, and Disposed*, 3 vols. (London, 1725), 1:iii. On Shaw's implicit critique of the "theocentric basis of Boyle's experimental program" as a reflection of early eighteenth-century ambivalence toward Newtonianism, see Markley, *Fallen Languages*, 217; and on Shaw's role in the transmission of Restoration chemistry into the eighteenth century, see Jan Golinski, "Peter Shaw: Chemistry and Communication in Augustan England," *Ambix* 30 (1983): 19–28.

28. On Boyle and romance, see Lawrence Principe, "Virtuous Romance and Romantic Virtuoso: The Shaping of Robert Boyle's Literary Style," *Journal of the History of Ideas* 56 (1995): 377–97.

29. Koen Vermeir, "The Magic of the Magic Lantern (1660–1700): An Analogical Demonstration of the Visualization of the Invisible," *British Journal for the History of Science* 38 (June 2006): 127–59. See also Keith Hutchison, "What Happened to Occult Qualities in the Scientific Revolution?" *Isis* 73 (1982): 233–53.

30. Robert Boyle, *Continuation of New Experiments Physico-Mechanical, Touching the Spring and Weight of the Air*, in *Works*, 3:178.

31. Since Aristotle, the "*aer*" that fills the terrestrial sphere had been understood as analogous to the "aither" that filled the celestial sphere that presumably surrounded it. The difficulty of telling them apart came of the analogy and consequent ease with which they could be taken for each other, as well as in *aither*'s supposed capacity to mix with *pneuma*, the animating force of all living things that is carried in air. G. N. Cantor and M. J. S. Hodge trace their interaction over a millennium of Western European science in *Conceptions of Ether: Studies in the History of Ether Theories, 1740–1900* (Cambridge: Cambridge University Press, 1981), 1–26.

32. Joseph Priestley, *The History and Present State of Electricity*, 2 vols. (London, 1775), 2:16.

33. Michael Hunter considers Boyle's bait of equivocation and its links to moral value in *Robert Boyle: Scrupulosity and Science*.

34. Rose-Mary Sargent, *The Diffident Naturalist: Robert Boyle and the Philosophy of Experiment* (Chicago: University of Chicago Press, 1995), 181.

35. Robert Boyle, letter to Samuel Hartlib, in Birch, *The Life of the Honourable Robert Boyle*, in *Works*, 1:221.

36. The method I have outlined here bears some resemblance to what Ronald Schleifer has recently termed "intangible materialism." Drawing on Bruno Latour's critique of the reductive and analogical turn in modern scientific discourse, Schleifer replaces analogy with homology to foreground the role that literary semiotics play in the conceptual laying of matter as an object of knowledge. See Schleifer, *Intangible Materialism: The Body, Scientific Knowledge, and the Power of Language* (Minneapolis: University of Minnesota Press, 2009), 1–34.

37. The other physico-mechanical property of air on which *New Experiments* touched, "weight," is difficult to separate from it in this regard. "The weight of the upper part of the air," Boyle wrote in *Continuation of New Experiments*, "does, if I may so speak, bend the spring of the lower" (*Continuation of New Experiments*, 3:176–77). Boyle's Law, meanwhile, expresses their relationship in mathematical terms, as the inverse ratio of air's volume to the pressure brought to bear on it.

38. Robert Boyle, *The Christian Virtuoso* (1690), in *Works*, 5:511.

39. As its title implies, a materialist interpretation of hair throughout Milton's poetry is offered in Stephen B. Dobranski, "Clustering Hair and Curling Locks: The Matter of Hair in *Paradise Lost*," *PMLA* 125 (2010): 337–53. Dobranski accounts for the "tangled syntax" (Dobranski, 345) in Milton's description of Eve's hair by stressing its analogy to her convoluted marriage to Adam.

40. Robert Boyle, "An Account of the Life of Philaretus during His Minority," reprinted in Hunter, *Robert Boyle by Himself*, 8. Boyle's "Account" also appears, spelling amended, in Birch's 1744 *Life of the Honourable Robert Boyle* and in R. E. W. Maddison, *The Life of the Honourable Robert Boyle, F.R.S.* (London: Taylor and Francis, 1969).

41. Contemporary portraits of Boyle are discussed in R. E. W. Maddison, "The Portraiture of the Honourable Robert Boyle," *Annals of Science* 15 (1959): 141–214.

42. John Evelyn to William Wotton, March 29, 1696, reprinted in Hunter, *Robert Boyle by Himself*, 89.

43. Theophrastus's *Characters* (319 BCE) is the ultimate source of eighteenth-century conceptions of character according to Deidre Lynch, *The Economy of Character: Novels, Market Culture, and the Business of Inner Meaning* (Chicago: University of Chicago Press, 1994), 54–55. Critical treatments of Theophrastus as a natural philosopher, however, open up more latitude within the concept of Theophrastian character, and, indeed, as Aaron Kunin has recently reminded us, even Theophrastus's formulation of character moves beyond impression. See Kunin, "The Failure of Portraiture in Richardson's Novels" (paper presented at the Philosophical Questions, Literary Practices Conference, University of California, Los Angeles, May 6, 2010). See also Richard Sorabji, "Theophrastus on Place," in *Theophrastian Studies on Natural Science, Physics and Metaphysics*, ed. William W. Fortenbaugh and Robert W. Sharples (New Brunswick, NJ: Transaction, 1988), 139–62.

44. Jean-Louis Barrault, *The Theatre of Jean-Louis Barrault*, trans. Joseph Chiari (New York: Hill and Wang, 1959), 60–61.

45. Steven Shapin sees this as a random instance of Boyle's impediment, whereas Michael Hunter reads it as a token of Boyle's "scrupulosity," the intense casuistical consciousness that Boyle did liken to an "infirmity or distemper of the mind" even as he identified it with his own persistently "anxious Doubts" and "disquieting Curiosity." See Shapin, *A Social History of Truth: Civility and Science in Restoration England* (Chicago: University of Chicago Press, 1994), 136–37, and Hunter, *Robert Boyle: Scrupulosity and Science*.

46. Gilbert Burnet, "A Sermon Preached at the Funeral of the Honourable Robert Boyle" (1691–92), reprinted in Hunter, *Robert Boyle by Himself*, 48.

47. Marc Shell only mentions Boyle in passing, but presents stuttering as intrinsically linked to language's figural maneuvers. He maintains, for example, that "verbal wit arises from the same involuntary blockages of means and shifting of terms that inform stuttering" (Shell, 3). See Shell, *Stutter* (Cambridge, MA: Harvard University Press, 2005).

48. Cited in W. E. K. Middleton, ed., *Lorenzo Magalotti at the Court of Charles II: His "Relazione d'Inghilterra" of 1668* (Waterloo, Ontario, Canada: Wilfred Laurier University Press, 1980), 135. Boyle's stutter has been psychoanalyzed by both the Freudians and the Jungians as a symptom of repressed grief over his mother's early death. See especially J. Clay, "Robert Boyle: A Jungian Perspective," *British Journal for the History of Science* 32 (1999): 285–98, and Brett Kahr, "Robert Boyle: A Freudian Perspective on an Eminent Scientist," *British Journal for the History of Science* 32 (1999): 277–84.

49. Description is subtly, wittily, and extensively investigated as a rhetorical "praxis" that reflects cultural devotion to the act of making visible throughout Cynthia Wall, *The Prose of Things: Transformations of Description in the Eighteenth Century* (Chicago: University of Chicago Press, 2006). On description within scientific discourses in the period, see Lorraine Daston, "Description by Omission: Nature Enlightened and Obscured," in John B. Bender and Michael Marrinan, *Regimes of Description: In the Archive of the Eighteenth Century* (Stanford, CA: Stanford University Press, 2005), 11–23.

50. Edmund Curll, *Rarities of Richmond: Being Exact Descriptions of the Royal Hermitage and Merlin's Cave* (London, 1736), 12.

51. *Gentleman's Magazine*, April 1733, 208. A poem on the "indigested Pile" of Caroline's grotto published earlier in 1733 targeted Boyle as one who "trac'd Nature to her dark Recess." See *Gentleman's Magazine*, January 1733, 41.

52. Cited in Maddison, "Portraiture of the Honourable Robert Boyle," 197.

Chapter Three

1. The title of Howard Erskine-Hill's *The Social Milieu of Alexander Pope* (New Haven, CT: Yale University Press, 1975) is thus representative. An especially subtle discussion of the question of Pope's imaginative relationship to "an increasingly recalcitrant external world" may be found in Leopold Damrosch, *The Imaginative World of Alexander Pope* (Berkeley: University of California Press, 1987). Definitive studies of Pope's physical "deformity" and its stylistic mobilization also locate him in a prefabricated world of determinate forms; see especially Helen Deutsch, *Resemblance and Disgrace: Alexander Pope and the Deformation of Culture* (Cambridge, MA: Harvard University Press, 1997).

2. William Hazlitt, "Dryden and Pope," in *Lectures on the English Poets* ([1818] London, 1849), 84.

3. Alexander Pope, *Windsor-Forest* (1713), in *The Poems of Alexander Pope*, ed. John Butt (New Haven, CT: Yale University Press, 1972). Unless otherwise noted, all quotations from Pope's poetry are taken from this one-volume edition of the Twickenham edition and are cited parenthetically in the text by canto and/or line number.

4. Edmund Bohun, *A Discourse concerning the Origine and Properties of Wind* (London, 1671), 26.

5. John Arbuthnot, *An Essay concerning the Effect of Air on Human Bodies* (London, 1733), vi.

6. Anthony Ashley Cooper, "Sensus Communis: An Essay on the Freedom of Wit and Humour," in *Characteristics of Men, Manners, Opinions, Times*, ed. Lawrence E. Klein ([1711] Cambridge: Cambridge University Press, 1999), 39–40. Klein's introduction outlines the ways that Shaftesbury's may be seen as "a project almost sociological in nature," though for Klein it is also a profoundly "philosophical" one (Klein, vii).

7. W. H. Auden, "Alexander Pope," in *Essential Articles for the Study of Alexander Pope*, ed. Maynard Mack (Hamden, CT: Archon, 1968), 22.

8. Edith Sitwell, *Alexander Pope* (London: Faber and Faber, 1930), 7.

9. Cleanth Brooks, "The Case of Miss Arabella Fermor," in *Essential Articles*, 12.

10. See Marjorie Hope Nicolson, *Newton Demands the Muse: Newton's Opticks and the Eighteenth-Century Poets* (Princeton, NJ: Princeton University Press, 1946). Ludmilla Jordanova succinctly revised Nicolson's view of science as merely a source of literary imagery and cognitive design, pointing out that the two are "united in their shared location within cultural history." See Jordanova, *Languages of Nature: Critical Essays on Science and Literature* (London: Free Association, 1986), 15.

11. Keith Hutchison, "What Happened to Occult Qualities in the Scientific Revolution?" *Isis* 73 (1982): 233–53.

12. Here a superficial comparison with contemporary microscopy seems to be invited. Where the "small things not ordinarily apprehended" that Robert Hooke's microscope made available to ordinary apprehension opened the secrets of a hitherto invisible world, the minute particulars that hovered in it made air visible only as a kind of negative space, thereby preserving a notional framework around sense perception. On the relationship between microscopy and an emergent aesthetics of narrative detail, see Tita Chico, "Minute Particulars: Microscopy and Eighteenth-Century Narrative," *Mosaic* 39 (2006): 143–62.

13. Hooke's "Method for Making a History of the Weather" was published in Thomas Sprat's *History of the Royal Society* in London in 1667. According to Richard Hamblyn, Hooke was only "briefly transfixed by the problem of recording the various transitions of the atmosphere," but his attempts to do so were an important model for the romantic codifier of the clouds, Luke Howard. See Hamblyn, *The Invention of Clouds: How an Amateur Meteorologist Forged the Language of the Skies* (New York: Picador, 2001), 94. The extent to which Hooke's weather records were equally a lived environment and a written one is apparent in the first ten months of entries in the journal he kept throughout the 1670s. Lotte Mulligan discusses Hooke's "self-recording" as "integral to his scientific purposes" (Mulligan, 48) in "Robert Hooke's Memoranda: Memory and Natural History," *Annals of Science* 49 (1987): 47–61.

14. Robert Burton, *Anatomy of Melancholy*, ed. William H. Gass (New York: New York Review Books, 2001), 237.

15. Andrea Rusnock argues that the first decades of the eighteenth century saw many concerted efforts to reconcile Hippocrates's "medical meteorology" with mechanistic physiological models in "Hippocrates, Bacon, and Medical Meteorology, 1700–1750," in David Cantor, *Reinventing Hippocrates* (Burlington, VT: Ashgate, 2001), 144–61.

16. George Cheyne, *The English Malady; or, A Treatise of Nervous Diseases of all Kinds, as Spleen, Vapours, and Lowness of Spirits* (London, 1733), 1.

17. Richard Blackmore, *A Treatise of the Spleen and Vapours* (London, 1729), A2r.

18. Mary Wortley Montagu, "The Court of Dulness" (1729), in *The Works of the Right Honorable Mary Wortley Montagu*, ed. J. P. Dove (London, 1825), 500. In fact, Pope's grotto was quite dry; a visit there in the summer of 2010 yielded little more than musk of fox.

19. Robert Boyle, *Essays of the Strange Subtility, Great Efficacy, and Determinate Nature of Effluviums*, in Thomas Birch, ed., *The Works of the Honourable Robert Boyle*, 6 vols. (London, 1772), 3:669.

20. John Floyer, *A Treatise of the Asthma* (London, 1707), 53, 60. Floyer's fascinating *Treatise* classifies both "Air" and "Changes in the Weather" among the "non-naturals" (Floyer, xxvii).

21. Bernard Annely, preface to *A Theory of the Air* (London, 1729), n.p.

22. Throughout the eighteenth century, airborne experiments tended both to aestheticize and gothicize "normal" science. Paul Keen, for example, observes the blurring of boundaries between a "carnivalesque world of parodic inversions and theatrical mock-heroes" and "the dignity of genuine scientific investigation" in the ballooning craze of the 1780s. See Keen, "Science and Spectacle in 1780s England," *Eighteenth-Century Studies* 39 (2006): 517.

23. See Tobias Smollett, *The Expedition of Humphry Clinker*, ed. Lewis M. Knapp and Paul Gabriel Boucé ([1771] Oxford: Oxford University Press, 1998), 95–96.

24. John Evelyn, *Fumifugium; of the Inconveniencie of the AER of London* (London, 1661), 12.

25. Peter Denton, "'Puffs of Smoke, Puffs of Praise': Reconsidering John Evelyn's *Fumifugium*," *Canadian Journal of History* 35 (December 2000): 11.

26. *Diary of John Evelyn*, ed. E. S. deBeer, 6 vols. (Oxford: Oxford University Press, 1955), 3:295.

27. Mark Jenner, "The Politics of London Air: John Evelyn's *Fumifugium* and the Restoration," *Historical Journal* 38 (1995): 535–51; see also Denton, "'Puffs of Smoke, Puffs of Praise,'" and Jeanne K. Welcher, *John Evelyn* (New York: Twayne, 1972), 44.

28. The first epistle of Pope's *Essay on Man* evokes a sense of "Touch" that is "tremblingly alive all o'er" and "quick effluvia darting through the brain" (1.197, 199).

29. G. K. Chalmers, "Effluvia: The History of a Metaphor," *PMLA* 52 (1937): 1031–50.

30. On Lucretius's related use in modern poetry as a way to uncover the lyric capacities of otherwise lumpish matter, see Daniel Tiffany, *Toy Medium: Materialism and Modern Lyric* (Berkeley: University of California Press, 2000), 201–2. Sophie Gee takes up Pope's imagery of effluvia as part of a larger cultural fascination with the ecology and economy of excess in *Making Waste: Leftovers and the Eighteenth-Century Imagination* (Princeton, NJ: Princeton University Press, 2009), 67–90.

31. Jonathan Swift materialized air as an inherently insignificant body throughout his satire, though see especially *The Mechanical Operation of the Spirit* (London, 1704) and *Gulliver's Travels* (London, 1726), pt. 3.

32. John Wilkins, *The Discovery of a New World in the Moone* ([1638] London, 1684), 41.

33. On Pope's medical condition and the asthma in particular, see Marjorie Hope Nicolson and George S. Rousseau, *"This Long Disease My Life": Alexander Pope and the Sciences* (Princeton, NJ: Princeton University Press, 1968). See also S. H. Monk, "Die of a Rose," *Huntington Library Quarterly* 21 (1958): 359–61.

34. Owen Ruffhead, *The Life of Alexander Pope* (London, 1769), 119.

35. *Correspondence of Alexander Pope*, ed. George Sherburn, 4 vols. (Oxford: Clarendon, 1956), 4:433.

36. Alexander Pope, *The Iliad of Homer*, ed. Steven Shankman ([1717] London: Penguin, 1996), 10.

37. John Dennis, *Remarks on Mr. Pope's Rape of the Lock* (London, 1728).

38. David Fairer, *Pope's Imagination* (Manchester, UK: Manchester University Press, 1984), 11.

39. Alexander Pope, "A Receit to Make an Epick Poem," reprinted in *The Rape of the Lock*, ed. Cynthia Wall (Boston: Bedford, 1998), 257.

40. The editor of Pope's *Correspondence*, George Sherburn, dismissed this letter as an "innocuous note," one "impossible to date . . . surely" because Pope was also "at Bath amidst congelation" the following December. It is, nonetheless, a classic example of Pope's lifelong preoccupation, epistolary and poetical, with the state of the air about him.

41. Virginia Woolf, *On Being Ill* ([1926] Ashfield, MA: Paris Press, 2006), 6.

42. Richard Kroll shows how integral such "metamorphic and fluid" dynamics were to Pope's engagement with the psychosomatic environment of contemporary pharmacology in "Pope and Drugs: The Pharmacology of *The Rape of the Lock*," *ELH* 67 (2000): 99–121. See also Reuben Brower, *Alexander Pope: The Poetry of Allusion* (Oxford: Clarendon, 1958–69), 26.

43. Joseph Warton, *An Essay on the Writings and Genius of Pope*, 2 vols. (London, 1756), 1:53.

44. Samuel Johnson, "Life of Pope," in *Lives of the Poets*, ed. Roger Lonsdale, 4 vols. (Oxford: Oxford University Press, 2006), 4:70.

45. William Hazlitt, *Lectures on the English Poets*, 86.

46. F. R. Leavis, "Pope," in *Essential Articles*, 8–9.

47. William Wimsatt, "One Relation of Rhyme to Reason: Alexander Pope," in *Essential Articles*, 54.

48. William Warburton, *Works of Pope* (London, 1751), 1:221.

49. Here Pope may have come closer to his French source of 1670 than is often conceded. Edmund N. Seeber notes that while *Le Comte de Gabalis* "became pleasurably entangled in what Diderot disdainfully termed '*ce tissu indigeste et ridicule de suppositions*,'" there was in fact plenty of room to "accept Count Gabalis as a real person" (Seeber, 76) who never intended to parody the Rosicrucian belief systems he appears to. Seeber even mentions the recommendation by early twentieth-century theosophist C. H. Bjerregard that the reader of Gabalis "leave the question [of its fictionality] open, and attend to the teachings of the book," which can "leave a subtle influence upon the mind and prepare it for a flight upwards" (Seeber, 76). See Seeber, "Sylphs and Other Elemental Beings in French Literature since *Le Comte de Gabalis*," *PMLA* 59 (1944): 71–83.

50. William Melmoth, "To Cleora," in *The Letters of Sir Thomas Fitzoborne* ([1748] London, 1750), 302.

51. Earl Wasserman, *The Subtler Language: Critical Readings of Romantic and Neoclassic Poems* (Baltimore: Johns Hopkins University Press, 1959), 174.

52. See John Sitter, "What the Sylphs Do," in *Approaches to Teaching Pope's Poetry*, ed. Wallace Jackson and R. Paul Yoder (New York: MLA, 1993), 132. See also Robin Grove, "Uniting Airy Substance: *The Rape of the Lock*, 1712–1736," in *The Art of Alexander Pope*, ed. Howard Erskine Hill (London: Vision), 52–88; and Patricia Bruckmann, "Virgins Visited by Angel Powers: *The Rape of the Lock*, Platonick Love, Sylphs and Some Mysticks," in *The Enduring Legacy: Alexander Pope Tercentary Essays*, ed. G. S. Rousseau and Pat Rogers (Cambridge: Cambridge University Press, 1988), 3–20.

53. This is a potential that lurks in Locke's own model of the "understanding," as Ernest Tuveson captures in *The Imagination as a Means of Grace: Locke and the Aesthetics of Romanticism* (Berkeley: University of California Press, 1960).

54. Francis Thomson, "Pope," *Academy* (July 3, 1897): 13.

55. Stephen Hales, *A Description of Ventilators, whereby Great Quantities of Fresh Air May with Ease be Conveyed into Mines, Gaols, Hospitals* (London, 1743), v.

56. Gotthold Ephraim Lessing, *Laocoön; An Essay on the Limits of Painting and Poetry*, trans. E. A. McCormick ([1766] Baltimore: Johns Hopkins University Press, 1962), 68.

57. David Wellbery, "Aesthetic Media: The Structure of Aesthetic Theory before Kant," in John B. Bender and Michael Marrinan, *Regimes of Description: In the Archive of the Eighteenth Century* (Stanford, CA: Stanford University Press, 2005), 202.

58. Stothard went on to train his attention closely on butterflies. See Robert Halsband, "*The Rape of the Lock*" *and Its Illustrations* (Oxford: Oxford University Press, 1980), 39. The full anecdote about Stothard and his butterfly sylphs may be found in Anna Eliza Bray, *Life of Thomas Stothard, R.A.*, 2 vols. (London: J. Murray, 1851), 1:31–32.

59. A valuable guide to the English fairy—charting, for example, its physical diminution over the course of the eighteenth century, and its increasing assimilation to forms found in nature like that of the butterfly and so its decreasing distinction from these same forms—is Katharine M. Briggs, *The Fairies in English Tradition and Literature* (London: Routledge, 1967).

60. Fuseli, we should note, rejected most poetry as nothing more than "metrical and rimed prose" but ironically found in *The Rape of the Lock* true "poetic genius," for "a Poet is an inventor; and what has Pope invented, except the Sylphs?" See John Knowles, *The Life and Writings of Henry Fuseli*, 3 vols. (London, 1831), 1:358.

61. *The Fan* (London, 1749), 10.

62. Jean-François Marmontel, *The Moral Tales of M. Marmontel*, trans. C. Denis and R. Lloyd (Dublin, 1764), 13.

63. Anne Francis, "The Sylph," in *Miscellaneous Poems, by a Lady* (Norwich, 1790), 268.

64. "The Hobgoblin, or the Amorous Sylph," in *The Gallant Companion; or, an Antidote for the Hyp. and Vapours* (London, 1746), 6.

65. Charlotte de Bressay, *Letters from Julia . . . to Ovid* (London, 1753), 102.

66. Georgiana Cavendish, *The Sylph*, ed. Jonathan Gross ([1779] Chicago: University of Chicago Press, 2007), 139.

67. On sylphs and readers, see Murray Cohen, "Versions of the Lock: Readers of *The Rape of the Lock*," *ELH* 43 (1976): 53–73.

Chapter Four

1. Daniel Defoe, *The Storm: or, A Collection of the most Remarkable Casualties and Disasters Which happen'd in the Late Dreadful Tempest*, ed. Richard Hamblyn ([1704] London: Penguin, 2004), 36. Future citations in this chapter will be to this edition and will appear parenthetically in the text by short title and page number.

2. Ian Watt's classic account of Defoe's fiction in *The Rise of the Novel* treats his "formal realism [as] an end rather than a means" (Watt, 117) and lies behind the many other recent accounts of his idiomatic commitment to the language of fact. See Watt, *Rise of the Novel* (Berkeley: University of California Press, 1957), 93–134, and Lennard Davis, *Factual Fictions: The Origins of the English Novel* (New York: Columbia University Press, 1983), 154–74. Defoe's relationship to the supernatural is an implicit concern throughout George A. Starr, *Defoe and Spiritual Autobiography* (Princeton, NJ: Princeton University Press, 1965), though for the standard (and only) book-length critical work on the subject, see Rodney M. Baine, *Daniel Defoe and the Supernatural* (Athens: University of Georgia Press, 1968).

3. John Bender, "The Novel as Modern Myth," in *Defoe's Footprints: Essays in Honour of Maximillian E. Novak* (Toronto: University of Toronto Press, 2006), 232; Walter Scott, "Daniel De Foe," in *Biographical Memoirs of Eminent Novelists*, 2 vols. ([1821–24] London, 1834), 2:274. Scott's exegesis of Defoe's "exact and circumstantial delineation" (Scott, "Daniel De Foe," 2:274) places it primarily in the context of his writing about the supernatural.

4. William Warburton, "The Life of Pope," in *Works of Pope*, 9 vols. (London, 1751), 1:221.

5. The one exception is Ilse Vickers, who in considering Defoe's relationship to the "Baconian legacy" of empirical observation systematized into theory touches on the mediating figure of Robert Hooke. See Vickers, *Defoe and the New Sciences* (Cambridge: Cambridge University Press, 1996), 108–10.

6. Influential genealogies of Defoe's fiction, their titles alone almost always revelatory, include Lincoln B. Faller, *Crime and Defoe: A New Kind of Writing* (Cambridge: Cambridge University Press, 1993); J. Paul Hunter, *The Reluctant Pilgrim: Defoe's Emblematic Method and Quest for Form in* Robinson Crusoe (Baltimore: Johns Hopkins University Press, 1966); and Maximillian E. Novak, *Economics and the Fiction of Daniel Defoe* (Berkeley: University of California Press, 1962).

7. See Maximillian E. Novak, *Realism, Myth and History in Defoe's Fiction* (Lincoln: University of Nebraska Press, 1983), 32. For the details of the "extratropical cyclone" that remains "the worst storm in British history" (p. x), see Richard Hamblyn, introduction to Defoe, *The Storm*, x–xxxix. For details of the storm, see also Jan Golinski, *British Weather and the Climate of Enlightenment* (Chicago: University of Chicago Press, 2007), 43–53.

8. On the generic promiscuity and fragmented forms of early meteorography, see Vladimir Janković, "Imperfect Mixtures," in *Reading the Skies: A Cultural History of the English Weather, 1650–1820* (Chicago: University of Chicago Press, 2000), 14–32, and Golinski, *British Weather*, 1–12.

9. Robert Markley, "'Casualties and Disasters': Defoe and the Interpretation of Climate Instability," *Journal for Early Modern Cultural Studies* 8 (2008): 117.

10. Daniel Defoe, *The Works of Daniel Defoe*, vol. 3, *Serious Reflections during the Life and Surprizing Adventures of Robinson Crusoe* (1720), ed. G. H. Maynadier (New York: George D. Sproul, 1903), ix.

11. Two classic readings of *Crusoe* that link the novel to earlier forms of spiritual inquiry are Hunter, *Reluctant Pilgrim*, and Starr, *Defoe and Spiritual Autobiography*. On the side of historical realism may be found Watt's *Rise of the Novel* and Karl Marx's identification of Crusoe's behavior as an extreme form of an emergent capitalist pattern in *Capital: A Critique of Political Economy*, ed. Frederick Engels, trans. Samuel Moore and Edward Aveling, 2 vols. ([1867–94] New York: International Publishers, 1967), 1:88. Notable mediators include Michael McKeon, *The Origins of the English Novel, 1600–1740* (Baltimore: Johns Hopkins University Press, 1987), and John Richetti's *Defoe's Narratives: Situations and Structures* (Oxford: Oxford University Press, 1975). In a more recent essay, Richetti offers an especially deft and instructive account of the critical "dilemmas" staged in and around *Crusoe* and imagines the novel in a more dialogic configuration than heretofore imagined. See Richetti, "Secular Crusoe," *Eighteenth-Century Genre and Culture: Serious Reflections on Occasional Forms in Honor of J. Paul Hunter*, ed. Dennis Todd and Cynthia Wall (Newark: University of Delaware Press, 2001), 58–78.

12. Leslie Stephen, *Hours in a Library* (1892), in *Defoe: The Critical Heritage*, ed. Pat Rogers (London: Routledge, 1972), 173.

13. Stories of shipwreck have often seemed to be unusually suited to this kind of representation. Thomas H. Uzzell's *Narrative Technique: A Practical Course in Literary Psychology* (New

York: Harcourt Brace, 1923), discussed in chap. 1, dwells at length on Robert Louis Stevenson's "The Merry Men" as an example of "the atmosphere story" because it interweaves a language of "effects" with that of "description" (Uzzell, 377). Walter de la Mare offers an unusually nuanced reading of *Robinson Crusoe* in *Desert Islands and Robinson Crusoe* (New York: Farrar and Rinehart, 1930).

14. Daniel Defoe, *Review* (April 11, 1710): 259, cited in Novak, *Realism, Myth and History*, 17.

15. Joseph Browne, *A Dialogue between Church and No-Church* (1706), cited in Novak, *Realism, Myth and History*, 17.

16. Daniel Defoe, *The Strange and Surprizing Adventures of Robinson Crusoe*, ed. Michael Shinagel ([1719] New York: Norton, 1996), 9. Future citations in this chapter will be to this edition and will appear parenthetically in the text by short title and page number.

17. David Marshall, "Autobiographical Acts in *Robinson Crusoe*," *ELH* 71 (2004): 899–920.

18. Thomas Appletree, journal, cited in Golinski, *British Weather*, 19. Both Golinski and Janković implicitly lament the disappearance of provincial, reflexive, anarchic, and individualistic meteorologies over the course of the eighteenth century.

19. August Kleinzahler, "The Exquisite Atmography of Thomas Appletree," *London Review of Books* (January 7, 2010), 22.

20. This reading is most vividly advanced in Janković, *Reading the Skies*, but see also McKeon, *Origins of the English Novel*; Lorraine Daston, "Marvelous Facts and Miraculous Evidence in Early Modern Europe," *Critical Inquiry* 18 (1991): 93–124; and Ann Geneva, *Astrology and the Seventeenth-Century Mind: William Lilly and the Language of the Stars* (Manchester, UK: Manchester University Press, 1995).

21. Arden Reed's perception that "meteorology produces signifiers" allows him to describe weather as preformal and open ended, but from the same perspective it could also be seen as hyperformal. See Reed, *Romantic Weather: The Climates of Coleridge and Baudelaire* (Hanover, NH: University Press of New England, 1983), 7, 11, and for the counterview, see Lucian Boia, *The Weather in the Imagination*, trans. Roger Leverdier (King's Lynn, UK: Reaktion, 2005), 41–65.

22. Aristotle, *Meteorologica*, trans. H. D. P. Lee (Cambridge, MA: Harvard University Press, 1962), 13.

23. Thomas Burnet, *The Sacred Theory of the Earth* [1681], ed. Basil Willey (Carbondale: Southern Illinois University Press, 1965), 173. The attentive reader of Stephen Jay Gould's *Time's Cycle, Time's Arrow: Myth and Metaphor in the Discovery of Geological Time* (Cambridge, MA: Harvard University Press, 1987) would expect Burnet to come up with this view. Gould tracks the conveyance of two different temporal models in Burney's scriptural geology and so uncovers his startling modernity.

24. See Janković, *Reading the Skies*, 79, and Mark Monmonier, *Air Apparent: How Meteorologists Learned to Map, Predict, and Dramatize the Weather* (Chicago: University of Chicago Press, 1999), 18–38.

25. Thomas Short, preface to *A Comparative History of the Increase and Decrease of Mankind in England and Several Countries Abroad*, ed. Richard Wall ([1767] Farnborough, UK: Gregg International, 1973), n.p.

26. Thomas Short, *A General Chronological History of the Air*, 2 vols. (London, 1749), 1:108, 363.

27. Lorraine Daston and Peter Galison argue that objectivity in the sense of a noninterpretive representation of facts is a nineteenth-century scientific ideal; the "collective empiricisms"

of the Enlightenment understand facts to be mixed with both representational and interpretive acts. See Daston and Galison, *Objectivity* (New York: Zone, 2007).

28. Hans Blumenberg captures this transaction as a defining feature of modernity and draws on a weather metaphor to explicate the secularizing trend that attends that modernity in his view: "The world that became even more worldly was a subject whose extension was about as obscure as that of the impersonal 'it' in the proposition 'It's raining.'" See Blumenberg, *The Legitimacy of the Modern Age*, trans. Robert M. Wallace ([1966] Cambridge, MA: MIT Press, 1983), 4.

29. Leo Spitzer, "Milieu and Ambiance," in *Essays in Historical Semantics* (New York: Russell and Russell, 1948), 189.

30. Scott here comments at length on "the air of truth about the delightful history of *Robinson Crusoe*."

31. See Maximillian E. Novak, "Picturing the Thing Itself, or Not: Defoe, Painting, Prose Fiction and the Arts of Describing," *Eighteenth-Century Fiction* 9 (1996): 1–20.

32. George Borrow, *Lavengro* (1851), in *Defoe: Critical Heritage*, 123–24. Borrow's remarkable account ends with the reader of *Crusoe* "cantering before a steady breeze over an ocean of enchantment" (Borrow, 125). Who could ask for anything more?

33. Samuel Taylor Coleridge, marginalia for *Robinson Crusoe*, in *Defoe: Critical Heritage*, 82.

34. Evelyn and Cotgrave are both cited in R. L. Chambers, "A History of the Umbrella," in Chambers, *Book of Days*, 2 vols. (London, 1865), 2:241–44.

Chapter Five

1. Lucinda Cole considers the "semiotic constellation among witches, decay, and plague" (Cole, 67–68) in relation to the common notion that all arise from bad—that is, wet and warm—air in "Of Mice and Moisture: Rats, Witches, Miasma, and Early Modern Theories of Contagion," *Journal for Early Modern Cultural Studies* 10 (2010): 65–84.

2. Daniel Defoe, *A Journal of the Plague Year*, ed. Louis A. Landa ([1722] Oxford: Oxford University Press, 1969), 22, 62, 61. Future citations in this chapter will be to this edition and will appear parenthetically in the text by short title and page number.

3. Walter Scott, "Daniel De Foe," in *Biographical Memoirs of Eminent Novelists*, 2 vols. ([1821–24] London, 1834), 2:252.

4. John Bender, "The Novel as Modern Myth," in *Defoe's Footprints: Essays in Honour of Maximillian E. Novak* (Toronto: University of Toronto Press, 2006), 227.

5. E. J. Clery explores the eighteenth-century ghost story's oft-posited travels first through oral culture (and often via the tales of illiterates, servants, and women) in *The Rise of Supernatural Fiction, 1762–1800* (Cambridge: Cambridge University Press, 1995), 2–5. On reader ambivalence as a function of textual markers in European literature of the supernatural later in the eighteenth century, see Dorothea von Mücke, *The Seduction of the Occult and the Rise of the Fantastic Tale* (Stanford, CA: Stanford University Press, 2003), 3–9.

6. W. E. H. Lecky was perhaps the first to notice that between 1660 and 1720, arguments against the reality of apparition were "extremely languid," while at least twenty-five works were published more or less on their behalf. See Lecky, *History of the Rise and Influence of the Spirit of Rationalism in Europe* ([1878] New York: Braziller, 1955), 70. Nonetheless, the now more authoritative work by Keith Thomas posits a radical decline in ghost belief, as does Stephen Greenblatt's influential study of ghosts in the English Renaissance, which does not imagine ghosts to have

survived Reginald Scot's 1584 *Discoverie of Witchcraft* as anything more than theatrical illusions. See Thomas, *Religion and the Decline of Magic: Studies in Popular Beliefs in Sixteenth and Seventeenth Century England* (Oxford: Oxford University Press, 1971), 605–6, and Greenblatt, *Hamlet in Purgatory* (Princeton, NJ: Princeton University Press, 2001), 258–62.

7. Rodney M. Baine, *Daniel Defoe and the Supernatural* (Athens: University of Georgia Press, 1968). Defoe's best-known piece of ghostwriting has long been the 1705 *A True Relation of the Apparition of One Mrs. Veal*, though his authorship of that interesting pamphlet has recently been challenged by George Starr in "Why Defoe Probably Did Not Write *The Apparition of Mrs. Veal*," *Eighteenth-Century Fiction* 15 (2003): 421–51. As both Starr and Baine note, many of Defoe's works engage the world of spirits, from *A Political History of the Devil* (1726) and *A System of Magick* (1726) to the brief "Vision of the Angelick World" found in *Serious Reflections of Robinson Crusoe* (1720).

8. Thus Leslie Stephen could find that the twelve-page *True Relation of the Apparition of One Mrs. Veal* (discussed below) so often attributed to Defoe "contains in a few lines all the essential peculiarities of his art." Stephen, *Hours in a Library*, cited in Manuel Schonhorn, introduction to *Accounts of the Apparition of Mrs. Veal* (Los Angeles: Augustan Reprint Society, 1965), i. On the ghost's permissibility in self-avowed art, and indeed its use in generating "enthusiastic terror," see also Clery, *Rise of Supernatural Fiction*, 33–44.

9. Robert Mayer analyzes the *Journal*'s reception history to establish it as a "pseudomorph" constructed out of a "nexus of historical and fictional discourse"—a nexus that becomes central to the reader's Jaussian "horizon of expectations" and ultimately to the novel genre itself. See Mayer, "The Reception of *A Journal of the Plague Year* and the Nexus of Fiction and History in the Novel," *ELH* 57 (1990): 542. Also relevant is Mayer, *History and the Early English Novel: Matters of Fact from Bacon to Defoe* (Cambridge: Cambridge University Press 1997). Weighing in on "the vexed question of whether [the *Journal*] is history or fiction" are, representatively, Louis Landa, introduction to *A Journal of the Plague Year*, xxxvii, and John Richetti, who decides that it is "pseudo-history in the interest of expert political propaganda." See Richetti, *Defoe's Narratives: Situations and Structures* (Oxford: Oxford University Press, 1975), 234.

10. Frank Ellis, "Defoe's *Journal of the Plague Year*," *Review of English Studies* 45 (1994): 79. See also Leonard Fabian Hirst, *The Conquest of Plague* (Oxford: Oxford University Press, 1953), 34.

11. This is the implication of Landa's introduction to the *Journal* and also underlies George Starr's now classic representation of H. F.'s "equivocation" as ultimately rhetorical in nature, a deliberate effort "to reconcile and combine . . . seemingly opposed outcomes" in order to present "a model of sustained moderation." See Starr, *Defoe and Casuistry* (Princeton, NJ: Princeton University Press, 1971), 81.

12. These are the terms in play in Michael McKeon, *The Origins of the English Novel, 1600–1740* (Baltimore: Johns Hopkins University Press, 1987), 315–37.

13. Daniel Defoe, *An Essay on the History and Reality of Apparitions* (London, 1727). (Future citations in this chapter will be to this edition and will appear parenthetically in the text by short title and page number.) For a historically contextualized discussion of how Defoe challenges "the primacy of narrative through this work's "fascinating and wide-ranging commentary on behaviour and belief" (p. 35), see George Starr's introduction to *An Essay on the History and Reality of Apparitions*, in *Satire, Fantasy and Writings on the Supernatural by Daniel Defoe*, ed. W. R. Owens and P. N. Furbank, 8 vols. (London: Pickering and Chatto, 2005), 8:1–35.

14. "The Matter of this Collection is beyond Humane reach," Aubrey wrote in continuation,

"we being miserably in the dark, as to the Oeconomie of the Invisible World, which knows what we do, or incline to, and works upon our Passions, and sometimes is so kind as to afford us a glimpse of its Prescience." See John Aubrey, dedication to James, Earl of Abingdon, in *Miscellanies* (London, 1696), n.p.

15. Richard Baxter, *The Certainty of the World of Spirits* (London, 1691), A4r.

16. McKeon examines some proto-novelistic features of these narratives (their doomed emphasis on eyewitness testimony in particular) in *Origins of the English Novel*, 83–89. Terry Castle presents these narratives as having been placed over the course of the eighteenth century by a "new technical language" supported by new visual technologies that made ghost sightings a matter of collective entertainment via ghost shows and other artificial phantasmagoria. See Castle, The *Female Thermometer: Eighteenth-Century Culture and the Invention of the Uncanny* (Oxford: Oxford University Press, 1995), 162.

17. Thomas Hobbes, *Leviathan; or, the Mater, Forme and Power of a Commonwealth, Ecclesiasticall and Civil* ([1651] New York: Penguin, 1962), 462, 482.

18. Stuart Clark, "The Reformation of the Eye: Apparitions and Optics in Sixteenth- and Seventeenth-Century England," *Journal of Religious History* 27 (2003): 143–60.

19. Or at least with the realist canvas as understood by Svetlana Alpers in her groundbreaking *The Art of Describing: Dutch Art and the Seventeenth Century* (Chicago: University of Chicago Press, 1983). Maximillian Novak notes the complicity between the methodology of this painting and English narrative techniques of the period and slightly later in *Realism, Myth and History in Defoe's Fiction* (Lincoln: University of Nebraska Press, 1983). Clery explores a different aspect of apparitions' continued viability with respect to art, noting a new "valorization of the supernatural as a source of aesthetic pleasure." See Clery, *Rise of Supernatural Fiction*, 35.

20. This complicates any role that apparitions might play in Simon During's account of the overhaul of "the structure of magic" into an eighteenth-century "language game" and may explain why they can find no real place in his argument. See During, *Modern Enchantments: The Cultural Power of Secular Magic* (Cambridge, MA: Harvard University Press, 2002), 17–25.

21. See Lawrence Principe, "Boyle's Alchemical Pursuits," in Michael Hunter, ed., *Robert Boyle Reconsidered* (Cambridge: Cambridge University Press, 1994), 91–102; John Henry, "Boyle and Cosmical Qualities," in *Robert Boyle Reconsidered*, 119–34; and Michael Hunter, ed., introduction to *The Occult Laboratory: Magic, Science and Second Sight in Late Seventeenth-Century Scotland* (Rochester, NY: Woodbridge, 2001), 1–32.

22. On Conway's circle, see Sarah Hutton, introduction to *The Conway Letters: The Correspondence of Anne, Viscountess Conway, Henry More and Their Friends, 1642–1684*, ed. Marjorie Hope Nicolson (New York: Oxford University Press, 1992), vii–xix.

23. Robert Boyle, letter to Henry Stubbe, in Thomas Birch, ed., *The Life of the Honourable Robert Boyle*, in *The Works of the Honourable Robert Boyle*, ed. Thomas Birch, 6 vols. (London, 1772), 1:48. Boyle's involvement in scriptural mediation is apparent in his funding of the "Translation and Impression of the New Testament into the *Malayan* Language, which he sent over all the *East Indies*." See Gilbert Burnet, "A Sermon Preached at the Funeral of the Honourable Robert Boyle" (1691–92), reprinted in Michael Hunter, ed., *Robert Boyle by Himself and His Friends* (Cambridge: Cambridge University Press, 1994), 49.

24. Daniel Defoe, *The Second-Sighted Highlander* (London, 1713), 15.

25. See, for example, the 1711 translation of Bordelon's cautionary *A History of the Ridiculous Extravagances of Mnsr. Oufle*; the anonymous *The Story of the St. Alban's Ghost, or the Apparition of Mother Haggy* (1712); Alexander Telfair, *A True Relation of an Apparition in the Parish of Ker-*

rick (1696); and the English translation of Pierre Du Moulin's *The Devil of Mascon*, which had gone into its fourth edition by 1669. Also of interest may be Isabel M. Westcott, ed., *Seventeenth-Century Tales of the Supernatural* (Los Angeles: Augustan Reprint Society, 1958).

26. The term "poltergeist" is anachronistic, having entered English only in the early nineteenth century with the work of Catherine Crowe, whose *The Night Side of Nature* (London, 1848) brought English ghost traditions into dialogue with their German counterparts.

27. Douglas Grant's *The Cock Lane Ghost* (New York: Macmillan, 1965), tells the story of the Cock Lane Ghost as a cultural phenomenon of the early 1760s. See also Clery, *Rise of Supernatural Fiction*, 13–32, and Jayne Lewis, "Sight Unseen: The Cock Lane Ghost among the Infidels," in *Sustaining Literature: Essays in Literature, History and Culture . . . Commemorating the Life and Work of Simon Varey*, ed. Greg Clingham (Lewisburg, PA: Bucknell University Press, 2007), 247–69.

28. Troy Boone tells how the supernatural found a home in eighteenth-century aesthetic theory in "Narrating the Apparition: Glanvill, Defoe, and the Rise of Gothic Fiction," *Eighteenth Century: Theory and Interpretation* 55 (1994): 173–89. Patricia Meyer Spacks's survey of supernatural referents in eighteenth-century poetry likewise assumes that this was the only place left for them after scientific skepticism had done its work. See Spacks, *The Insistence of Horror: Aspects of the Supernatural in Eighteenth-Century Poetry* (Cambridge, MA: Harvard University Press, 1962), 7–28.

29. Johnson's ideas about the second sight are usefully summarized in Thomas Jemielity, "Samuel Johnson, the Second Sight, and Its Sources," *Studies in English Literature* 14 (1974): 403–20.

30. John Frazer, *Deuteroskopia; or, A Brief Discourse concerning the Second Sight*, in Hunter, *Occult Laboratory*, 197.

31. George Hickes to Samuel Pepys, June 19, 1700, reprinted in Hunter, *Occult Laboratory*, 172.

32. John Beaumont, *Historical, Physiological, and Theological Treatise of Spirits* (London, 1705), 9.

33. Troy Boone's cogent assessment of apparition narratives by Defoe and Glanville comes to a conclusion close to McKeon's. Boone maintains that both Glanvill and "his most popular successor," Defoe, resisted "rationalistic deprivileging of the supernatural," but did so under a rationalist rubric that ultimately undermined their efforts, sped cultural flight from McKeon's register of the "metaphysical," and helped to create the refuge of Gothic fiction. See Boone, "Narrating the Apparition," 173.

34. Daniel Defoe, *An Essay upon Literature* (London, 1726), 92.

35. Beaumont, dedication to the Earl of Carbury, in *Historical, Physiological, and Theological Treatise*, n.p.

36. Henry More, preface to *Saducismus Triumphatus; or, Full and Plain Evidence concerning Witches and Apparitions* (1681), in *Collected Works of Joseph Glanvill*, ed. Bernhard Fabian (Hildesheim, Germany: Georg Olms Verlag, 1978), 9:n.p.

37. Joseph Glanvill, *Saducismus Triumphatus; or, Full and Plain Evidence concerning Witches and Apparitions* (1681), in *Collected Works*, F4r.

38. Daniel Defoe, *A True Relation of the Apparition of One Mrs. Veal*, in *Accounts of the Apparition of Mrs. Veal*, 3.

39. Hogarth's print is discussed in rich detail in Ronald Paulson, *Hogarth*, 3 vols. (New Brunswick, NJ,: Rutgers University Press, 1993), 3:317, and in Jenny Uglow, *Hogarth: A Life and a World* (London: Faber and Faber. 1997), 651–57.

40. Starr marshals an impressive array of evidence from Defoe's supernatural writings, along with his likely contempt for ghost-believing Anglicans such as Joseph Horneck, to conclude that the attribution of *The Apparition of Mrs. Veal* that dates from the "List of Writings" George Chalmers printed along with his *Life of De Foe* (London, 1790) is false and that the text "should return to the limbo of anonymity" (p. 450). But if we see Defoe's conception of apparition as linked to his conception of literature, the attribution could still stand. The text might—as Starr leaves us room to believe—be a parody of contemporary accounts of its ilk, or it might simply be that ghost belief in and of itself is simply irrelevant to the other kinds of credit that the text accumulates.

41. On the "crisis of repetition" played out in the *Journal*'s linguistic field, see Ellis, "Defoe's *Journal*," 78.

42. A summary of the Henry Foe thesis may be found in Maximillian E. Novak, *Daniel Defoe: Master of Fictions* (Oxford: Oxford University Press, 2001), 605.

43. The problem of form in and for the *Journal* is central to virtually all recent readings of it. Scott J. Juengel incorporates this problem into his interpretation of the work's "narrative vagrancy" as a mirror for its overarching lack of "epistemological stability," as well as for "the narrator's perambulations" and for "the plague's undisciplined progress." See Juengel, "Writing Decomposition: Defoe and the Corpse," *Journal of Narrative Technique* 25 (1995): 143.

44. In the not irrelevant early seventies, for example, Everett Zimmerman engaged Rodney Baine in an influential debate over the extent to which H. F. is "a character, not a lesson." See Zimmerman, "Henry Foe's Meditations: *A Journal of the Plague Year*," *PMLA* 87 (1972): 417, and Baine, *Daniel Defoe and the Supernatural*, 6–7.

Chapter Six

1. Vladimir Janković shows how the earthquakes "quickly became a commodity used by competing groups" to display their own interpretive power. See Janković, *Reading the Skies: A Cultural History of the English Weather, 1650–1820* (Chicago: University of Chicago Press, 2000), 22. A similar perspective is adopted by Lorraine Daston and Katharine Park in *Wonders and the Order of Nature, 1150–1750* (Cambridge, MA: MIT Press, 1995).

2. Thomas Sherlock, *A Letter from the Lord Bishop of London on the Occasion of the Late Earthquakes* (London, 1751), 4, 8.

3. On the relationship between modern geology and textual materiality, see Noah Heringman, *Romantic Rocks, Aesthetic Geology* (Ithaca, NY: Cornell University Press, 2004), 12–21.

4. *An Account of Explosions in the Atmosphere, or Airquakes* (London, 1750), 7.

5. Seneca, *Naturales quaestiones*, trans. Thomas H. Corcoran, 2 vols. (Cambridge, MA: Harvard University Press, 1972), 2:163. As borrowed from Pliny, Seneca's idea was that, like the human body, the earth was shot through with air passages. When obstructed, the air—naturally loving freedom—"rages and is driven around this way and that, hurling some obstacles aside, cutting through others. . . . It pushes through obstructions, however large, and whatever it enters it splits by its force" (Seneca, 2:171). Seneca, who himself suffered from an upper respiratory disorder, imagines terrestrial air as both a kind of glue "by which [the earth] holds itself together and joins the parts of itself" and a force that "is vital and active, nourishing all things" (Seneca, 2:173).

6. Seventeenth-century naturalist John Woodward traced earthquakes to "sulphureous vapours" exhaled from the earth—vapors that substantively identified it with the upper air. Thus aerial events such as the aurora borealis or even lightning storms could foretell seismic activity

here below. See Woodward, *An Essay towards a Natural History of the Earth and Terrestrial Bodies* (London, 1695), 18. Woodward's theory is discussed in Janković, *Reading the Skies*, 27, and in Jan Golinski, *British Weather and the Climate of Enlightenment* (Chicago: University of Chicago Press, 2007), 66.

7. Kenneth R. Olwig, "All That Is Landscape Is Melted into Air: The 'Aerography' of Ethereal Space," *Environment and Planning D: Society and Space* (2011): 526.

8. Arthur Eddington, *New Pathways in Science* (New York: Macmillan, 1935), 40; cited also in Olwig, "All That Is Landscape Is Melted into Air," 526.

9. The author of *Airquakes* makes virtually no distinction between the matter and the media—the representations—of atmospheric events: "In all Aeril Concussions, the Shock and the Sound are in Reality the same Thing, and must be perceived in the same Instant" (*Airquakes*, 47).

10. John Flamsteed, *A Letter concerning Earthquakes* ([1693] London, 1750), 29, 31, 26. Flamsteed was also an astrologer.

11. Walter Scott perfectly captured Fielding's Englishness in the first sentence of his memoir of the writer: "Of all the works of imagination, to which English genius has given origin, the writings of Henry Fielding are, perhaps, most decidedly and exclusively her own." Scott added that, in *Tom Jones*, for example, "the persons of the story live in England, travel in England, quarrel and fight in England; and scarce an incident occurs, without its being marked by something which could not well have happened in any other country." See Scott, "Henry Fielding," in Scott, *Biographical Memoirs of Eminent Novelists*, 2 vols. ([1821–24] London, 1834), 1:77.

12. William Empson, "*Tom Jones*," in *Modern Critical Views: Henry Fielding*, ed. Harold Bloom (Philadelphia: Chelsea House, 1984), 29.

13. See Jill Campbell, *Natural Masques: Gender Identity in Fielding's Plays and Novels* (Stanford, CA: Stanford University Press, 1995).

14. This perspective is implicit in Jill Campbell's evocative essay on Fielding's style, which she describes as something that cannot "be separated from meaning" but can be "scored, in quasi-musical notation as a pleasurable rhythm of intakes of breath" (Campbell, 419). See Campbell, "Fielding's Style," *ELH* 72 (2005): 407–28.

15. Thomas Wilkes, *A General View of the Stage* (London, 1759), 63. Wilkes's comments on acting as the assuming of an "air" (Wilkes, 92) are anticipated in Aaron Hill's *An Essay on the Art of Acting* ([1753] London, 1779), which pegs theatrical emotions from joy ("a vivacity in look, air and accent" to grief ("a careless air") and anger ("a threatening air") to specific "airs." (Wilkes, 12, 20, 22)

16. "Verses to the Memory of Mr. Garrick," in *Poetical Works of David Garrick, Esq.*, ed. George Kearsley, 2 vols. (London, 1785), 1:lviii.

17. Joseph R. Roach brilliantly analyzes Garrick's body language in this scene in "Garrick, the Ghost, and the Machine," *Theatre Journal* 34 (1982): 31–40.

18. Samuel Taylor Coleridge, "Notes on *Tom Jones*," in *Literary Remains of Samuel Taylor Coleridge*, ed. Henry Nelson Coleridge, 4 vols. ([1836] New York: AMS Press, 1967), 2:374.

19. William Makepeace Thackeray, *The English Humourists of the Eighteenth Century* ([1853] London: Grey Walls Press, 1949), 171.

20. Betty Rizzo, "The Devil in *Tom Jones*," *1650–1850: Ideas, Aesthetics and Inquiries in the Early Modern Era* 10 (2004): 103–27. See also Robert Wess, "The Probable and the Marvelous in *Tom Jones*," *Modern Philology* 68 (1970): 32–45.

21. James Boswell, *The Life of Johnson* ([1791] Oxford: Oxford University Press, 1980), 389. Fielding's "natural pictures" have supported a heavy phalanx of historical readings, exemplified

in Martin Battestin's general introduction to the two-volume Wesleyan *Tom Jones* (Middletown, CT: Wesleyan University Press, 1975), 1:xiii–xxx; J. Paul Hunter, *Occasional Form: Henry Fielding and the Chains of Circumstance* (Baltimore: Johns Hopkins University Press, 1975), 13; and Ronald Paulson, *Satire and the Novel in Eighteenth-Century England* (New Haven, CT: Yale University Press, 1967). The opposing view of *Tom Jones* as primarily concerned with fictive form is developed in Wayne Booth, *The Rhetoric of Fiction* (Chicago: University of Chicago Press, 1961), and Robert Alter, *Fielding and the Nature of the Novel* (Cambridge, MA: Harvard University Press, 1961). Homer Brown mediates between these positions, treating history as itself a tropology, in *Institutions of the English Novel from Defoe to Scott* (Philadelphia: University of Pennsylvania Press, 1997). A long and useful note on the history versus fiction debate may be found on pp. 223–24.

22. William Hazlitt, "Trifles Light as Air," in *Henry Fielding: A Critical Anthology*, ed. Claude Rawson (Harmondsworth, UK: Penguin, 1973), 255.

23. Henry James, preface to *The Princess Casamassima*, in *Works of Henry James* (New York: Charles Scribner's Sons, 1908), 5:xiv. On James's own sense of atmosphere, see Thomas J. Otten, *A Superficial Reading of Henry James: Preoccupations with the Material World* (Columbus: Ohio State University Press, 2006).

24. Henry Fielding, *The History of Tom Jones, A Foundling*, ed. Fredson Bowers and Martin Battestin, 2 vols. (Middletown, CT: Wesleyan University Press, 1975), 1.4.42. Future citations in this chapter will be to these volumes and will appear parenthetically in the text by short title and book, chapter, and page number.

25. Samuel Taylor Coleridge, *Table Talk* (July 5, 1834), in *Specimens of the Table Talk of the Late Samuel Taylor Coleridge*, ed. Henry Nelson Coleridge, 2 vols. (New York: Harper and Brothers, 1835), 2:171.

26. Austin Dobson, *Fielding* (London, 1833), 125.

27. Another remark of Coleridge is equally apposite: "To take [Fielding] up after Richardson is like emerging from a sick room heated by stones into an open lawn on a breezy day in May." See Coleridge, *Table Talk* (July 5, 1834), in *Specimens*, 2:171.

28. The metaphor of shape has often lent itself to some of recent modern criticism's best attempts to comprehend and describe Fielding's fiction, as witness Sheldon Sacks, *Fiction and the Shape of Belief* (Berkeley: University of California Press, 1964).

29. Dorothy Van Ghent's penetrating commentary on *Tom Jones* is well worth citing here: "We may think of *Tom Jones* as a complete architectural figure, a Palladian palace perhaps. . . . The structure is all but in the light of intelligibility: air circulates around and over it and through it." See Van Ghent, *The English Novel: Form and Function* (New York: Harper, 1953), 80. Finally, Fielding's nondepiction of Paradise Hall recalls Vitruvius's celebrated *De Architectura* (25 BCE), which conceives of a building not merely as a shelter of human design but as the expressive obverse of the airspace around it. In each of its many translations, reprintings, and citations throughout the first half of the eighteenth century, the ancient Roman's *Theory and Practice of Architecture* urges the aspiring architect first to "distinguish the Qualities of the Air" in the place he would build. Vitruvius takes up the question of "the Convenience of Fabrics" by announcing that foremost "the Wholesomeness of the Air must be regarded. And to this end we must chuse a high situated place, that it may be less subject to Fogs and mists; it must be likewise far from all Morasses, because the Corruption that may be occasion'd by the infectious Breath of Venomous Creatures . . . makes the place very unwholesome." See Vitruvius Pollio, *The Theory and Practice of Architecture . . . carfully* [*sic*] *done into English* (London, 1703), 17.

30. An admirably straightforward account of Fielding's concept of character assessment

may be found in Arlene Fish Wilner, "Henry Fielding and the Knowledge of Character," *Modern Language Studies* 18 (1988): 181–94.

31. Claude Rawson discusses "Professor Empson's *Tom Jones*" in his essay of that title in *Tom Jones: A Casebook*, ed. Neil Compton (London: Macmillan, 1970), 173–81.

32. Fielding's "providential theme" is linked to his narrative structure through rational religious argument in Richard A. Rosengarten, *Henry Fielding and the Narration of Providence* (New York: Palgrave, 2000). Most critics, as Rosengarten points out, see a thoroughgoing embrace of the providential to have been forestalled by Fielding's commitment to "historical" narration. See in particular Leopold Damrosch Jr., *God's Plot and Man's Stories: Studies in the Fictional Imagination from Milton to Fielding* (Chicago: University of Chicago Press, 1985).

33. William Empson, *Seven Types of Ambiguity* ([1930] Harmondsworth, UK: Penguin, 1961), 17.

34. Arthur Murphy, "An Essay on the Life and Genius of Henry Fielding, Esq." (1762), in *Henry Fielding: The Critical Heritage*, ed. Ronald Paulson and Thomas Lockwood (London: Routledge, 1969), 425.

35. For a biographical treatment of Fielding that is sensitive to the folkloric and even mystical influences on his literary imagination, see Donald Thomas, *Henry Fielding: A Life* (New York: St. Martin's, 1991).

36. Martin Battestin and Ruthe Battestin, *Henry Fielding: A Life* (London: Routledge, 1989), 5.

37. "Aretine," *Old England*, in *Henry Fielding: Critical Heritage*, 167. J. Paul Hunter makes sense of Fielding's peculiar geography by proposing his affiliation with "the country of the mind and spirit" whose "closest approximation in reality was Somersetshire." See Hunter, *Occasional Form*, 13.

38. Samuel Richardson, letter to Astraea and Minerva Hill (August 4, 1749), in *Henry Fielding: Critical Heritage*, 174.

39. Kenneth R. Olwig, "Performance, *Aetherial* Space and the Practice of Landscape/Architecture: The Case of the Missing Mask," *Social and Cultural Geography* 12 (2011): 315–18.

40. Nicholas Hudson evocatively captures the improbable "physical expansion of this scene" in "*Tom Jones*," in *The Cambridge Companion to Henry Fielding*, ed. Claude Rawson (Cambridge: Cambridge University Press, 2007), 91.

41. Wolfgang Iser, *The Implied Reader: Patterns of Communication in Prose Fiction from Bunyan to Beckett* (Baltimore: Johns Hopkins University Press, 1974), 48.

42. On Fielding's one foray into contemporary natural history—via a "parodic amalgam" in which he has an ant develop a quasi-meteorological analysis of "the infinite Number of Drops of Water . . . released by the Wind" when a cow urinates on its anthill—see Bertram Goldgar, "Fielding, the Flood Makers, and Natural Philosophy: *Covent Garden Journal* 70," *Modern Philology* 80 (1982): 136–44.

43. As Rizzo also notes, Fielding's relationship to the supernatural is remarkably uncharted territory. See Rizzo, "Devil in *Tom Jones*," 103–4. L. F. Sells takes Fielding's skepticism about it for granted even as he catalogues its narrative value in "Fielding and the Supernatural: *The Journal of a Voyage to Lisbon*," *Susquehanna University Studies* 11 (1981): 117–22.

44. "Orbilius," *An Examen of the History of Tom Jones* (1750), in *Henry Fielding: Critical Heritage*, 199.

45. An especially subtle analysis of the ways in which Bridget Allworthy makes Fielding's habit of characterization visible is Sheridan Baker's "Bridget Allworthy: The Creative Pressures of Fielding's Plot," in *Tom Jones*, ed. Baker (New York: Norton, 1973), 906–16. The "pressures"

Baker has in mind are the conflicting ones of treating Bridget as both a figure of depth and a comic type; her full "personality," Baker suggests, "emerges from the conflicting evidence" (Baker, 908) and, perhaps more important, is responsible for the crucial dynamic of concealment in Fielding's plot. I analyze Bridget's atmospheric role in the novel in "The Air of *Tom Jones*; or, What Rose from the Novel," *Eighteenth Century: Theory and Interpretation* 52 (2011): 303–21.

46. I borrow the term "vivacity" from Elaine Scarry, who shows how the sense of life "comes about by reproducing the deep structure of perception" in aesthetic and specifically literary experience. See Scarry, *Dreaming by the Book* (Princeton, NJ: Princeton University Press, 1999), 9.

47. On the role of spectral evidence in these trials, with a brilliant exposition of the concept, see Victoria Silver, "'Wonders of the Invisible World': The Trial of the Lowestoft Witches," in *Sir Thomas Browne: The World Proposed*, ed. Reid Barbour and Claire Preston (Oxford: Oxford University Press, 2010), 118–45.

48. John Stevenson is one of few critics to recognize the complexity of Partridge's relationship to Fielding. Without denying Fielding's satiric distance from Partridge, Stevenson nonetheless proposes that he often stands as "a witness for his creator, and thus . . . a reminder of the novelist's capacity to be in at least two places at the same time." See Stevenson, *The Real History of Tom Jones* (New York: Palgrave, 2005), 182–83. The critical interpretations of *Tom Jones* sympathetic to romance often result in more lenient perspectives on Partridge. See especially Henry Knight Miller, *Henry Fielding's "Tom Jones" and the Romance Tradition* (Victoria, Canada: University of Victoria Press, 1976), and, by contrast, J. Paul Hunter, *Before Novels: The Cultural Contexts of Eighteenth-Century English Fiction* (New York: Norton, 1990).

49. On the moon's consistent record of inconsistent meaning, see Timothy Harley, *Moon Lore* (Detroit: Singing Tree Press, 1969).

50. For an illustration of the eighteenth-century provenance of the term "celestial science," see Ebenezer Sibly, *An Illustration of the Celestial Science of Astrology* (London, 1776). Sibly famously cast the horoscope of the United States in 1787.

51. John Booker, draft (1644), cited in Ann Geneva, *Astrology and the Seventeenth-Century Mind: William Lilly and the Language of the Stars* (Manchester, UK: Manchester University Press, 1995), 10.

52. Astrology's interface with contemporary science is discussed at length in Bernard Capp's indispensable *Astrology and the Popular Press: English Almanacs, 1500–1800* (London: Faber, 1979), 180–214.

53. See, for example, the title page of *Merlinus Liberatus* (London, 1739), which describes the year 1739 as "the 50th of our Deliverance by K. *William*, from Popery, and Arbitrary Government."

54. Most scholars of astrology stress that it was a system of thought adjacent to both science and religion; natural philosophers from Boyle to Hooke were far from rejecting it, and a number of late Stuart astrologers produced tables and compiled data in a manner they deemed no less "experimental" than what one might find in the transactions of the Royal Society. Capp thus suggests that astrology aimed not to discern the "nature of stellar inference" from the top down but to accumulate "data about its effects, with a comparison of known events with the astrological circumstances" (*Astrology and the Popular Press*, 184).

55. On Fielding's use of the almanac in the service of formal realism, see Ian Watt, *The Rise of the Novel* (Berkeley: University of California Press, 1957), 25.

56. These positions are taken up, respectively, by Capp, Keith Thomas, and Frank Palmeri. While his student Maureen Perkins contests the fact itself, Capp attributes the almanac's "decay" not just to astrology's "reduced intellectual and thence standing" but to their increasingly "vague

and repetitive" quality (Capp, 270, 238, 239). Besides astrology's lack of "the essential quality of a science—the capacity for demonstration," Thomas stresses a supposed disconnect between astrology's deterministic assumptions and the Puritan—later Protestant—doctrine of free will, pointing out also that it was most threatening because it actually mimed Christian Providence. See Thomas, *Religion and the Decline of Magic: Studies in Popular Beliefs in Sixteenth and Seventeenth Century England* (Oxford: Oxford University Press, 1971), 352, and Maureen Perkins, *Visions of the Future: Almanacs, Time, and Cultural Change, 1775–1870* (Oxford: Oxford University Press, 1996). Speaking of mime and parody, Palmeri traces the almanac's degeneration into an object of ridicule in "History, Nation and the Satiric Almanac, 1660–1760," *Criticism* 40 (1998): 377–408. See also Patrick Curry, "Astrology in Early Modern England: The Making of a Vulgar Knowledge," in *Science, Culture and Popular Belief in Renaissance Europe*, ed. Stephen Pumfrey et al. (Manchester, UK: Manchester University Press, 1989), 274–91.

57. Partridge, *Merlinus Liberatus* (London, 1751), n.p.; Partridge, *Merlinus Liberatus* (London, 1742), n.p.

58. Jonathan Swift, "Predictions for the Year 1708" (1707–8), in *Writings of Jonathan Swift*, ed. Robert A. Greenberg and William B. Piper (New York: Norton, 1973), 427.

59. William Ingpen, *The Secrets of Numbers* (1624), cited in Thomas, *Religion and the Decline of Magic*, 323.

60. Francis Moore, "The Use and Explanation of the Almanack," in *Vox Stellarum, Being an Almanack for . . . 1703* (London, 1703), n.p.; Edward Smith, entries for November 12 and December 29, *Annual Advertiser* (London, 1739), n.p.

61. Geneva proposes that astrology was done in partly by the rise of referential languages; astrology belongs with universal language schemes such as those of John Wilkins, which likewise "assumed isomorphism between words and what they signified" (Geneva, *Astrology*, 273).

62. Alison Chapman, "Marking Time: Astrology, Almanacs, and English Protestantism," *Renaissance Quarterly* 60 (2007): 257–90. This function may be seen to have been taken over by the English novel as it stabilized points of temporal and spatial reference by taking them out of the real world altogether.

63. Thomas Sprat, *History of the Royal Society* (London, 1667), 362.

64. Both Perkins and Geneva identify astrology with mentality rather than as a sphere of knowledge: Perkins calls it "the way people thought" (Perkins, *Visions of the Future*, 3), while Geneva, as her title promises, explores the way it was integral to "the seventeenth-century mind."

65. John Dove, *Speculum Anni* (London, 1653), C6r. Of course, a still subtler medium was supposed to integrate planetary events and earthly ones. In 1674's *Suspitions about Some Hidden Qualities in the Air*, Boyle proposed "subtil, but corporeal, emanations" from lunar and sidereal bodies (Boyle, in Thomas Birch, ed., *The Works of the Honourable Robert Boyle*, 6 vols. [London, 1772], 4:85).

66. Fielding's own classical education would have equipped him with other astrological schemas; as Geneva suggests, one of the Stoic schools gave each of the earth's climates "its own planet which determined the characteristics of its inhabitants." See Geneva, *Astrology*, 123. This conception of character introduces something far more complex than the environmental conditioning that Montesquieu and Hume both contemplated.

67. Cited in Geneva, *Astrology*, 123.

68. On Jacobitism and the cryptic, see Paul Kléber Monod, *Jacobitism and the English People, 1688–1788* (Cambridge: Cambridge University Press, 1989).

69. Martin Battestin, "Tom Jones and 'His Egyptian Majesty,'" *PMLA* 82 (1967): 76. For the

most erudite analysis of this scene as political allegory, see Robert Folkenflik, "Tom Jones, the Gypsies, and the Masquerade," *University of Toronto Quarterly* 44 (1975): 244–57.

70. Manuel Schonhorn, "Fielding's Ecphrastic Moment: Tom Jones and His Egyptian Majesty," *Studies in Philology* (1981): 305–6.

71. John Stevenson, "Fielding's Mousetrap: Partridge, Hamlet, and the '45," *Studies in English Literature* (1997): 553–71.

Chapter Seven

1. Henry Fielding, review of *The Female Quixote, Covent-Garden Journal* 24 (March 24, 1752), in *The Covent-Garden Journal*, ed. Gerard Edward Jensen, 2 vols. (New York: Russell and Russell, 1964), 1:280.

2. Georg Simmel, "Sociability," in *On Individuality and Social Forms*, ed. Donald N. Levine (Chicago: University of Chicago Press, 1971), 139.

3. Catherine Gallagher, *Nobody's Story: The Vanishing Acts of Women Writers in the Marketplace, 1670–1820* (Berkeley: University of California Press, 1994), 178.

4. This debate has shaped recent criticism of Lennox's novel. The opposing positions are set out by Laurie Langbauer in *Women and Romance: The Consolations of Gender in the English Novel* (Ithaca, NY: Cornell University Press, 1990), 78–85, and Margaret Anne Doody, introduction to *The Female Quixote* (Oxford: Oxford University Press, 1989), xx–xxii. Langbauer thinks that romance has been expunged from Lennox's fiction; Doody insists that it persists.

5. Charlotte Lennox, *The Female Quixote; or, The Adventures of Arabella*, ed. Margaret Dalziel ([1752] Oxford: Oxford University Press, 1989), 382, 305, 368. Future citations in this chapter will be to this edition and will appear parenthetically in the text by short title and page number.

6. Miriam Rossiter Small, *Charlotte Ramsay Lennox* ([1935] New York: Yale University Press, 1935), 74.

7. Samuel Richardson, *Clarissa; or, the History of a Young Lady*, ed. Angus Ross ([1747–48] London: Penguin, 2004,) 145. It is Richardson's Lovelace who characterizes Clarissa as a "frost-piece" and then plays libertine meteorologist as he tries to determine "whether her frost is frost indeed" (p. 878).

8. *Correspondence of Samuel Richardson*, ed. Anna Laetitia Barbauld, 6 vols. (London, 1804), 6:243.

9. *The Autobiography and Correspondence of Mary Glanville, Mrs. Delany*, ed. Lady Llanover (London, 1861), 3:116.

10. *A Blow against Modern Sadducism* was the title of the 1668 version of what became Glanvill's *Saducismus Triumphatus*, discussed at length in chapter 5 and in this chapter (below).

11. Joseph Glanvill, *Saducismus Triumphatus; or, Full and Plain Evidence concerning Witches and Apparitions* (1681), in *Collected Works of Joseph Glanvill*, ed. Bernhard Fabian (Hildesheim, Germany: Georg Olms Verlag, 1978), 9:15.

12. Charlotte Lennox, *The Lady's Museum*, 2 vols. (London, 1762), 2:722.

13. Clara Reeve, *The Progress of Romance through Times, Countries and Manners*, 2 vols. (Colchester, UK, 1785), 2:6.

14. Lennox's Grace Pythoness—of *Lady's Museum* fame—likewise identifies superstition and witchcraft with bygone days: "My mother and my grandmother have often informed me of many wonderful noises, apparitions, and visions that have been seen and heard in our family. My honoured parents were not only pious matrons but great believers," Grace declares, "and

shall I degenerate?" Her sorrow is that she "foresee[s] that with me must die the usual family veneration for supernatural causes" for her own granddaughters are modern young women who "give each other knives and scissars [*sic*], without considering the consequence, that such kind of instruments invisibly cut love and affection," and the elder one "never fails to quit the room as soon as I begin to read my fate in coffeegrounds." It is impossible to tell whether we are finally meant to laugh at the wittily named "Grace," who longs for an era when "dread and reverence were paid to omens, dreams, visions, blue burning candles, knives and forks across each other, salt spilt by aukwardness [*sic*], and every kind of prognostic that led into the avenues of fate." See Lennox, *Lady's Museum*, 2:721.

15. Joseph Glanvill, *Scepsis Scientifica; or Confest Ignorance the Way to Science* (London, 1665), 68, 82–83.

16. Harry Price, *Poltergeist over England: Three Centuries of Mischievous Ghosts* (London: Country Live, 1945). See also H. Stanley Redgrove and I. M. L. Redgrove, *Joseph Glanvill and Psychical Research in the Seventeenth Century* (London: W. Rider, 1921). The Redgroves credit Glanvill with having introduced into "psychical research" methodological consciousness of the observer flaws that distort information, such as fancy, precipitancy, and haste. Meanwhile, many Glanvillean premises inform *Phantasms of the Living* (1886), the Victorian era's most substantive inquiry into the "paradox" of phantasmic experience. Commissioned by the Council of the Society for Psychical Research, this two-volume compendium of spectral sightings includes a long philosophical meditation titled "Collective Cases." It thus focuses on the words that groups use to describe the objectification of supernatural presence: "Gaseous and vaporous ghosts are, I imagine, quite at a discount; but the word 'ether' seems sometimes to be used as a way out of the difficulty. For many ears the word has, no doubt, a convenient vagueness, . . . [but] even if ethereal ghosts could be seen, the auditory phenomena would remain a hopeless obstacle to a satisfactory physical explanation of them" (*Phantasms*, 2:168–69). See Edmund Gurney, Frederic W. H. Myers, and Frank Podmore, *Phantasms of the Living* ([1886] Gainesville, FL.: Scholars' Facsimiles, 1970), 2:168–69.

17. Anthony Horneck, preface to *Some Discourses, Sermons, and Remains of the Rev. Joseph Glanvill* (London, 1681), 21. See too Ferris Greenslet, *Joseph Glanvill: A Study in English Thought and Letters in the Seventeenth Century* (New York: Columbia University Press, 1900), 144.

18. Glanvill, epistle dedicatory to *Saducismus Triumphatus*, n.p.

19. Henry More, "The Letter of Dr. H. M.," in Glanvill, *Saducismus Triumphatus*, 12.

20. Samuel Pepys, December 25, 1667, in *The Diary of Samuel Pepys*, vol. 8, *1667*, ed. Robert Latham and William Matthews (Berkeley: University of California Press, 1971), 589.

21. Pepys, April 16, 1663, in *Diary*, vol. 4, *1663*, 185–86.

22. The jury stays out on the ghostwriting question: in 1843, the Anglican divine John Mitford first advanced the theory that Johnson wrote this pivotal exchange, and scholars have remained divided ever since. See Mitford, *Gentleman's Magazine*, n.s., 20 (1843): 132–33. See also Patricia Meyer Spacks, *Desire and Truth: Functions of Truth in Eighteenth-Century Novels* (Chicago: University of Chicago Press, 1990), 15; Duncan Isles, "Johnson, Richardson, and *The Female Quixote*," appendix to Lennox, *Female Quixote*, 419–28; O. M. Brack Jr. and Susan Carlisle, "Samuel Johnson's Contributions to Charlotte Lennox's *The Female Quixote*," *Yale University Library Gazette* 77, 3–4 (2003): 165–70; and David Marshall, who surveys this critical archive in *The Frame of Art: Fictions of Aesthetic Experience, 1750–1815* (Baltimore: Johns Hopkins University Press, 2005), 239–40.

23. Richardson enumerated some of the "Lovely Visionary's" "Absurdities" in an admiring

letter to Lennox. See Duncan Isles, "The Lennox Collection A," *Harvard Library Bulletin* 18 (1970): 338.

24. Joseph Glanvill, *Essays on Several Important Subjects in Philosophy and Religion* (1675), in Glanvill, *Collected Works* 4:4. This position has given Glanvill a significant place in the same history of skepticism that includes Hume. See Richard Popkin, "Joseph Glanvill: A Precursor of David Hume," *Journal of the History of Ideas* 14 (1953): 292–303.

25. Contrast Dorcas Sheldon, the heroine of Tabitha Tenney's 1801 novel, *Female Quixotism*: "She was of middling stature, a little embonpoint, but neither elegant nor clumsy. Her complexion was rather dark; her skin somewhat rough; and features remarkable for neither beauty nor deformity." Tenney, *Female Quixotism: Exhibited in the Romantic Adventures of Dorcasina Sheldon*, ed. Jean Nienkamp and Andrea Collins (New York: Oxford University Press, 1992), 5.

26. See Deborah Ross, "Mirror, Mirror: The Didactic Dilemma of *The Female Quixote*," *Studies in English Literature* 27 (1987): 455–73, and Marshall, *Frame of Art*, 147–76.

27. Leo Spitzer, "Milieu and Ambiance," in *Essays in Historical Semantics* (New York: Russell and Russell, 1948), 260.

28. Lorraine Daston, "The Cold Light of Fact and the Fact of Cold Light: Luminescence and the Transformation of Scientific Fact," in *Signs of Early Modern France, 1600–1750*, ed. David Rubin (Charlottesville: University of Virginia Press, 1997), 17–44.

29. Terry Castle, *The Female Thermometer: Eighteenth-Century Culture and the Invention of the Uncanny* (Oxford: Oxford University Press, 1995), 22.

30. John Hawkins, *Life of Samuel Johnson* (London, 1787), 285–86. Hawkins's figurative language is especially telling in light of the possible factitiousness of much of this account, which is considered in James L. Clifford, "Johnson's First Club," in *Evidence in Literary Scholarship: Essays in Memory of James Marshall*, ed. René Wellek and Alvaro Ribeiro (New York: Oxford University Press, 1979), 197–213.

31. Duncan Isles, "The Lennox Collection B," *Harvard Library Bulletin* 19 (1970): 46–47.

32. Daniel Defoe, *The Political History of the Devil*, ed. John Mullan ([1726] London: Pickering and Chatto, 2005), 89. Dorothea von Mücke observes a similar phenomenon in Jacques Cazotte's 1772–76 *The Devil in Love*, a literary work that diabolically creates "a mixed semiotic universe that provokes a state of uncertainty as to how to discern the markers of the real from the markers of the imaginary." See von Mücke, *The Seduction of the Occult and the Rise of the Fantastic Tale* (Stanford, CA: Stanford University Press, 2003), 6.

33. Michael Oakeshott, *On Human Conduct* (Oxford: Oxford University Press, 1975), 74.

34. Richard Sennett, *The Fall of Public Man* (New York: Norton, 1974), 38–39.

Chapter Eight

1. Even Michel Foucault put Radcliffe among the "initiators of discursive practices" who set down "the possibility and the rules of formation of other texts." See Foucault, *Language Counter-Memory, Practice: Selected Essays and Interviews*, trans. Donald F. Bouchard and Sherry Simon (Ithaca, NY: Cornell University Press, 1977), 131–32. There has been little critical commentary in the last two hundred years that has not seen Radcliffe to have minted many of the conventions that define gothic writing, but see Eve Kosofsky Sedgwick, *The Coherence of Gothic Conventions* ([1976] New York: Arno, 1980), and Marshall Brown, *The Gothic Text* (Stanford, CA: Stanford University Press, 2005) for summaries of the ways in which Radcliffe is understood as an originator. An unusually lucid and compelling commentary on Foucault's remark (in light

of her own thesis that what Radcliffe pioneered was "a commentary on the psychodynamics of narrative form") may be found in Margaret Russett, "Narrative as Enchantment in *The Mysteries of Udolpho*," *ELH* 65 (1998): 159–86.

2. Deidre Lynch, "Gothic Fiction," in *The Cambridge Companion to British Romantic Fiction*, ed. Katie Trumpener and Richard Maxwell (Cambridge: Cambridge University Press, 2008), 47.

3. Jane Austen, *Northanger Abbey*, ed. John Davie and Claudia L. Johnson ([1817] Oxford: Oxford University Press, 2003), 82. On atmosphere as gothic trope, see Jayne Lewis, "'No Colour of Language': Radcliffe's Aesthetic Unbound," *Eighteenth-Century Studies* 39 (2006): 377–90.

4. Paul Gilmore considers the figure of electricity as a signature of romantic aesthetics in a nineteenth-century American context in his "Romantic Electricity, or the Materiality of Aesthetics," *American Literature* 76 (2004): 467–94.

5. Without addressing the terminology of atmosphere, Russett identifies the "indiscreet 'charm' of gothic genre" as "its power to attract and absorb." See Russett, "Narrative as Enchantment," 160.

6. Some of the most fruitful readings of Radcliffe's work stress its suspense between neoclassical and romantic aesthetic styles. See especially Barbara M. Benedict, "Pictures of Conformity: Sentiment and Structure in Ann Radcliffe's Style," *Philological Quarterly* 68 (1989): 363–72; Elizabeth Napier, *The Failure of Gothic: Problems of Disjunction* (Oxford: Oxford University Press, 1987); and Wylie Sypher, "Social Ambiguity in a Gothic Novel," *Partisan Review* 12 (1945): 50–60.

7. The panorama as a virtual reality technology relevant to literary experiment in the period is explored in Markman Ellis, "Spectacles within Doors: Panoramas of London in the 1790s," *Romanticism* 14 (2008): 133–48, and Jay Martin, "The Panorama: History of a Mass Medium," *ArtForum*, March 1998. Jennifer B. Johnson explores the intersection of "sublimity, virtuality, materiality" in spaces such as the eidophusikon in "Absorbing Hesitation: Wordsworth and the Theory of the Panorama," *Studies in Romanticism* 45 (2006): 357–75.

8. Benjamin Wilson, *An Account of Experiments Made at the Pantheon, on the Nature and Use of Conductors* (London, 1778), 84. For a discussion of the politics of Wilson's attempt, as a member of an elite scientific coterie with a background in theater, to promote the technology of rounded lightning rods over that of pointed rods advocated by his ostensibly more democratic rival, Benjamin Franklin, see Troy A. Mitchell, "The Politics of Experiment in the Eighteenth Century: The Pursuit of Audience and the Manipulation of Consensus in the Debate over Lighting Rods," *Eighteenth-Century Studies* 31 (1998): 37–51. On exhibition science more generally, see the essays in Bernadette Bensoude-Vincent and Christine Blondel, eds., *Science and Spectacle in the Enlightenment* (New York: Ashgate, 2008).

9. Ann Radcliffe, *The Mysteries of Udolpho*, ed. Terry Castle and Bonamy Dobrée ([1794] Oxford: Oxford University Press, 1998), 3; Ann Radcliffe, *The Italian*, ed. Frederic Garber and E. J. Clery ([1797] Oxford: Oxford University Press, 1998), 413. Future references in this chapter will be to these editions and will appear parenthetically in the text.

10. Samuel Taylor Coleridge, "Notes on *Tom Jones*," in *Literary Remains of Samuel Taylor Coleridge*, ed. Henry Nelson Coleridge, 4 vols. ([1836] New York: AMS Press, 1967), 2:374.

11. Walter Scott, "Mrs. Ann Radcliffe," in *Biographical Memoirs of Eminent Novelists*, 2 vols. ([1821–24] London, 1834), 1:379.

12. Mary Wollstonecraft, review of *The Mysteries of Udolpho*, *Analytical Review* 25 (May 1797).

13. Ann Radcliffe, *A Journey Made in the Summer of 1795, through Holland and the Western Frontier of Germany* (London, 1795), 456.

14. "To the wild landscape of Salvator Rosa," wrote one contemporary, she "added the softer graces of a Claude." See Nathan Drake, "On the Objects of Terror," in *Literary Hours; or Sketches, Critical, Narrative and Poetical*, 3 vols. ([1798] London, 1820), 1:270.

15. Deidre Lynch reads the gothic fiction of Radcliffe's period, especially Radcliffe's own, as an exercise in ambivalence toward the British literary canon. See Lynch, "Gothic Libraries and National Subjects," *Studies in Romanticism* 4 (2003): 29–48.

16. James Boswell, *Life of Johnson* ([1791] Oxford: Oxford University Press, 1980), 441.

17. Rictor Norton, *Mistress of Udolpho: The Life of Ann Radcliffe* (London: Leicester University Press, 1999), 68. Radcliffe's landscape aesthetic in relation to Burke's sublime is discussed at length by Elizabeth Bohls in *Women Travel Writers and the Language of Aesthetics, 1716–1818* (Cambridge: Cambridge University Press, 1995), 216–18, and by Anne K. Mellor in *Romanticism and Gender* (London: Routledge, 1993), 95.

18. Joseph Priestley, preface to *The Rudiments of English Grammar* (London, 1761), vi. Jane Hodson considers the changes between Priestley's two (1761 and 1768) versions of this text in "Joseph Priestley's Two *Rudiments of English Grammar*: 1761 and 1768," in *Grammars, Grammarians and Grammar-Writing in Eighteenth-Century England*, ed. Ingrid Tieken-Boon van Ostade (Berlin: de Gruyter, 2008), 177–89.

19. On the transformation of van Helmont's "chaos" into "gas," see Jenny Uglow, *The Lunar Men: Five Friends Whose Curiosity Changed the World* (New York: Farrar, Straus and Giroux, 2002), 231.

20. Steven Johnson, *The Invention of Air: A Story of Science, Faith, Revolution, and the Birth of America* (New York: Riverhead, 2008), 70.

21. Elizabeth Inchbald's 1784 play, *The Mogul Tale; or, The Descent of the Balloon* (a British balloon, into India) epitomizes, if it also ironizes, the giddily liberal and universal conception of air that developed in the chemical revolution and was realized in newly discovered "airs" such as helium. On the "ballonomania" more generally, see Paul Keen, "Science and Spectacle in 1780s England," *Eighteenth-Century Studies* 39 (2006): 517.

22. According to Isaac Kramnick, Priestley supplied "a metaphorical model" for his radical vision of the universal and indiscriminate "unfolding [of] truth and perfection with the American and French revolutions." See Kramnick, "Eighteenth-Century Science and Radical Social Theory: The Case of Joseph Priestley's Scientific Liberalism," *Journal of British Studies* 25 (1986): 3.

23. Joseph Priestley, *Experiments and Observations on Different Kinds of Air*, 3 vols. ([1774] Birmingham, UK, 1790), 1:xxiii. Future citations in this chapter will be to these volumes and will appear parenthetically in the text by short title and page number. Other writings by Priestley will be cited by short title and page number after their first mention in the text.

24. Joseph Priestley, "The Importance and Ethics of Free Inquiry," in Priestley, *Discourses on Various Subjects* (London, 1787), 184.

25. Edmund Burke, *Reflections on the Revolution in France*, ed. C. C. O'Brien ([1790] Harmondsworth, UK: Penguin, 1968), 90.

26. Stephen Hales's *Vegetable Staticks; or an Account of Some Statickal Experiments on the Sap of Vegetables* (London, 1721) contrasts with Priestley's pneumatic writing partly because of its preoccupation with statistics. Black, a Regius professor at the University of Glasgow, coined the term "fixed air" in 1754.

27. Clara Reeve, *The Progress of Romance through Times, Manners, and Countries* (Colchester, UK, 1785).

28. Joseph Priestley, *A Course of Lectures on Oratory and Criticism* (London, 1777), 80–81.

29. Joseph Priestley, *Memoirs of Dr. Joseph Priestley, Written by Himself*, in *The Autobiography of Joseph Priestley*, ed. John T. Boyer (Teaneck, NJ: Fairleigh Dickinson University Press, 1970), 74.

30. See Priestley, *Experiments*, 2:162: "Who can tell but that, in time, this pure air may become a fashionable article in luxury?"

31. Alain Corbin documents the rise of a "new calculus" based on codification and centered on smell via, for example, the so-called eudiometer. See Corbin, *The Foul and the Fragrant: Odor and the French Social Imagination* (Cambridge, MA: Harvard University Press, 1986), 15–16.

32. In an enlightening series of essays, James L. McEvoy considers Priestley's nomenclature partly in light of the ways it signaled subject position in relation to objective events, tracks its resistance to hypothesis, and draws attention to its peculiarities. See McEvoy, "Joseph Priestley, 'Aerial Philosopher,' Metaphysics and Methodology in Priestley's Chemical Thought from 1767–81," *Ambix* 25 (1978): 1–55, 113–16, 153–75, and *Ambix* 26 (1979): 11–37. See also Jan Golinski, "The Chemical Revolution and the Politics of Language," *Eighteenth Century: Theory and Interpretation* 33 (1992): 238–57.

33. An especially crisp analysis of Priestley's resistance to a shifting paradigm that was generating new "tickets" of discursive belonging may be found in Gilman McCann, *Chemistry Transformed: The Paradigmatic Shift from Phlogiston to Oxygen* (Norwood, NJ: Ablex, 1978). Against the standard view of scientific revolutions put forth in Thomas Kuhn's classic *The Structure of Scientific Revolutions* (Chicago: University of Chicago Press, 1962), Steven Johnson discusses Priestley's resistance to acknowledging the internal flaws that motivate paradigm shifts and intriguingly proposes that thought paradigms also exist in fundamentally "ecological" relationships to conditions outside them. See Johnson, *Invention of Air*, 48–55.

34. Robert Schofield, *The Enlightened Joseph Priestley: A Study of His Life and Work from 1733 to 1783* (University Park: Pennsylvania State University Press, 1997), 274. See also Dan Eshet, "Rereading Priestley," *History of Science* 39 (2001): 127–59.

35. Wilda Anderson, *Between the Library and the Laboratory: The Language of Chemistry in Eighteenth-Century France* (Baltimore: Johns Hopkins University Press, 1984), 3–4.

36. McEvoy notes that the verbs that Priestley used to capture the behavior of his airs automatically invoked the observer's relationship to them. See McEvoy, "Joseph Priestley, 'Aerial Philosopher,'" 113–16.

37. Anna Laetitia Barbauld, "An Inventory of the Furniture in Dr. Priestley's Study," in *A Selection from the Poems and Prose Writings of Mrs. Anna Laetitia Barbauld* (Boston, 1874), 33–34.

38. Joseph Priestley, *The History and Present State of Electricity*, 3 vols. (London, 1767), 1:45.

39. See Uglow, *Lunar Men*, and Robert M. Schofield, *The Lunar Society of Birmingham: A Social History of Provincial Science in Eighteenth-Century England* (Oxford: Oxford University Press, 1963). Priestley's interest in materializing immaterial dimensions, including language, provoked a poetic diatribe against him in the anonymous *The Sadducee* (London, 1778).

40. Joseph Priestley, *A Course of Lectures on the Theory of Language and Universal Grammar* (London, 1762), 12, 14.

41. Priestley found that, as time passed and languages fell into ruins, their harsher sounds faded away but that the "easy and agreeable" ones could be resurrected into new languages. See Priestley, *Universal Grammar*, 195.

42. Robert Alexander, "The Language of 'Naked Facts': Joseph Priestley and the Apocalypse of Language," *Language and Communication* 28 (2008): 21–35.

43. Stephen K. Land outlines a broader shift in language theory over the long eighteenth century, from a preoccupation with discrete units of language to an interest in the forms that relate them to one another. Priestley clearly belongs toward the far end off this trajectory. See Land, *From Signs to Propositions: The Concept of Form in Eighteenth-Century Semantic Theory* (London: Longman, 1974).

44. Jane Hodson defends Priestley's language theory against charges of prescription and casts him as an unusually innovative descriptivist in "The Problem of Joseph Priestley's Descriptivism," *Historiographia Linguistica* 33 (2006): 57–84.

45. Tobias Smollett, *The Expedition of Humphry Clinker*, ed. Lewis M. Knapp and Paul Gabriel Boucé ([1771] Oxford: Oxford University Press, 1998), 25, 95–96.

46. Edmund Burke, *Letters on a Regicide Peace*, in *Select Works of Edmund Burke*, ed. Francis Canavan, 3 vols. ([1796] Indianapolis: Liberty, 1999), 3:102.

47. Joseph Priestley, *A Familiar Introduction to the Theory and Practice of Perspective* (London, 1770), iii.

48. Joseph Priestley, *A Description of a New Chart of Biography* (Warrington, 1765), 24–25.

49. Daniel Rosenberg, "Joseph Priestley and the Graphic Invention of Modern Time," *Studies in Eighteenth-Century Culture* 36 (2007): 55–103. See also John McLachlan, "Joseph Priestley and the Study of History," in *Transactions of the Unitarian Historical Society* 19 (1987–89): 252–63, and Arthur Sheps, "Joseph Priestley's Time Charts: The Use and Teaching of History in Late Eighteenth-Century England," *Lumen* 18 (1999): 135–54.

50. Rictor Norton notes that Priestley's sublime not only attended to "novels, fictions, and feigned tragedies" as sources of this order of experience but was based on an inherently writerly perspective. See Norton, *Mistress of Udolpho*, 68–69.

51. Priestley's failure to identify oxygen as such has been the subject of a play—Carl Djerassi's *Oxygen* (Cambridge: Wiley, 2001). It has stuck in the craws of many historians of science, though see especially Joe Jackson, *A World on Fire: A Heretic, an Aristocrat, and the Race to Discover Oxygen* (New York: Viking, 2005); Schofield, *Enlightened Joseph Priestley*; and J. B. Conant, ed., "The Overthrow of the Phlogiston Theory: The Chemical Revolution of 1778–89," in *Harvard Case Histories in Experimental Science* (Cambridge, MA: Harvard University Press, 1950).

52. On Priestley's difference from Lavoisier in a linguistic context, see James L. McEvoy, "Priestley Responds to Lavoisier's Nomenclature: Language, Liberty and Chemistry in the English Enlightenment," in *Lavoisier in European Context: Negotiating a New Language for Chemistry*, ed. Bernadette Bensaude-Vincent and Ferdinando Abbri (Canton, MA: Science History Publications, 1995), 123–42.

Chapter Nine

1. Julia Kavanaugh, *English Women of Letters: Biographical Sketches*. 2 vols. ([1862] London, 1863), 1:251.

2. Walter Scott, "Mrs. Ann Radcliffe," in *Biographical Memoirs of Eminent Novelists*, 2 vols. ([1821–24] London, 1834), 1:379.

3. On the consolidation of French chemical knowledge, see Isaac Kramnick, "Eighteenth-Century Science and Radical Social Theory: The Case of Joseph Priestley's Scientific Liberalism," *Journal of British Studies* 25 (1986), 3; and Wilda Anderson, *Between the Library and the*

Laboratory: The Language of Chemistry in Eighteenth-Century France (Baltimore: Johns Hopkins University Press, 1984).

4. Joseph Priestley, *Experiments and Observations on Different Kinds of Air*, 3 vols. ([1774] Birmingham, UK, 1790), 1:62.

5. Ann Radcliffe, *A Journey Made in the Summer of 1794 through Holland and the Western Frontiers of Germany* (London, 1795), 176. Future citations in this chapter will be to this edition and will appear parenthetically in the text by short title and page number.

6. For an illuminating reading of the model of difference written into Priestley's description of carbonated water in relation to the poetry of Robert Burns, see Celeste Langan, "Scotch Drinks and Irish Harps: Mediations of the National Air," in *The Figure of Music in Nineteenth-Century British Poetry*, ed. Phyllis Weliver (Aldershot, UK: Ashgate, 2005), 25–48.

7. On helium and the intersection of transportation and chemistry, see Robert Schofield, *The Lunar Society of Birmingham: A Social History of Provincial Science in Eighteenth-Century England* (Oxford: Oxford University Press, 1963), 250–54, and Paul Keen, "Science and Spectacle in 1780s England," *Eighteenth-Century Studies* 39 (2006): 517.

8. Mark Monmonier discusses the intriguing question of why the weather map was not invented until 1816 or so, noting that previously "no one thought that cartographic snapshots of barometric pressure and wind might prove revealing" and citing an interest only "in describing climate, not forecasting weather" and a focus "on celestial influences as well as on local links among weather, magnetism, insects, and health." See Monmonier, *Air Apparent: How Meteorologists Learned to Map, Predict, and Dramatize the Weather* (Chicago: University of Chicago Press, 1999), 18. On the militarized imagery of cloud fronts and other affinities between war and weather, see Monmonier, 78; and on this conjunction as it fed an "air-borne sense of suffering," see Mary Favret, *War at a Distance: Romanticism and the Making of Modern Wartime* (Princeton, NJ: Princeton University Press, 2010), 145.

9. Mark Harrison considers the erosion of difference between such far-flung entities as England and India in *Climates and Constitutions: Health, Race, Environment and British Imperialism in India, 1600–1850* (Oxford: Oxford University Press, 1999), 18–20; see also Rajani Sudan, "Chilling Alabahad: Climate Control and the Production of Anglicized Weather in Early Modern India," *Journal for Early Modern Cultural Studies* 8 (2008): 56–73; and Dror Wahrman, *The Making of the Modern Self: Identity and Culture in Eighteenth-Century England* (New Haven, CT: Yale University Press, 2004), 83–126. From a feminist perspective, Clare Brant explores the new relativity of directions and figures of national difference (such as sangfroid), direction (north and south), and temperature (hot and cold) in the waning decades of the eighteenth century. See Brant, "Climates of Gender," in *Romantic Geographies*, ed. Amanda Gilroy (Manchester, UK: Manchester University Press, 2000), 129–49.

10. This view diverges from the one that Dorothy McMillan advances when she sees *A Journey* to break with the stylized pictorialism of Radcliffe's novels. McMillan speculates that "while . . . Radcliffe's contemporaries anxiously wanted to believe that the special characteristics of her novels were born out of her experience of real places, . . . [t]he experience of real places is what may well have put an end to her fiction." See McMillan, "The Secrets of Ann Radcliffe's English Travels," in *Romantic Geographies*, 63. Elizabeth Bohls similarly considers Radcliffe to have skewed the aesthetic aspects of her travel writing toward the political and historical, and George Dekker sees *A Journey* as the source of the travelogues embedded in her fiction. See Bohls, *Women Travel Writers and the Language of Aesthetics, 1716–1818* (Cambridge: Cambridge University Press, 1995), 209–29, and Dekker, *The Fictions of Romantic Tourism: Radcliffe, Scott, and Mary Shelley* (Stanford, CA: Stanford University Press, 2005), 71–123.

11. Ann Radcliffe, *The Italian*, ed. Frederic Garber and E. J. Clery ([1797] Oxford: Oxford University Press, 1998); Ann Radcliffe, *The Romance of the Forest*, ed. Chloe Chard ([1792] Oxford: Oxford University Press, 1986), 13; Ann Radcliffe, *The Mysteries of Udolpho*, ed. Terry Castle and Bonamy Dobrée ([1794] Oxford: Oxford University Press, 1998). Future citations in this chapter will be to these editions and will appear parenthetically in the text by short title and page number.

12. A pertinent discussion of the sense of motion gothic novels like this create through scenes like this one may be found in Mark Blackwell, "The Gothic: Moving in the World of Novels," in *A Concise Companion to the Restoration and Eighteenth Century*, ed. Cynthia Wall (Maldon, MA: Blackwell, 2005), 144–61.

13. Air's movement, via Adeline's voice, between national registers anticipates nineteenth-century writers' use of the gothic form to define and undermine national boundary, as described in Cannon Schmitt, *Alien Nation: Gothic Fictions and National Identity* (Philadelphia: University of Pennsylvania Press, 1997).

14. On Radcliffe's music, see Noelle Chao, "Musical Letters: Eighteenth-Century Writings on Music and the Fictions of Burney, Radcliffe, Scott" (PhD diss., University of California, Los Angeles, 2008).

15. Terry Castle, "The Spectralization of the Other in *The Mysteries of Udolpho*," in Castle, *The Female Thermometer: Eighteenth-Century Culture and the Invention of the Uncanny* (Oxford: Oxford University Press, 1995), 120–39.

16. Ann Radcliffe, "The Supernatural in Poetry," in *New Monthly Magazine and Literary Journal* (London, 1826): 145–52.

17. William Hazlitt, "On the English Novelists," in *Lectures on the English Comic Writers* ([1818] London: Taylor and Hessey, 1815), 250; Hazlitt, "Why the Heroes of Romance Are Insipid," in *Sketches and Essays* (London: John Templeman, 1839), 267, reprinted in *The Critical Response to Ann Radcliffe*, ed. Deborah Rogers (Westport, CT: Greenwood, 1994), 154. Robert Mayer analyzes Hazlitt's "defensiveness" as an effect of Radcliffe having "conditioned the way in which he sees the world." See Mayer, *Ann Radcliffe: The Great Enchantress* (Manchester, UK: Manchester University Press, 1995), 3.

18. See Leo Spitzer, "Milieu and Ambiance," in *Essays in Historical Semantics* (New York: Russell and Russell, 1948), 206–7.

19. Gabriele Schwab, "Words and Moods: The Transference of Literary Knowledge," *SubStance* 26 (1997): 107.

20. Amy May King, *Bloom: The Botanical Vernacular of the English Novel* (New York: Oxford University Press, 2003).

21. Claudia Johnson, *Equivocal Beings: Politics, Gender, and Sentimentality in the 1790s* (Chicago: University of Chicago Press, 1995), 99–101.

22. Robert Harbison, *Eccentric Spaces* (Cambridge, MA: MIT Press, 1977), 85. Harbison links this ambiguity to the reader's own "prison of subjectivism" in which "our thoughts happen to us, we are spectators of moods which appear and disappear like characters we do not understand" (p. 86).

23. Thomas Noon Talfourd, "Memoir of the Life and Writings of Mrs. Ann Radcliffe," in Radcliffe, *Gaston de Blondeville; or The Court of Henry III*, 4 vols. (London, 1826), 1:100. Scott mentions this same preference in his biography of Radcliffe.

24. Elihu Hubbard Smith, *Diary of Elihu Hubbard Smith* (1796), ed. James E. Cronin (Philadelphia: American Philosophical Society, 1973), reprinted in Rogers, *Critical Response*, 26.

25. *Monthly Review* 18 (1795): 241.

26. Virginia Woolf, "Gothic Romance," *Times Literary Supplement*, May 5, 1921, 288.

27. Thomas Beddoes, *Outline of a Plan for Determining the Medicinal Powers of Factitious Airs* (London, 1795), 1. See Mike Jay, *The Atmosphere of Heaven: The Unnatural Experiments of Dr. Beddoes and the Sons of Genius* (New Haven, CT: Yale University Press, 2009).

28. Thomas Beddoes, *Considerations on the Medicinal Use and on the Production of Factitious Airs*, 3 vols. (London, 1795), 1:9.

29. As Alain Corbin has shown, French chemical revolutionaries of the 1790s were increasingly preoccupied with "bad air" (Lavoisier himself had studied the "mephitic air" that bred in French prisons); to them air itself had become a "laboratory of decomposition." In a heavily Foucauldian argument, Corbin reports a shift in interest in the 1790s from air as an external entity to "qualities of the confined, enclosed area of everyday life, the aerial envelope and the atmosphere of bodies." See Corbin, *The Fragrant and the Foul: Odor and the French Social Imagination* (Cambridge, MA: Harvard University Press, 1986), 16, 20.

30. Letter to Humphry Davy, April 3, 1794, cited in Jay, *Atmosphere of Heaven*, 97.

31. See Jay, *Atmosphere of Heaven*, 233. Jan Golinski offers a fascinating reading of Davy's reputation in relation to his pneumatic researches in "Humphry Davy's Sexual Chemistry," *Configurations* 7 (1999): 15–41.

32. Thomas Beddoes. *Notice of Some Observations Made at the Medical Pneumatic Institution* (London, 1799), 9, 11.

33. Matthew Lewis, *The Monk*, ed. Christopher Maclachlan ([1796] London: Penguin, 1998), 38, 20.

34. Matthew Lewis, *Journal of a West India Proprietor*, ed. Judith Terry ([1834] Oxford: Oxford University Press, 1999), 28.

35. William Godwin, "The Composition of *Caleb Williams*," in *Caleb Williams; or, Things as They Are*, ed. Pamela Clement ([1794] Oxford: Oxford University Press, 2009), 349.

36. Charles Dickens, *Bleak House*, ed. Stephen Gill ([1852–53] Oxford: Oxford University Press, 1999), 95.

37. Though we may not take them for granted for long: see Justine Pizzo, "Atmospheres in the Nineteenth-Century Novel" (PhD diss., University of California, Los Angeles, in progress).

38. Karl Marx, *Capital: A Critique of Political Economy*, ed. Frederick Engels, trans. Samuel Moore and Edward Aveling, 2 vols. ([1867–94] New York: International Publishers, 1967), 1:533.

39. Charlotte Brontë, *Jane Eyre*, ed. Sally Shuttleworth ([1847] Oxford: Oxford University Press, 2009), 42.

40. Mark Twain, *Pudd'nhead Wilson and Those Extraordinary Twins*, ed. E. Berger (New York: Norton, 2005), appendix 1, 235.

41. Jane Austen, *Northanger Abbey*, ed. John Davie and Claudia L. Johnson ([1817] Oxford: Oxford University Press, 2003), 82.

Index